INTERFACIAL ENZYME KINETICS

INTERFACIAL ENZYME KINETICS

Otto G. Berg

Uppsala University, Sweden

Mahendra Kumar Jain

University of Delaware, USA

JOHN WILEY & SONS, LTD

Other Wiley Editorial Offices

John Wiley & Sons, Inc., 605 Third Avenue,
New York, NY 10158-0012, USA

Wiley-VCH Verlag GmbH, Pappelallee 3,
D-69469 Weinheim, Germany

John Wiley & Sons Australia Ltd, 33 Park Road, Milton,
Queensland 4064, Australia

John Wiley & Sons (Asia) Pte Ltd, 2 Clementi Loop #02-01,
Jin Xing Distripark, Singapore 0512

John Wiley & Sons (Canada) Ltd, 22 Worcester Road,
Rexdale, Ontario M9W 1L1, Canada

Library of Congress Cataloguing-in-Publication Data
Jain, Mahendra Kumar.
    Interfacial enzyme kinetics / Mahendra Kumar Jain, Otto G. Berg.
        p. cm.
    Includes bibliographical references and index.
    ISBN 0-471-49304-X (alk. paper)
    1. Enzyme kinetics. 2. Biological interfaces. I. Berg, Otto G. II Title.

QP601.2.J35 2001
572′.7 – dc21

                                                                2001035960

British Library Cataloguing in Publication Data

A catalogue record for this book is available from the British Library

ISBN 0-471-49304-X

Typeset in 10.5/12.5pt Palatino by Laserwords Pvt. Ltd., Chennai, India.
Printed and bound in Great Britain by Antony Rowe, Chippenham, Wiltshire.
This book is printed on acid-free paper responsibly manufactured from sustainable forestry, in which at least two trees are planted for each one used for paper production.

To Barbara who made it all seem worthwhile

and

to Preeti who learned to ask questions and understand the

answers

Contents

Preface   xvii

Theory Boxes   xxi

List of Symbols   xxiii

1 Why Interfacial Enzymes?   1

1.1 Structural Diversity of Nonpolar and Amphiphilic Solutes   3

1.2 To Be or Not to Be in Water   5
1.2.1   Solubility and partitioning   7
1.2.2   Amphiphiles   7
1.2.3   Phospholipids   9

1.3 The Hydrophobic Effect   10
1.3.1   Hydrophobic drive and energetic balance   10

1.4 Knotty Issues of Interfacial Enzymology   11
1.4.1   Problems of accessibility of lipid substrates   12
1.4.2   Understanding the variables   13
1.4.3   The implicit variables in a kinetic scheme   14
1.4.4   Substrate accessibility and other variables in an ensemble of
        dispersed particles   15
1.4.5   Concerns for the interfacial variables in the cellular environment   15

1.5 Pathophysiology of Lipolytic Enzymes   16
1.5.1   Arachidonate from phospholipids   17

1.6 Secreted PLA2 (sPLA2) Prototype for Interfacial Enzymology   18
1.6.1   Molecular characteristics of the sPLA2 family   18
1.6.2   The catalytic site versus the interface binding region
        (i-face) of sPLA2   19

1.7   **Summary and Outlook: Towards the Paradigm for Interfacial
       Enzymology**   19

       Further Reading   21

2   **Interface Phenomena: Accessibility and Exchange**   23

2.1   **Aggregates and Dispersions**   25
       2.1.1   Stability of dispersions   25

2.2   **Organized Micellar Aggregates in Aqueous Dispersions**   26
       2.2.1   Onset of micelle formation   27
       2.2.2   The CMC, critical micellization concentration   28
       2.2.3   Significance of the CMC   30
       2.2.4   Nonidealities in micellization   31
       2.2.5   Structural requirements for micellization   31
       2.2.6   Factors affecting the monomer–micelle equilibrium   32

2.3   **Amphiphile Organization and the Monomer–Micelle Equilibrium
       and Exchange**   33
       2.3.1   Cooperativity of micellization   34

2.4   **The Concentration Issues in the Monomer–Aggregate Equilibria**   34
       2.4.1   The monomer residence time in micelles   35

2.5   **Exchange of Amphiphiles between Organized Interfaces**   37
       2.5.1   No exchange between the bilayer-enclosed vesicles   39
       2.5.2   Trough geometry determines the exchange in monolayers at the
                air–water interface   40
       2.5.3   Mixing by fusion–fission of mixed micelles   40

2.6   **Dispersed Phases**   41
       2.6.1   Stable emulsions   41
       2.6.2   Serum lipoproteins   42

2.7   **Choice of Interface for Study of an Interfacial Enzyme**   42

2.8   **Summary: Interface Determines the Accessibility and
       Replenishment Rate**   44

       Further Reading   45

3 To Be or Not To Be: Dilemma for the Substrate in Solution

47

3.1 Constrained Reaction Path for the Turnover in the Aqueous Phase
50
3.1.1 Assumptions, valid constraints and boundary conditions 50
3.1.2 The steady state condition for flux 51

3.2 Analysis of the Steady State Turnover Cycle in the Aqueous Phase
52
3.2.1 The substrate concentration dependence of the flux 53
3.2.2 Time-dependent change in the rate 53
3.2.3 More assumptions 55
3.2.4 Significance of the primary kinetic parameters 56

3.3 Ascertaining the Reaction Path in the Aqueous Phase
56
3.3.1 Premicellar aggregates 59

3.4 PAF-Acetylhydrolase Hydrolyzes the Monodisperse Substrate
60
3.4.1 Effects of the interface in a matrix mechanism 61
3.4.2 Partitioning of the monodisperse substrate 63
3.4.3 Nonequilibrium effects 63
3.4.4 Teleology 66

3.5 Equilibrium Partitioning and Binding to Aggregates
66
3.5.1 Neutral diluent 67
3.5.2 Defining equilibrium partitioning as the dissociation constant, K'_L 67
3.5.3 Nonidealities 68
3.5.4 Ideal partitioning of a solute in micelles 68
3.5.5 Effect of equilibrium partitioning of the substrate on the steady state rate 71

3.6 Summary: The Equilibrium Dilemma
73

Further Reading
75

4 Interfacial Processivity: Ensemble Behavior in the Scooting Mode

77

4.1 Conceptualizing Interfacial Processivity
78
4.1.1 Characteristics of the reaction progress in the scooting mode 78
4.1.2 Consequences of undefined processivity 80

4.2   The Fourfold Meaning of the Substrate Concentration   81
   4.2.1   Variables change with the quality of the interface   82

4.3   Constraints for Defining the Variables for the Reaction Progress in the Scooting Mode   83

4.4   Integrity of Vesicles during the Scooting Mode Reaction Progress   86
   4.4.1   Stability of bilayer vesicles   86
   4.4.2   Hydration of phospholipids   86
   4.4.3   Size and dispersity   87
   4.4.4   Conformational and motional constraints within the bilayer   88
   4.4.5   Organizational polymorphism and metastability   88

4.5   Tests for the Scooting Mode Reaction Progress by PLA2   89
   4.5.1   Extent of hydrolysis increases with the vesicle size   89
   4.5.2   Product and substrate do not exchange   90
   4.5.3   Vesicle integrity is maintained   90
   4.5.4   Binding to the interface as a pre-steady state event   91
   4.5.5   The order of addition and the distribution of the components   91
   4.5.6   Substrate replenishment conditions   92

4.6   Ensemble Behavior of the Binding of the Enzyme to Vesicles   92
   4.6.1   Counting the catalytically active enzyme species   92
   4.6.2   Poisson distribution of enzyme over vesicles   92
   4.6.3   Substrate replenishment through polymyxin B (PxB) induced vesicle-to-vesicle contacts   93
   4.6.4   Apparent activation due to substrate replenishment   94

4.7   Summary: Ensemble Behavior in a Microscopically Heterogeneous Environment   95

5   Analysis of the Processive Reaction Progress

                                                   97

5.1   Michaelis–Menten Equation for the Turnover in the Interface   97
   5.1.1   The initial rate conditions with microscopic steady state   101

5.2   Catalytic Parameters from the Integrated Processive Reaction Progress 101
   5.2.1   Integrated rate equation   102
   5.2.2   Relations in the reaction progress   104
   5.2.3   Extended initial rate by rapid substrate replenishment   105

5.3 Additional Constraints for the Analysis of the Interfacial Turnover Events for PLA2 106
  5.3.1 The Michaelis parameters 107
  5.3.2 Product inhibition 109

5.4 Uses of the Primary Catalytic Parameters 110
  5.4.1 The chemical step is rate-limiting 110
  5.4.2 The carbonyl isotope effect 110
  5.4.3 The solvent isotope effect 110
  5.4.4 Substrate specificity and substrate conformation 111

5.5 Catalytic Mechanism of PLA2 112
  5.5.1 Catalytic triad versus the calcium-coordinated oxyanion mechanism for the chemical step 112

5.6 The Quality-of-Interface Effects 114
  5.6.1 Lateral phase separation in the bilayer 114
  5.6.2 Lateral diffusion and the target search 116
  5.6.3 Changes in the rate-limiting step 117

5.7 Apparent Interfacial Activation 117
  5.7.1 Reaction progress with slow enzyme exchange 117

5.8 Hopping and the Fast Enzyme Exchange Limit 120
  5.8.1 Reaction progress with fast enzyme exchange 121
  5.8.2 Not all enzymes are in the interface 121
  5.8.3 Reaction progress with fast exchange and not all enzyme in the interface 123

5.9 Summary: Variations in the Processivity 123

Further Reading 124

6 Detailed Balance Conditions for Interfacial Equilibria 125

6.1 Binding of Ions to the Interface 126
  6.1.1 Electrical double layer 127
  6.1.2 Electrostatic binding of cationic peptides to anionic interfaces 128
  6.1.3 Selective partitioning of anions into zwitterionic interfaces 128
  6.1.4 Specific interactions at close contacts 129

6.2   Equilibria for the Binding of the Enzyme to the Interface   129
6.2.1   The detailed-balance conditions   130

6.3   Effective Equilibrium Constants   132
6.3.1   Effective interface binding of the enzyme   133
6.3.2   Effective catalytic site binding of the ligand(s)   134

6.4   The Cofactor Binding Obligatory for the Substrate Binding   135
6.4.1   Detailed balance condition for the cofactor binding   136
6.4.2   Effect of cofactor partitioning   136

6.5   Detailed Balance Conditions and Local Concentrations for Effective Ligand Binding   137

6.6   Resolution of the Interfacial Constants for PLA2   138
6.6.1   The high-affinity binding of PLA2 to the anionic interface   139
6.6.2   Dissecting the active site events with the protection method   142
6.6.3   The dissociation constant for E* or E*L from the interface   144
6.6.4   The dissociation constant for L* from E*L   145

6.7   Detailed Balance Conditions for PLA2   146
6.7.1 K_L^* activation Factor   146

6.8   Summary: Primary Equilibrium Parameters for the Kinetic Path   147

7   Rapid Substrate Replenishment in the Quasi-Scooting Mode   149

7.1   Interfacial Catalytic Cycle Turnover in the Quasi-Scooting Mode   150

7.2   Sparingly Soluble Substrates   154
7.2.1   Effect of solubility limit and other nonidealities on the partitioning   154

7.3   Reaction Rate with the Partitioned Substrate   158

7.4   Interfacial Turnover by Triglyceride Lipase   158
7.4.1   Hydrolysis of PNPB in POPG vesicles by tl-lipase from *Thermomyces lanuginosa*   160
7.4.2   Assumption of tl-lipase as an interfacial enzyme   160
7.4.3   Partitioning of PNPB into POPG-suv   161
7.4.4   Substrate dilution on POPG-suv   161
7.4.5   The substrate concentration dependence   163

7.5 **Lid on tl-Lipase Active Site** 164
7.5.1 Relationship of the 'lid' to the active site 165
7.5.2 Cationic residues as the 'hinge' for the interfacial activation by the anionic charge 165

7.6 **Interfacial Allostery for the Quality-of-Interface Effects** 167
7.6.1 Defining the problem of lid opening versus interfacial activation 167

7.7 **Motifs for Close Contact of Proteins with the Interface** 168
7.7.1 Hydrophobic residues 168
7.7.2 Hydrophobic segments 169
7.7.3 Amphipathic helix 170
7.7.4 Membrane anchors 170
7.7.5 Putative membrane-binding domains 170
7.7.6 Colipases 171

7.8 **Summary: Multiple States of the Enzyme at the Interface** 171

8 **Interfacial Allostery** 173

8.1 **Interfacial Catalytic Turnover in the Quasi-Scooting Mode** 176

8.2 **The Apparent Rate Parameters for the Pig Pancreatic PLA2** 179
8.2.1 Resolution of the anomalous contribution of the monomer rate 180
8.2.2 Hydrolysis of monodisperse anionic substrate via premicellar aggregate 182
8.2.3 [S*] dependence due to fusion–fission of micelles 183

8.3 K_S^* **Allosteric Effects of the Interface** 183
8.3.1 Quasi-equilibrium for the catalytic turnover in micelles 183

8.4 **Analysis of the Interfacial Anionic Charge Preference: k_{catS}^* Activation** 184
8.4.1 Possible consequences of product accumulation in zwitterionic micelles 185
8.4.2 Activation by added NaCl 186

8.5 **The Structural Basis for the Anionic Interface Preference** 187
8.5.1 Additivity of the incremental effect on the primary parameters 187

8.6 **Modeling the i-face of PLA2** 191
8.6.1 Conceptualization of the i-face versus the catalytic site 192
8.6.2 The anion-assisted dimer of PLA2 has five coplanar anions 193

8.7 **Site-Directed Mutagenesis to Discern Interactions Along the i-face** 194

8.7.1 The Trp-3 environment 195
8.7.2 Energetics of the substitutions at the N terminus 195
8.7.3 Charge compensation versus surface electrostatics 196
8.7.4 Conformational reciprocity 196
8.7.5 Multiple conformational changes 197

8.8 **Summary: Residues Involved in Charge Compensation** 197

9 **Inhibition: Specific or Nonspecific** 199

9.1 **Specific Inhibitors** 200
9.1.1 Concerns for the interfacial inhibitor design 201
9.1.2 Active-site-directed competitive inhibitors of secreted PLA2 202
9.1.3 Detailed balance condition for the active site binding and stability 203
9.1.4 Sulfated glycoconjugates bind to the E form of PLA2 203
9.1.5 Covalent modifiers of PLA2 204

9.2 **Kinetic Effects of Nonspecific Inhibitors** 205
9.2.1 Substrate dilution and phase change 205
9.2.2 Nonspecific inhibition of PLA2 by alkanols 206

9.3 **Kinetic Effects of the Interface-Based Competitive Inhibitors** 208
9.3.1 The inhibitor concentration dependence of the apparent rate 210
9.3.2 The Z-factor for inhibitor potency 212

9.4 **Influence of Cofactor on the Scooting Kinetics** 213

9.5 **Effects of the Interface-Based Inhibitor on the Integrated Rate Equation in the Scooting Mode** 214

9.6 **Partitioning of the Inhibitor and Substrate between the Interface and the Aqueous Phase** 215
9.6.1 Effect of the chain length on the inhibitor potency 219
9.6.2 Solution-based inhibitor 222
9.6.3 Apparent noncompetitive behavior 225

9.7 **Summary: Multiple Pathways for Reduction of the Observed Rate** 225

Further Reading 226

10 The Delay to the Steady State in the Reaction Progress — 227

10.1 Effects of the Accumulated Products in Zwitterionic Bilayers — 228
10.1.1 Effect of the gel–fluid phase transition temperature — 230
10.1.2 Origin of the anomalous effect of the gel–fluid phase transition — 230

10.2 Model for the Delay Due to the Product Accumulation during the Reaction Progress on Phosphatidylcholine Vesicles — 232

10.3 Effect of the Accumulated Products on the Delay in the Monolayer Reaction Progress — 235
10.3.1 Trough geometry and unstirred layer — 239
10.3.2 Diffusion through the unstirred layer — 239
10.3.3 Time for diffusion to the monolayer versus the equilibration time through the unstirred layer — 241
10.3.4 Product accumulation and the build-up of the anionic charge at the interface — 242
10.3.5 Reaction progress from the model — 245

10.4 Summary: Activation by the Anionic Charge Induced by the Product Accumulation — 245

11 Nonidealities of the Dispersed Phases — 249

11.1 The Exchange Limit — 249

11.2 Exchange-Limited Kinetics of PLA2 in Detergent-Dispersed Mixed Micelles of Long Chain Phospholipids — 253
11.2.1 The intermicellar concentration (IMC) — 253
11.2.2 The exchange-limited interfacial turnover — 254
11.2.3 Replenishment by fusion–fission — 255

11.3 Effect of Bile Salts on the Pancreatic PLA2 Catalyzed Hydrolysis of Phosphatidylcholines — 257

11.4 Kinetic Concerns for Interfaces of the Dispersed Phase — 260
11.4.1 Surface dilution in detergent dispersions — 260
11.4.2 Polymorphism, metastability and lyotropic behavior — 261

11.5 **The Nonideality Factor** 262
11.5.1 Effect of nonideal partitioning on substrate–enzyme binding 263

11.6 **Summary: Nonidealities for Replenishment and Binding** 265

12 Effects of Reduction of Dimensionality 267

12.1 **Dissection of the Entropy Loss on Interface Binding** 268
12.1.1 Motional constraints and accessible volume in the interface 268
12.1.2 Free energy differences 271

12.2 **Synergistic Effects of the Interface on Enzyme–Substrate Binding, Local Concentration and Scaffolding** 274

12.3 **Diffusion Times in 1D, 2D and 3D** 276
12.3.1 Diffusion with mixed dimensionality 279
12.3.2 Effective association rates in 2D and 3D 280

12.4 **Rate Enhancement by Facilitated Diffusion in 2D** 282

12.5 **Summary: Dimensionality Effects on the Equilibrium and Diffusion** 285

References 287

Index 297

Preface

The perception of increasing validity comes from directed interactions with the awareness of a priori knowledge.

Shatkhandagam (ca. 100 AD) by Pushpadant and Bhutbali

Interfacial enzymes have evolved to deal with the biophysical realities of interfaces. About half the protein in cells is membrane-associated and must function within the interfacial kinetic constraints for substrate accessibility, distribution, orientation, partitioning and exchange. Such processes ultimately control the processivity of the interfacial turnover cycles and the equilibria that feed into the events of the turnover cycle. Thus the binding of the enzyme to the interface and the substrate accessibility and replenishment in the microenvironment of the bound enzyme determine the microscopic steady state condition for the catalytic turnover. An understanding of the interfacial processes during the reaction progress also permits an unequivocal identification of the kinetic path for the analysis of the steady state turnover events of the Michaelis–Menten cycle in the aqueous phase or processively at the interface.

In this book we outline the principles, rules and analytical protocols for the kinetic analysis of the catalytic function of enzymes at interfaces. The interfacial kinetic paradigm has emerged from the prototypical studies with secreted phospholipase A_2. With an appropriate description of underlying variables, analysis of the steady state turnover kinetics in terms of a defined path provides the primary rate and equilibrium parameters. Under suitable conditions such parameters relate to the elemental events defined by the thermodynamic and kinetic variables with their origins in the molecular structures and organization. Starting from a well-defined kinetic path under the assay conditions, such relationships permit quantitative insights into the inhibitor and substrate specificities, the mechanism for the chemical step and the allosteric effects due to cross-talk between the distal interactions. The underlying parameters have well-established structural and functional significance. Ultimately the parameters are useful for the interpretation and prediction of the enzyme function under virtually all assay conditions, including the biological *milieux*. Of course, this is possible only if the kinetic path and the underlying variables are unequivocally defined in the microscopically heterogeneous environment encountered by an interfacial enzyme during the steady state turnover.

Study of the interfacial rate and equilibrium processes is a discipline in its own right because the interplay of elementary processes under the dimensionally constrained conditions pose conceptual challenges that begin with the definition of the variables rooted in the model. As developed in this book, for the analysis of the steady state kinetic analysis of the interfacial turnover it is necessary to consider the variables that control the interfacial processivity of the turnover cycles, as well as the substrate accessibility and replenishment rates. Thus the challenge of interfacial enzymology is to define and characterize the steady state condition for the variables of the elementary events that make up the steady state catalytic turnover. This is critical for ensemble- and time-averaging of all the events that occur during the observed reaction progress. If the microscopic steady state condition in a constrained microscopically heterogeneous system can be defined, an analysis of the observed rate and equilibrium processes provides the primary rate and equilibrium parameters for the interfacial binding of the enzyme and the catalytic turnover cycle of the bound enzyme. Such considerations of the microscopically constrained environment are not intuitive in classical enzymology where only the turnover events in a homogeneous solution environment are considered.

As a primer for a potentially vast field, in this book we outline the basic concepts that set the foundation for the kinetic analysis of the catalytic behavior of enzymes within the interfacial constraints (Chapters 1 and 2). Starting with a consistent set of basic assumptions to extend the classical Michaelis–Menten paradigm (Chapter 3), we develop the principles and formalism to obtain the rate and equilibrium parameters for the primary processes at the interfaces (Chapters 4 to 8). Such parameters have mechanistic significance and thus correlate with the structural bases for the less constrained conditions (Chapters 9 to 11). In the last chapter, we consider some of the dimensionality effects on the variables and constants of the interfacial kinetic paradigm in terms of the fundamentals of thermodynamics (Chapter 12).

A well-grounded quantitative description of the observed kinetic behavior raises issues of concern for researchers in cell biology, biochemistry, pharmacology, and colloid and interface sciences. We start with fundamentals of chemistry and physics. The 400 level college courses in biochemistry and in physical chemistry should generally be an adequate prerequisite for following the arguments developed in this book. The descriptive theme of the text is based on key concepts that are sequentially developed with explicit assumptions to arrive at more complex kinetic conditions. The analytical description of the defined kinetic models is developed in separate sub-sections. The theory boxes deal with the more elaborate concepts that form the basis for certain critical assumptions used for the analysis. Such quantitative relations encourage thinking about the real-world problems and concerns in terms of the well-defined parameters. The models, with their assumptions and limitations, have been chosen as representative of many of the situations that we have encountered. The possible variations are virtually limitless, and it is our hope that by understanding the structure of these models it will be easier to extend the description to other situations as well.

We thank Drs Apitz-Castro, Cajal, Homan, Romsted and Wilton for useful suggestions with earlier drafts of some chapters. As noted in the text, the work of our collaborators, Drs Cajal, Jahagirdar, Maliwal, Rogers, Upreti and Yu, has provided many of the critical results that have proven useful for the strategies and analytic relations developed in this book. The work from the authors' laboratories was funded by NIH, Swedish Natural Science Research Council and Sterling Inc. Finally, we acknowledge extensive fruitful discussions over the last three decades with Drs Cordes, de Haas, Gelb, Tsai, and Verheij, which ultimately led to the formulation of the interfacial kinetic paradigm developed in this book.

Otto G. Berg
Uppsala University, Sweden

Mahendra Kumar Jain
University of Delaware, USA

Theory Boxes

2.1   Diffusion-limited rates of exchange between aggregates   36
3.1   Desorption of an amphiphile from the interface at the steady state   65
3.2   Choice of concentration units in the interface   69
4.1   Interfacial processivity   80
4.2   Poissonian enzyme distribution over vesicles: ensemble-averaging   85
5.1   Steady state flux in the interfacial reaction loop   99
5.2   Time scales for target search by lateral diffusion   115
5.3   Slow enzyme exchange   120
6.1   Statistical weights of enzyme states   131
6.2   Excluded surface effects in the equilibrium binding of enzyme to
      large vesicles   140
7.1   Requirements for rapid substrate replenishment   153
7.2   Molecular area effects on partitioning and binding   157
8.1   Exchange-limited kinetics: activation by product desorption   177
8.2   Two-state model for k^*_{catS} activation in the interface   190
9.1   Partitioning equilibria for a three-component ideal interfacial
      mixture   221
10.1 Model for the time delay at zwitterionic vesicles   234
10.2 Stationary diffusion to the monolayer through the unstirred layer   240
10.3 Delay to steady state in monolayer kinetics   243
11.1 Substrate replenishment by free monomer diffusion between
      particles   251
11.2 Substrate replenishment through fusion–fission of mixed micelles   255
11.3 Thermodynamics of nonideal partitioning   264
12.1 Entropy change on translational confinement   270
12.2 Mean-time calculations   277

List of Symbols

A. Concentrations

[..]	Square brackets always denote a concentration in moles per bulk solution volume; examples follow:	M	
$[A_T]$	Total concentration of A added to the solution	M	Figure 2.6; Table 2.2
$[A^*]$	Concentration of A in the accessible interface	M	Figure 2.6; Table 2.2
$[M^*]$	Concentration of all molecules in the accessible interface	M	Equation (3.16)
$[A]$	Concentration of monomer A in solution	M	Figure 2.6
$\{A^*\}$	Local concentration of A in the interface	M	Equation (12.3)
CMC	Critical micellization concentration	M	Figure 2.6; Equations (2.3) to (2.7)
X_A^*	Mole fraction of A in the interface	Mole fraction	Equation (3.10); Theory Box 3.2
$[I_T(50)]$	Total inhibitor concentration that reduces enzyme rate j_0 by 50 %	M	Equation (9.8)
$X_I^*(50)$	Inhibitor concentration in the interface that reduces j_0 by 50 %	Mole fraction	Equation (9.9)
$n_I^*(50)$	Inhibitor concentration in the interface that reduces enzyme rate k_i by 50 %	Mole fraction	Equation (9.20)

B. Geometric Factors for Aggregates

N_T	Total number of molecules in the accessible interface of an aggregate		Theory Box 2.1
N_S^0	Total number of substrate molecules initially in the accessible interface of an aggregate		Theory Box 4.2; Equations (5.5) and (5.6)
a_A	Interface area occupied by one A molecule	Å^2, dm^2	Theory Boxes 2.1 and 3.2
b	Radius for the interface area (approximated as a circular disc) for one molecule	Å, dm	Theory Boxes 2.1 and 5.2
φ	Area correction factor		Theory Boxes 3.2 and 7.2
n_E	Number of enzymes bound to one aggregate		Theory Box 4.2; Equations (5.5) and (5.6)

C. Equilibrium Constants

K_A'	Dissociation constant for A from the interface	M or M per mole fraction	Equations (2.7) and (3.10); Figures 3.1 and 6.1
K_d	Dissociation constant $E^* \rightarrow E$	M	Theory Box 2.1; Figure 3.1
K_d^L	Dissociation constant $E^*L \rightarrow EL$	M	Figures 3.1 and 6.1
K_L	Dissociation constant $EL \rightarrow E + L$	M	Figures 3.1 and 6.1
K_L^*	Dissociation constant $E^*L \rightarrow E^* + L^*$	Mole fraction	Figures 3.1 and 6.1
$K^{\#}$	Interfacial activation constant		Figures 7.12 and 8.7; Theory Box 8.2

D. Rate Constants

k_a	Association rate constant for amphiphile to the interface	$M^{-1}\,s^{-1}$	Equations (2.6) and (2.7); Theory Box 2.1
k_d	Dissociation rate constant for amphiphile from interface	s^{-1}	Equations (2.6) and (2.7); Theory Box 2.1

k_{on}	Association rate constant for enzyme to the interface	$M^{-1} s^{-1}$	Figure 1.8; Theory Box 2.1
k_{off}	Dissociation rate constant for enzyme from interface	s^{-1}	Figure 1.8; Theory Box 2.1
k_{exch}	Exchange rate of enzyme (or amphiphile) between aggregates	s^{-1}	Theory Box 2.1; Equation (5.19)
k_1, k_{-3}	Association rate constants for $E + S \rightarrow ES$ and $E + P \rightarrow EP$	$M^{-1} s^{-1}$	Figure 3.1
k_{-1}, k_3	Dissociation rate constants for $ES \rightarrow E + S$ and $EP \rightarrow E + P$	s^{-1}	Figure 3.1
k_2, k_{-2}	Rate constants for chemical change $ES \rightarrow EP$ and $EP \rightarrow ES$	s^{-1}	Figure 3.1
k_1^*, k_{-3}^*	Association rate constants for $E^* + S^* \rightarrow E^*S$ and $E^* + P^* \rightarrow E^*P$	(Mole fraction)$^{-1}s^{-1}$	Figures 3.1 and 5.2; Theory Box 5.1
k_{-1}^*, k_3^*	Dissociation rate constants for $E^*S \rightarrow E^* + S^*$ and $E^*P \rightarrow E^* + P^*$	s^{-1}	Figures 3.1 and 5.2; Theory Box 5.1
k_2^*, k_{-2}^*	Rate constants for chemical change $E^*S \rightarrow E^*P$ and $E^*P \rightarrow E^*S$	s^{-1}	Figures 3.1 and 5.2; Theory Box 5.1

E. Enzyme Flux and MM Parameters

v	Product flux per solution volume	$M s^{-1}$	Equation (3.2)
j	Product flux per enzyme molecule	Molecules/s	Equation (3.5)
j_0	Initial product flux per enzyme	Molecules/s	Equations (5.3) and (5.5)
j_{ss}	Stationary state (apparent initial) flux per enzyme	Molecules/s	Theory Box 8.1
k_{cat}	Catalytic rate for enzyme in solution	s^{-1}	Equations (3.2) and (3.3)
K_M	Michaelis constant for enzyme in solution	M	Equations (3.2) and (3.4)
k_{catS}^*	Catalytic rate for enzyme in the interface	s^{-1}	Theory Box 5.1
K_{MS}^*	Michaelis constant for substrate and enzyme in the interface	Mole fraction	Theory Box 5.1
K_{MP}^*	Michaelis constant for product and enzyme in the interface	Mole fraction	Theory Box 5.1
k_i	Relaxation rate for the integrated MM equation	s^{-1}	Equations (5.8), (5.9) and (5.10)

k_{cat}^{app}	Apparent catalytic rate	s^{-1}	Equations (5.24) and (5.25)
K_M^{app}	Apparent Michaelis constant	M	Equations (5.24) and (5.26)
B	Effective inhibitor strength		Equations (9.5) and (9.9)

1 Why Interfacial Enzymes?

It is a miracle of the shared knowledge that I have an idea translated into symbols on a page and someone can read those and have the same idea appear in their minds. Yet the writing condition is a strange one. Its unique excellence is at the same time its tragic flaw. Even a positive thing casts a shadow as coherence is imposed on raw phenomenology to make the world amenable. It is necessary to impose a linear order on the multi-dimensional world through discrete concepts with boundaries that we may share but not perceive in the same way.

Collective Wisdom

A conceptual need for boundaries and interfaces is intrinsic in all attempts to impose order. The discreteness and identity of a space are provided by boundaries and dividing lines. Such interfaces are also critical and obligatory at all hierarchical levels of organization and function in the biosphere. Membranes and interfaces are the organizing principles to contrast with the aqueous environment. As a unit of living organisms, the very existence of a cell is based on membranes and interfaces that identify it as a morphological entity distinct from the rest of the universe.

The subcellular interfaces

If life is impossible without water, it is inconceivable without membrane–water interfaces. The aqueous *milieu* in organisms is compartmentalized by membranes. Tissues are made up of cells, and the cytoplasmic space in cells is interrupted by organelles. Virtually all processes that sustain an organism, ranging from the morphology to specialized functions, depend on discrete functional identities provided to the space by membranes and interfaces. These hydrophobic barriers of molecular dimensions are less than 10 nm thick compared to the 1000 to 50 000 nm diameter of most cells. An appreciation of the role of interfaces in the form and function of a cell is probably best gained by examining the compartments of a cell as in Figure 1.1. Here all the dark lines represent the bilayer interfaces. All such membrane-bounded structures enclose an aqueous compartment that is distinct in composition from the surrounding medium. Noteworthy features of the heterogeneity in the cellular milieu include:

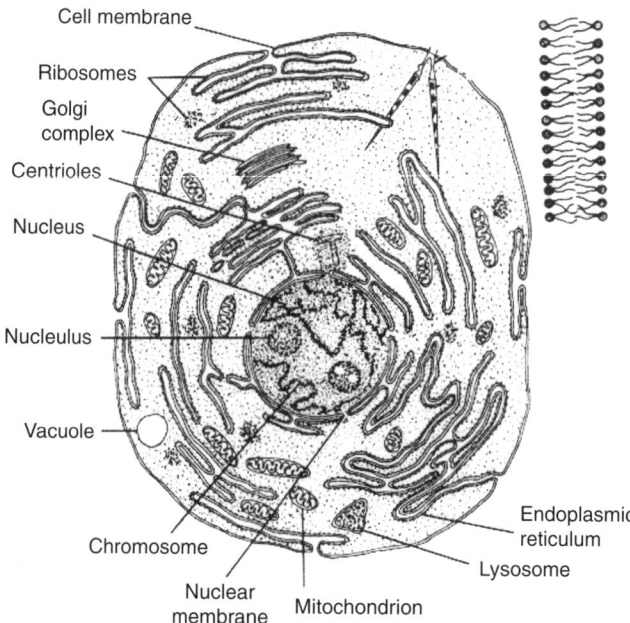

Figure 1.1 Membrane-bounded compartments (organelles) in a typical animal cell. The thick lines represent a bilayer matrix (upper right insert) with a hydrophobic core. Thus vesiculated bilayers compartmentalize the cytoplasmic space and sequester the organelle components

(a) An eukaryotic cell typically contains one to several thousand copies each of a variety of organelles. Although in metabolic communication, organelles retain their characteristic morphology, composition (lipids, proteins, and other macromolecules) and specialized functions.

(b) The organizational matrix of membranes (the thick lines in Figure 1.1) is a phospholipid bilayer with a hydrophobic interior sandwiched between two polar interfaces in contact with the aqueous phases.

(c) Virtually all phospholipid molecules in a cell are present in the bilayer of membranes. The interface area is twice the area of the bilayer. With a cross-sectional area of about 100 $\text{Å}^2 (= 1 \text{ nm}^2)$ per phospholipid molecule, one mole of phospholipid dispersed as bilayer will have the interface of 600 000 m^2 (about a quarter of a square mile). Consider the consequences of the fact that the total phospholipid concentration in tissues like brain, kidney, liver and heart is typically 50 mM. Therefore, 1 cm^3 of a tissue has 30 m^2 of the phospholipid interface. In other words, with 0.05 millimole of the phospholipid packed within a 1 cm cube, equivalent to 50 mM total concentration, the average thickness of the interspersed aqueous compartments separating 1 cm^2 slices of the interfaces would be approximately 300 Å with 50 Å thickness for the phospholipid bilayer. (Note that 1 cm = 0.01 m = 10^7 nm = 10^8 Å.)

(d) The bilayer interface spans a broad range of polarity within a thickness of less than 1 nm. The dielectric constant is 80 in the aqueous phase compared to less than 2 in the hydrophobic interior.

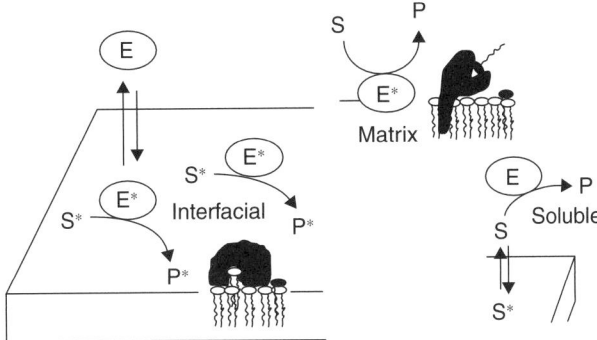

Figure 1.2 A view of bilayer interface to emphasize the operational distinction between interfacial versus matrix enzymes. Although both are membrane-localized, only the interfacial enzymes access the substrate directly from the interface (S*). In contrast, matrix enzymes, like the soluble enzymes, access their substrate from the aqueous phase (S)

(e)   In terms of weight, number or genome length, about half of the protein in cells is membrane-associated.

(f)   Membranes act as a workbench for about half of the cellular protein. Membrane enzymes are operationally (Figure 1.2) either *matrix enzymes* that access their substrate from the bulk aqueous phase or *interfacial enzymes* that directly access the membrane-localized substrate. The *soluble enzymes* in the aqueous phase access their substrate directly from the aqueous phase.

1.1 Structural Diversity of Nonpolar and Amphiphilic Solutes

The primary functional distinction between interfacial versus matrix or soluble enzymes comes from the relative tendency of the substrate to be in the aqueous phase or the interface. Such effects assume added importance in the microscopically heterogeneous cellular environment, where the thickness of the aqueous compartment between the interfaces may rarely exceed more than a few hundred Å. In effect, the cytoplasmic aqueous phase is interspersed by stable membranes with a rather steep polarity gradient that spans from the hydrophobic interior of the bilayer to the water-compatible polar surface. Much of the polarity change occurs within 10 Å interface near the bilayer surface. As developed in this book, mechanisms and biophysical strategies that have evolved to deal with the physical reality of the substrate solubility and accessibility are beginning to be understood. As implicit in Figure 1.2, the partitioning (S to S*) and enzyme binding (E to E*) equilibria determine the kinetic path and influence the turnover rate.

The evolutionary significance of biochemical processes at or across interfaces is best appreciated by examining molecular structures of some of the nonpolar and amphiphilic solutes with nonpolar functional groups with or without polar substituents. For example, galactosyl lipids (Figure 1.3) are the most abundant lipids

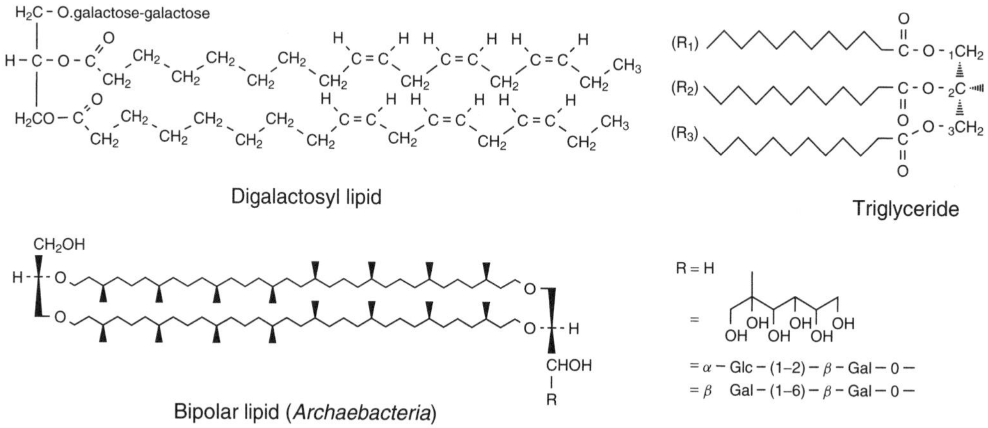

Figure 1.3 Structures of digalactosyl lipid, triglyceride (triacylglycerol) and bipolar lipid. The balance between the polar and nonpolar substituents in these molecules is the basis for their behavior in an aqueous environment

in the biosphere. With a polar polyhydroxylic moiety at one end, the glycolipids provide the bilayer matrix for virtually all the plant membranes. The membrane of *Archaebacteria* from volcanic springs contains a unique class of bipolar lipids, possibly the rarest of the membrane lipids. Bipolar lipids contain two polar head groups attached at the two ends of the intervening isoprene chains. Thus the 'monolayer' of a bipolar lipid molecule has two polar interfaces as the structural basis for the stability of a membrane that functions at the near-boiling temperatures of volcanic springs. Triglycerides (Figure 1.3) are the predominant components of dietary fat. They are among the most hydrophobic of the naturally occurring lipids. Virtually devoid of a polar group, acylglycerols do not dissolve or disperse in water. Glycerides with long acyl chains are stored as phase-separated droplets in plant seeds to provide food and fuel during germination. In addition to the survival value during starvation, the fat deposits in adipose tissues of animals also provide insulating layers. Properly distributed body fat has also inspired changing conceptions of beauty and health.

In addition to the food and fiber dependence on plants for the survival of all organisms, there is an even broader symbiosis between the phototrophs and chemotrophs. The sustainability of the biosphere at such a fundamental level is based on the ability of plants to use sunlight to fix carbon dioxide into carbohydrates that are used as metabolic fuel and building blocks by virtually all other organisms. At the heart of this cosmic machine driven by the sunlight are the oriented molecules of chlorophyll and other pigments (Figure 1.4) in the chromophores of the photosystem P680 and P700 complex. Such photosynthesis complexes are localized in a virtually crystalline lattice in the thylakoid membrane of chloroplasts and other plastids. They capture photons and transduce a part of their energy to generate the proton concentration (pH) gradient across the membrane. The pH gradient drives the reduction of NADP to NADPH and also the synthesis of ATP. Both of these high-energy products are used for the incorporation of CO_2 into ribulose-1,5-bisphosphate. Through such pathways

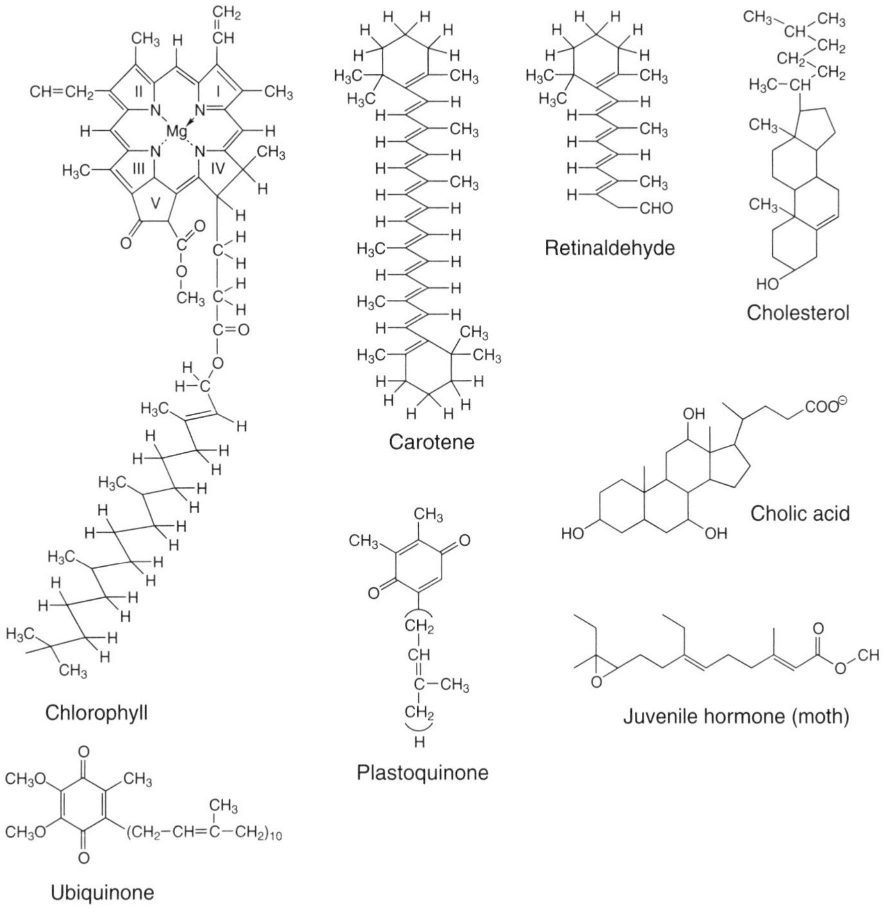

Figure 1.4 Structures of nonpolar molecules with a diverse range of biological functions

CO_2 is ultimately 'fixed' in glucose and its polymers such as starch (food) and cellulose (fiber).

In addition to their role in the biosynthesis of food and fiber, the roles of hydrophobic solutes as trace components are also impressive. Structurally diverse minor nonpolar metabolites include pigments, steroid hormones, pheromones, eicosanoids and other signal molecules. Of course, their behavioral, developmental and pathophysiological effects are the end result of the exquisite control they exert on processes ranging from macromolecular syntheses to osmoregulation and ion metabolism. Obviously, metabolic pathways have evolved to deal with a large polarity spectrum of biomolecules, where the water-soluble molecules represent only one extreme.

1.2 To Be or Not to Be in Water

Depending on the energetic balance of the interactions between the polar and nonpolar groups of a solute with each other and with water molecules, solutes in the aqueous phase exhibit wide-ranging solution characteristics. One extreme is the ideal solution with molecularly monodispersed solute in the bulk solvent without any solute–solute interactions. At the other extreme, virtually insoluble nonpolar solutes avoid any interaction with the aqueous phase by forming a separate phase. As developed in Chapter 2, complex phase behavior is observed in dispersions of amphipathic solutes with polar as well as nonpolar hydrophobic functional groups in the same molecule.

Accommodation of a solute between the organized water molecules in the aqueous phase depends on the ability of the functional groups of the solute to disrupt water–water interactions by promoting favorable water–solute interactions (Figure 1.5). Ions and most of the oxygen- and nitrogen-containing functional groups are capable of such favorable interactions with the polarized $H^{\delta+}-O^{2\delta-}-H^{\delta+}$ molecule that has a fractional negative charge on the electronegative oxygen and fractional positive charges on both hydrogens. As a consequence of the electronegativity of oxygen and nitrogen, solutes with fractional charges on polarized O- and N-containing functional groups have water–solute interactions that are energetically more favorable than the water–water interactions. For the same reasons, solute–solute and solute–water interactions are also at the heart of the interface phenomena as seen in the organization and exchange properties of the

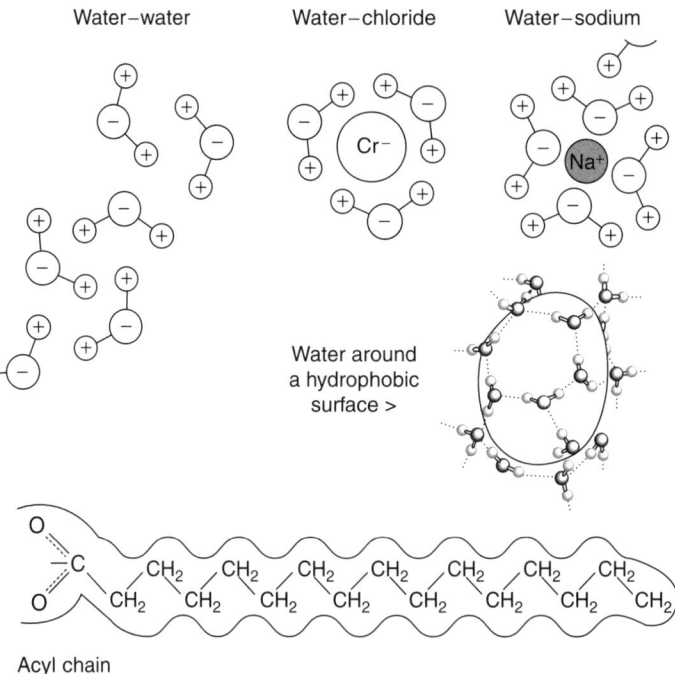

Figure 1.5 Interactions of water (from upper left) with other water molecules, with chloride or sodium ions, or the clatharate organization of water molecules around the surface of a caged nonpolar group or solute such as the acyl chain of a fatty acid salt (soap). The + and − shown on the water molecule represent only the partial charges resulting from polarization of the O–H bond

nonpolar and amphipathic solutes in aqueous dispersions. Such interactions are critical for understanding the effect of substrate accessibility on the enzyme-catalyzed turnover in the presence of dispersed phases.

1.2.1 Solubility and partitioning

Nonpolar solutes, with fewer polar groups capable of interacting with water, are less likely to compete in the water–water interactions. As a result, the dominant hydrophobic effect leads to lower solubility in the aqueous phase, higher partitioning in a nonpolar phase and a nonideal solution behavior due to the formation of aggregates. For example, above their solubility limit in the aqueous phase, nonpolar solutes phase separate and precipitate. As the basis for the S to S* equilibrium (Figure 1.2) the molecularly dispersed nonpolar solutes in solution favorably partition into nonpolar phases. The *equilibrium concentration* of a monodisperse nonpolar solute in the aqueous phase, as well as the concentration of the partitioned solute, does not change with *total concentration* of a solute above its solubility limit (developed further in Chapters 2 to 4).

The solution behavior and the partitioning characteristics of a nonpolar or amphiphilic substrate in a heterogeneous environment introduce the challenge of defining, establishing and evaluating the consequences of the substrate concentration variable for an enzyme-catalyzed reaction. It appears that several strategies have evolved to deal with the physical reality of low solubility and aggregation behavior. For example, sparingly soluble long chain fatty acids in the extracellular milieu bind to serum albumins that 'ferry' nonpolar solutes from tissue to tissue. Such binding proteins also act as 'buffers' by reducing the aqueous phase concentration of nonpolar solutes. Another strategy is that of acyl-coenzyme A derivatives with higher aqueous phase solubility compared to the free fatty acids. Thus the enzyme affinity well above the solubility limit makes the acyl-CoA suitable soluble substrates for fatty acid metabolism within the cell. Both of these mechanisms also solve potential problems associated with the detergent action of ionized free fatty acids (soaps).

1.2.2 Amphiphiles

The term refers to the ambivalence of the aqueous phase for the accommodation with polar and nonpolar regions of a solute molecule such as an ionized fatty acid (Figure 1.4), glycolipids (Figure 1.3) and phospholipids (Figure 1.6). Amphiphiles have a polar (*head*) region that is separated from the nonpolar (*tail*) polymethylene chain, and share a common tendency to form organized phases and dispersions. The molecular diversity of amphiphiles is large because specific interactions are not required for the aggregate formation or the partitioning of the solute driven by the hydrophobic effect. In addition to the acyl and alkyl chains of the hydrophobic tail there may be a steroid nucleus or an aromatic hydrocarbon moiety. The hydrophilic character of the head group may come from virtually any hydrogen-bonding or ionic group that interacts with water. The minimum polarity required of a

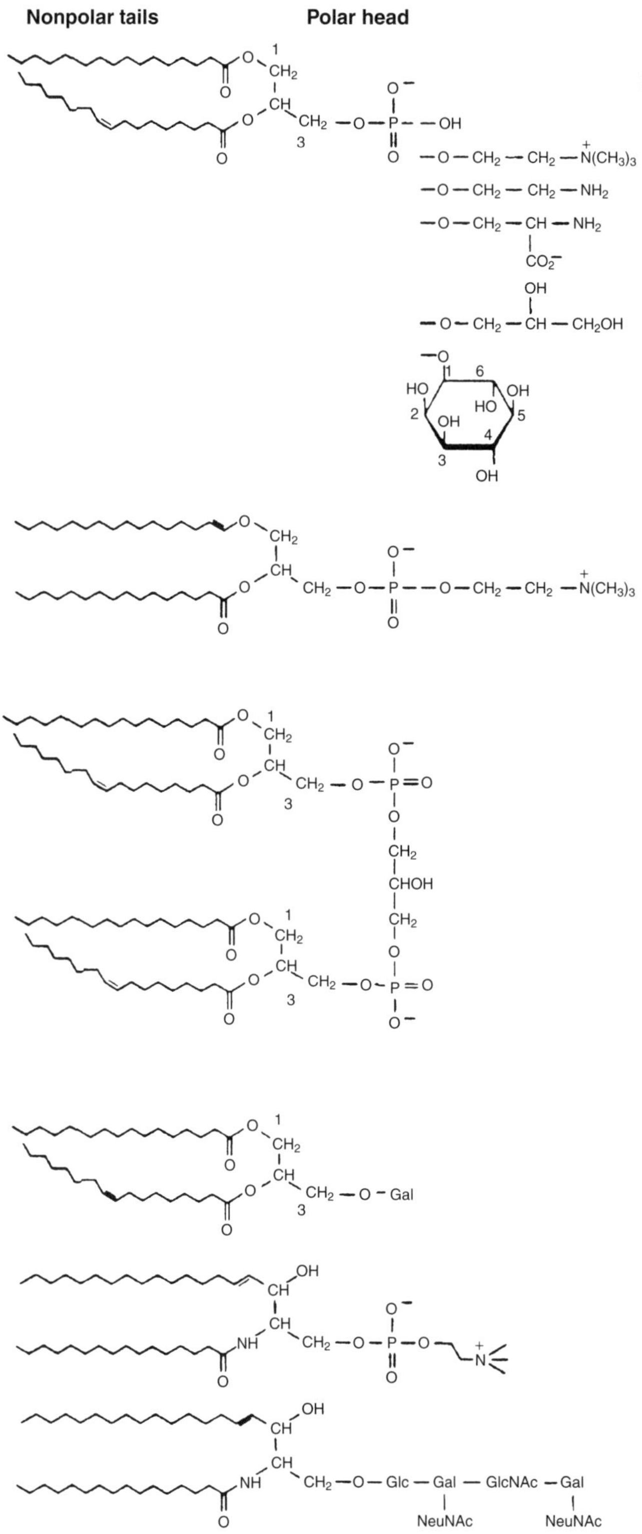

Nonpolar tails   **Polar head**

Phospholipids **PL**   (abbrev.)

Phosphatidic acid **PA**

Phosphatidylcholine **PC**

Phosphatidylethanolamine **PE**
Phosphatidylserine **PS**

Phosphatidylglycerol **PG**

Phosphatidylinositol **PI**

Choline plasmalogen

Cardiolipin **CL**

Glycosyl diacylglycerol

Sphingomyelin **SM**

Ganglioside **GM**

head group of an amphiphile appears to be between a hydroxyl ($-CH_2OH$) and a carboxylate ($-COO^-$) group. Amphiphiles with weaker polarity form microcrystals above their solubility limit in water. Organized structures, such as bilayers or micelles, are formed in dispersions of amphiphiles with stronger polar groups (sulfate, sulfonate, phosphate, ammonium, carboxylate) or bulky oxygen-containing uncharged groups (amine oxide, polyoxyethylene, phosphine oxide, sugar residue) capable of extensive hydrogen bonding and dipolar interactions with water. As the basis for the interface phenomena, the organizational constraints for an amphiphile in aggregates depend on the polar and nonpolar groups.

1.2.3 Phospholipids

Glycolipid (Figure 1.3) and phospholipid (Figure 1.6) make up the bilayer matrix of membranes (Figure 1.1) in virtually all organisms. Such amphiphiles have a common structural motif: cylindrical shape with a strongly polar head group at one end of the strongly hydrophobic diacylglycerol moiety. Phospholipids are the diglycerides with a polar *sn*-3-glycero phosphodiester substituent. Typically, the acyl chains contain 6 to 20 carbon atoms and 0 to 6 double bonds in the *sn*-2-chain of membrane phospholipids. Each type of membrane in a cell has a characteristic composition and a complement of the minor lipids. Such characteristic components include sterols of plasma membranes, cardiolipins of the mitochondrial inner membrane, sphingolipids on the outer surface of red cell and epithelial tissues, sulfolipids of sperm cells, gangliosides and cerebrosides of nerve cells. The relative composition of membrane amphiphiles, as well as the pattern of unsaturation, length and branching of their apolar chains, is precisely controlled, apparently in relation to the growth conditions. The head group and chain composition of phospholipids change with changes in the physiological and pathological environment of the organism.

Membrane lipids with other structural variations are also found. Such changes are common in organisms from unusual ecological niches, e.g. the bipolar lipids of *Archaebacteria* (Figure 1.3). Among the minor amphiphilic components of membranes are the acylated proteins containing specific nonpolar chains, apparently necessary for their specialized regulatory functions (Low, 1989). In addition, complex structures derived from mycolic acid, lipid A, lipid I and II are specific to bacterial membranes, where they act as antigenic markers, as well as the receptors for natural antibacterial peptides. In short, both the bilayer matrix function and the specific metabolic regulatory roles of membrane amphiphiles appear to be controlled through subtle changes in the composition as well as through structural modification.

The interfacial propensity of strongly amphiphilic phospholipids is also seen in their ability to codisperse insoluble solutes such as triglycerides, sterols and certain proteins. For example, phospholipids are also present at the aqueous interfaces of the

Figure 1.6 Major amphiphilic components of eukaryotic membranes. Note that the hydrophobic nonpolar polymethylene chains (tail) are attached to the polar charged moieties (head) at one end. The abbreviations refer to the head group (PC, etc.) with the chain identifiers (P = palmitoyl, M = myristoyl, O = oleoyl) for the acyl chains in the *sn*-1- and *sn*-2-positions. For example, POPC is 1-palmitoyl-2-oleoyl-*sn*-glycero-3-phosphocholine. The symmetrical saturated phospholipids are abbreviated as DC_nPL

fat deposits and droplets, serum lipoproteins, lamellar bodies, lung surfactant, milk and gastrointestinal emulsions with bile salts. The structure and mole fraction of the codispersed components control the phase properties of the dispersed particles, including the size, shape, dispersity, packing and the exchange behavior between the dispersed particles.

1.3   The Hydrophobic Effect

Phospholipids spontaneously organize as vesiculated bilayers in aqueous dispersions. The aggregation tendency of an amphiphile in the aqueous phase and the thermodynamic stability of the aggregate is a measure of the balance between the hydrophobic and hydrophilic character of the functional groups in the amphiphile. As developed in subsequent chapters, the hydrophobic drive is also at work for the partitioning of solutes into bilayer, or for the aggregation above the solubility limit or the micellization concentration, or for the exchange of solutes through the aqueous phase which is limited by its desorption rate from the interface.

1.3.1   Hydrophobic drive and energetic balance

Hydrophobicity and hydrophilicity of a solute are the consequences of the unique properties of water that come into play when a solute molecule is introduced in the aqueous phase. The hydrophobic drive for amphiphile aggregation in the aqueous phase demonstrates itself as the nonpolar chains are segregated away from the aqueous phase and the polar region remains at the aggregate interface in contact with water. The hydrophobic effect is *entropically driven*. The presence of the molecular clusters in liquid water (Figure 1.5) means that a solute can be included in bulk water only if solute–solute interactions and the hydrogen-bonding tendency of liquid water can be energetically compensated by competing interactions between water and solute. Water molecules interact favorably with most functional groups containing electronegative atoms, such as hydroxyl, carbonyl, carboxyl, phosphate and ammonium groups. On the other hand, as shown in Figure 1.5, nonpolar groups force an extra ordering on the neighboring water molecules, which increase their bonding with each other to compensate for the loss of hydrogen bonding in the direction of caged solute. To avoid the entropy decrease involved in this ordering, nonpolar moieties are squeezed out and segregated to form a separate 'phase' where contacts with water molecules are minimized.

Note that the hydrophobic drive is essentially a consequence of a lack of hydrophilic interactions. The energetic balance of the interactions in the hydrophilic and hydrophobic regions determines the net balance of the hydrophobic drive. Attempts to quantify such factors are intrinsic in the concept of the mass-based hydrophil–lipophil balance of an amphiphile molecule or the relationship of the curvature of the particle to the tangential pressure component due to the close packing of the polar groups at the interface. For the interfaces of single amphiphiles, such geometrical and organizational factors are defined in terms of the features of the effective shape of the amphiphile by taking the molecular motions and dynamics into consideration. In effect, the packing constraints that depend on the relationship of the polar and nonpolar regions of an amphiphile determine the shape, size and dispersity of the particles formed in aqueous dispersions.

As a first approximation, the free energy of transfer of a hydrocarbon from water to a hydrocarbon solvent is taken as the strength of the hydrophobic effect. The ΔG values for the transfer of a solute are obtained from the solubility of hydrocarbons in water or from the distribution coefficient of hydrocarbons between water and a water-immiscible organic solvent. The incremental free energy gain ($\delta \Delta G = \delta \Delta H - T \delta \Delta S$) for transferring a methylene residue in homologous hydrocarbons, alkanols or fatty acids from water to bulk hydrocarbon solvents, i.e. from water to a medium of low polarity, is about -0.8 to -0.6 kcal/mole. This drive amounts to a 10-fold change in the underlying property from a change of two methylene residues in the solute.

The free energy change for the transfer of a methylene residue away from water is largely due to an increase in entropy ($\delta \Delta S$) resulting from the release of the ordered water molecule surrounding the nonpolar solute. The ordered water forms a cage or clathrate structure (Figure 1.5) with distorted hydrogen bonds around a polymethylene chain that results in the loss of configurational entropy of water molecules. The enthalpic ($\delta \Delta H$) contribution is small because the number of water–water hydrogen bonds are not necessarily decreased to form the cage. The change in the van der Waals and dispersion forces due to the inter-chain interactions within the organized interfaces is also negligible. These results suggest that the entropy effect associated with the disruption and ordering of the water–water hydrogen bonds around the hydrocarbon chain dominates the distortion effect of the methylene residues on the surrounding water.

1.4   Knotty Issues of Interfacial Enzymology

Interfacial and matrix enzymes mediate pathways for the synthesis, degradation and tailoring of nonpolar and amphiphilic solutes. The better characterized enzymes of lipid metabolism (Table 1.1) include triglyceride hydrolases (lipases), phospholipid hydrolases (phospholipases), acylglycerol kinases, acylases and acyltransferases. Relatively few inherited metabolic defects of lipolytic enzymes are known, although certain disorders of the central nervous system are known to be associated with the enzymes of sphingolipid and glycolipid metabolism.

Interfacial enzymes are also implicated in the synthesis of potent regulatory and signal molecules (Table 1.2), such as steroidal hormones and eicosanoids. A wide range of metabolic and control processes occur at membrane interfaces and many of these trigger transmembrane signals transduced across membranes by other regulatory metabolic, biosynthetic and transport events. Some of the signal transduction events require exceedingly selective modification of certain membrane proteins by myristoylation, prenylation, geranylation or attachment of phosphatidyl inositol. Such modifications, as well as hydrolysis, glycosidation, phosphorylation and dephosphorylation of membrane components, would certainly require membrane-associated enzymes. It appears that such fine-tuned post-translational modifications at the interface are critical for the control of cell cycle and differentiation.

Table 1.1   Membrane lipid substrates for the interfacial enzymes

Substrate	Enzyme	Product
Phospholipid	Phospholipase A_2	1-Acyl-lysophospholipid + fatty acid
Phospholipid	Phospholipase C	Diglyceride + phosphoryl base
Phospholipid	Phospholipase D	Phosphatidic acid + base
Phospholipid	Acyltransferase	Glycerides and cholesterol esters
Phosphatidyl inositol (PI)	PI-PLC	Diacylglycerol + inositolphosphate(s)
Phosphatidyl inositol	PI-kinases	Phospho-PI
Sphingomyelin	Sphingomyelinase	Ceramide + phosphocholine
Ceramide	Cermaidase	Sphingosine + fatty acid
Gangliosides	Neuraminidase	Ganglioside + neuraminic acid
Ganglioside	Galactosidase	Ganglioside + galactose
Ganglioside	Sialidase	Ganglioside + sialic acid
Ganglioside	Galactosyltransferase	Ganglioside
Lysophospholipid	Lysophospholipase	Fatty acid + glycerolphosphoester
Lysophospholipid	Acylase	Phospholipids
Glycerides	Lipases	Fatty acid + glyceride or glycerol
Glycerides	Diglyceride kinase	Phosphatidic acid
Glycerides	Acyltransferase	Glycerides, cholesterol esters
Cholesterol esters	Cholesterol esterase	Cholesterol
Cholesterol	Cholesterol oxidase	Cholestanone
Protein factors	Kinases, acylase	GPI-anchor, G-proteins

Table 1.2   Regulatory products from phospholipids

Product	Precursor for	Function/response
Fatty acids	Eicosanoids	Chemotaxis, inflammation
Lysophospholipid	Platelet-activating factor	Inflammation
Diacylglycerol	Phosphatidic acid	Protein-kinase activation
Inositolphosphate(s)	PI-polyphosphate	Calcium mobilization
Phosphatidic acid	Phospholipid synthesis	Cell division
Ceramide	Sphingomyelin, gangliosides	Apoptosis

1.4.1   Problems of accessibility of lipid substrates

The importance of lipid metabolism is not in dispute. Even the tabloid press addresses the health concerns from fat and cholesterol levels. Obesity is a disorder of triglyceride metabolism made worse by the dietary intake of animal products containing 35 to 80 % triglyceride. Such foods also provide virtually all the dietary cholesterol, and the balance of the cholesterol level is typically controlled in equal parts through *de novo* synthesis and excretion. The lipase inhibitors are being explored as potential drug targets as the ever-increasing caloric intake coupled with a sedentary life style has become a health problem of lipid malnutrition afflicting over 30 % of the US population.

For an appreciation of the role of the lipolytic enzymes in the gastrointestinal tract, consider the fact that phospholipids provide less than 5 % of the dietary fatty acids and the balance comes from the triglycerides hydrolyzed by lipases. Gastric

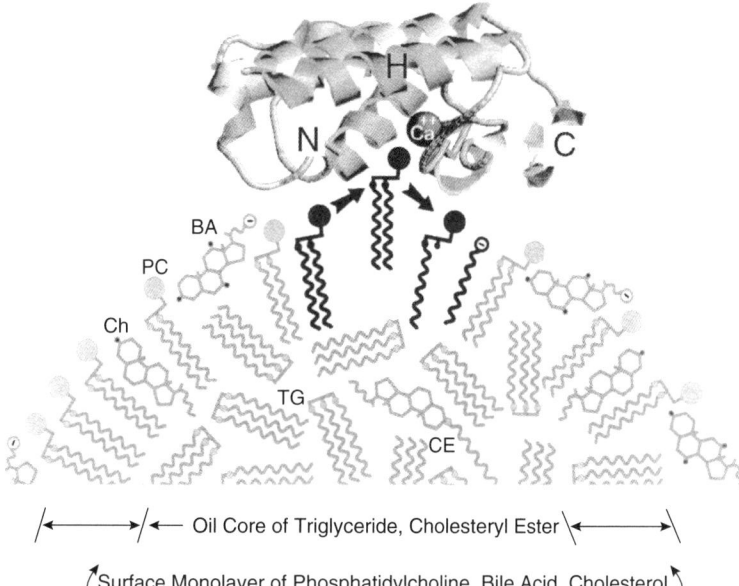

Figure 1.7 A cartoon of pancreatic PLA2 molecule on an emulsion particle of dietary lipids with the composition that predominates in the gastrointestinal tract. The protein structure shows the N and C termini, calcium binding site and the catalytic His-48 residue (H). The contact of the protein with the interface is along the i-face, the putative interface binding site or region. Note that the active site accesses the substrate from the interface along the i-face. TG, triglycerides; CE, cholesterol esters; BA, bile acids; PC, phosphatidyl choline; Ch, cholesterol. Reprinted from Homan and Jain (2000). Reproduced by permission of Kluwer Academic/Plenum Publishers

and milk lipases act on glycerides emulsified with bile salts and phospholipids (Figure 1.7). These multicomponent emulsions are also sites of action of a variety of digestive enzymes. Since the properties of the interface are also affected by other enzymes acting on the emulsion particles, there is always the possibility of cross-talk between the actions of the various enzymes acting on the same interface. Similar considerations for the accessibility of water-insoluble dispersed substrates come into play whether one considers use of lipases for the removal of grease stains from laundry, or for the control of the absorption of fat in adults, or for formulation of a lipid-based dietary supplement for infants. In all such cases, the size, dispersity and stability of the substrate interface available to the enzyme determine the fraction of the enzyme at the interface. Of course, additional variables control the events of the turnover cycle at the interface.

1.4.2 Understanding the variables

Operationally, as shown in Figure 1.8, the enzyme in the aqueous phase accesses the substrate as a two-step process: the E to E* step for the binding of the enzyme to the interface followed by the formation of the E*S complex at the interface for the catalytic turnover cycle. Two independent variables control the two independent steps. The *concentration of the interface* determines the fraction of the enzyme that

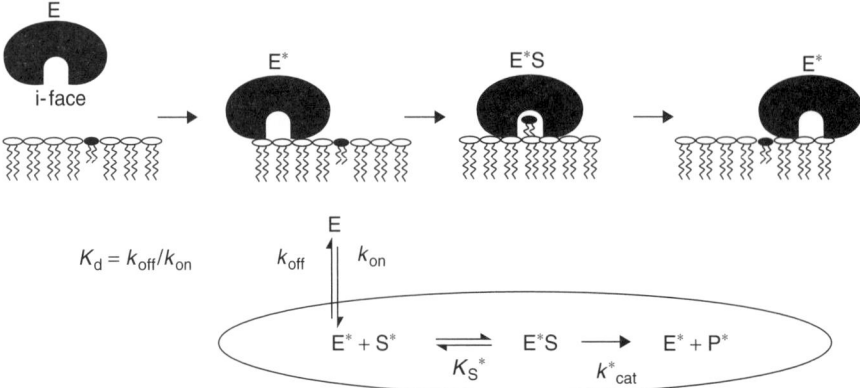

Figure 1.8 The catalytic turnover cycle of interfacial enzymes is mediated by the species marked with an asterisk (shown in the oval). According to this interfacial kinetic paradigm, if the enzyme is present in the aqueous phase (E), its binding to the interface precedes the binding of the substrate to the active site of E*. Depending on the implicit variables, numerous kinetic paths for the steady state kinetic analysis are possible within this paradigm. For example, the processivity of the interfacial turnover cycles changes with the residence time of E*. Similarly, the rate of substrate replenishment on the enzyme-containing particles determines the substrate concentration that the bound enzyme 'sees' at the interface

is bound to the interface, i.e. a larger accessible interface area from the aqueous phase ensures a higher fraction of the total enzyme at the interface. In addition, one must also consider the microscopic environment of the bound enzyme where catalytic turnover occurs, i.e. the fraction of the bound enzyme in the E*S form as controlled by the *substrate concentration* in the interface that E* 'sees'. Together, these two independent variables are explicitly considered in Figure 1.8. As we will see in subsequent chapters, analyses of the kinetic path explicitly defined by these two variables are possible only under certain conditions where other variables implicit in Figure 1.8 are suitably constrained.

1.4.3   The implicit variables in a kinetic scheme

The description and interpretation of the functional behavior of a protein in a microscopically heterogeneous environment requires an understanding of the microscopic relationship of the protein with the ensemble of the particles as well as with the biophysical properties of their interfaces. Both in the cellular milieu, as well as for *in vitro* assays, the functional analyses require identification of the kinetic path of the catalytic turnover events by delineating the kinetic contributions of parallel processes. The equilibrium and exchange processes implicit in the scheme in Figure 1.8 include partitioning, binding, exchange and transfer of substrate and products between the coexisting interfaces. Without an appreciation of their steady state kinetic contribution it is not possible to define the kinetic path for the interpretation of the kinetic behavior of interfacial enzymes.

The properties of interfaces determine the microenvironment of membrane-associated enzymes by controlling the underlying variables. The quantitative consequences of such effects constitute the main theme of this book, but a qualitative appreciation can be summarized at this stage. For the application of the paradigm in Figure 1.8, the fraction of the total enzyme bound to the interface is proportional to the *accessible surface area*, whereas the number density of the substrate molecules at the interface controls the substrate concentration dependence of the interfacial turnover cycle. In addition, the residence time of the bound enzyme determines the *processivity* of the successive turnover cycles. Also depending on the intrinsic catalytic rate, the *substrate replenishment rate* in the microenvironment of the bound enzyme contributes to the observed rate.

1.4.4 Substrate accessibility and other variables in an ensemble of dispersed particles

For an appreciation of the complexities encountered during the steady state reaction progress in the real world, consider an ensemble of the multicomponent emulsion particles. Such particles (Figure 1.7) provide the substrate interface for the action of digestive hydrolases including the lipases. As the major amphiphilic component of these particles, phospholipids play a critical role in the dispersal of dietary glycerides, sterols and other hydrophobic substrates. The polar regions of phospholipids and bile salts make up the interface between the aqueous phase, and the microdroplets of nonpolar glycerides and cholesterol esters form the core of the emulsion particle. Based on simple geometrical considerations, the effective surface area in such emulsions depends on the ratio of the polar amphiphiles at the interface to the nonpolar solutes that form a microdroplet as the core of the particle. Degradation and uptake of lipidic components in an emulsion also influence the organization and exchange of the components between emulsion particles. The proportion of the components at the interface will change not only the lateral number density of each of the components but also the number, size and the effective exchange rate between the particles that ultimately control the substrate accessibility. In addition, the accumulated products of reaction at the interface could influence the interfacial rate and equilibria of other enzymes at the interface. Possibly for such indirect reasons, inhibition of pancreatic PLA2 activity lowers cholesterol uptake.

1.4.5 Concerns for the interfacial variables in the cellular environment

Having outlined the interfacial kinetic paradigm and some of the underlying variables, it is worth reflecting on issues of metabolic fluxes of lipidic substrates within the confines of the coexisting interfaces in a cell (Figure 1.1). Such concerns are described by generic terms such as 'trafficking', 'sorting' and 'tailoring' of the membrane components between the cellular compartments. At best such terms refer to the phenomenology without providing insights into the underlying equilibria and exchange processes. The variables underlying the phenomenology are fundamentally the same as outlined above, i.e. the distribution and replenishment of the substrate and product between the coexisting interfaces in an ensemble. Such problems of distribution and exchange at interfaces are adequately understood and quantitatively described in isolated systems. For evaluating the function of an enzyme in a cellular environment one is concerned not only with the total product concentration but also with the integrity of the interfaces where the products are distributed. Considering the difficulties intrinsic in quantifying the heterogeneity in the cellular environment, it is certainly not trivial to interpret and identify the origins of the metabolic pools of the substrate, let alone predict consequences of transcriptional over-expression of an interfacial enzyme in intact cells.

1.5 Pathophysiology of Lipolytic Enzymes

Several phospholipid hydrolyzing enzymes are bacterial virulence factors (Table 1.3). As a part of the broader defense mechanism, other enzymes trigger the pathogenesis of inflammation, allergenic and immune response. At least some, if not most, of these effects are attributed to the products of catalytic activity and the downstream metabolites. For example, phospholipase A_2 activity is implicated as the rate-limiting enzyme in the control of the eicosanoid pathophysiology. Many of these enzymes are expressed in response to proinflammatory agonists with high tissue selectivity. The interfacial kinetic paradigm has evolved from the detailed characterization of the 14 to 18 kDa secreted PLA2s (sPLA2). Since our focus is

Table 1.3 Toxic and allergenic effects of interfacial enzymes

Enzyme	Effect
Phospholipase A2	Allergen, toxic, antibacterial
Phospholipase C	α-toxin (Gram +)
Phospholipase C (PI specific)	Virulence factor (Gram +)
Sphingomyelinase	β-hemolysin (Gram +, sea anemone)
Cholesterol oxidase	Insecticide

Table 1.4 Members of the calcium-dependent secreted (sPLA2) PLA2 family

Group	MW (kDa)	S-S number	Source	Chromosome number[b]
IA	13–14	7	Cobra and krait venom	
IB	14	7	Pancreas	12
IIA	14–15	7	Tears, synovial exudates	1
IIB	14	6	Gabon viper venom	1
IIC	15	8	Mouse and rat testes	1
IID	14	7	Mouse and human tissues	1
IIE	14	7	Mouse and human tissues	1
IIF	17	7	Mouse and human tissues	1
III	18	5	Bee and lizard venoms	22
V	14	6	Mouse and human tissues	1
IX	18[a]	?	Cone snails	
X	14	8	Mouse and human tissues	16
XI			Plants	
XII	14	7	Mouse and human tissues	

[a]It is a disulfide-linked hetero-dimer.
[b]In humans.

also on sPLA2, we close this chapter by introducing the basic properties of the members of the sPLA2 family (Table 1.4).

1.5.1 Arachidonate from phospholipids

Phospholipase A_2 (PLA2) hydrolyzes the *sn*-2-ester bond in 1,2-diacyl- or 1-alkyl-2-acyl-glycero-*sn*-3-phospholipids (Figure 1.9). Both the products, lysophospholipid and fatty acid, directly or through downstream metabolites, are known to induce a range of regulatory responses in cells and tissues. For example, 1-alkyl-2-acetyl-glycero-*sn*-3-phosphocholine, known as the platelet activating factor (PAF), mediates a variety of cellular changes. Similarly, polyunsaturated fatty acids are the metabolic precursors for exceedingly potent, selective and often short-lived regulators (Smith, 1989). Virtually the entire cellular pool of polyunsaturated fatty acids, especially the arachidonate, is found as the *sn*-2-acyl substituent in the membrane phospholipids. The release of arachidonate, the C_{20} fatty acid with four double bonds in positions 5, 8, 11 and 14, is often the rate-limiting step for the pathways (Figure 1.9) leading to major groups of eicosanoids (prostaglandin, thromboxane, leukotrienes and lipoxins). The 85 kDa cytoplasmic phospholipase A_2 appears to control such arachidonate release (Leslie, 1997). In addition to their well-known effects on smooth muscle functions, eicosanoids have been implicated in a wide range of processes leading to inflammation, arthritis, asthma, ischemia, toxic shock, psoriasis, pancreatitis and burn trauma. In addition, specific inhibitors of pancreatic

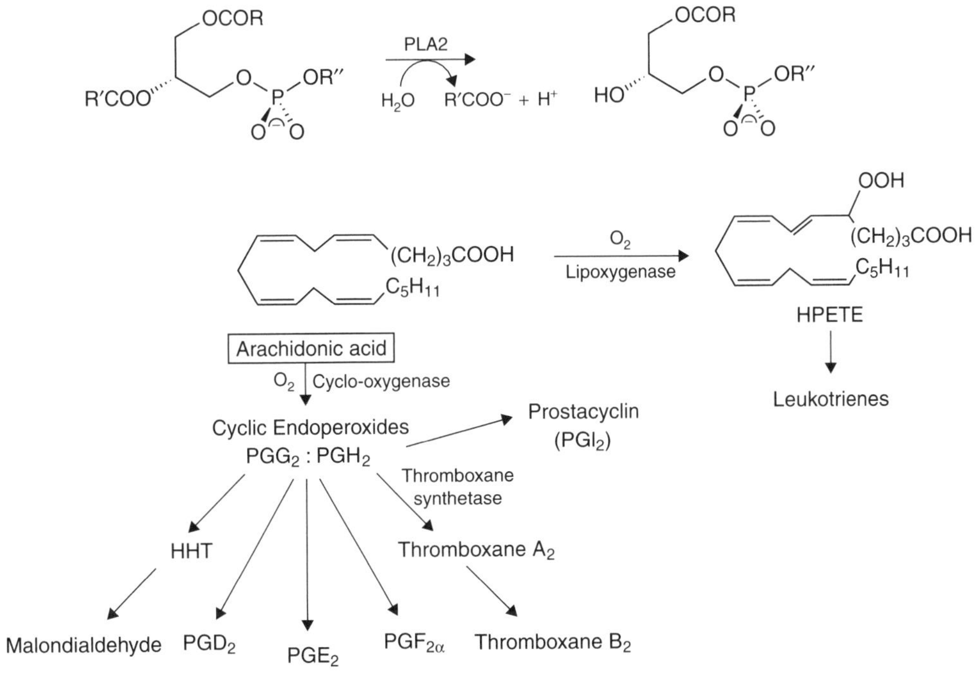

Figure 1.9 The phospholipase A2 catalyzed hydrolysis of phospholipids (top) provide the arachidonic acid precursor for the synthesis of eicosanoids (bottom)

PLA2 decrease degradation of lung surfactant, decrease uptake of dietary cholesterol, alter permeability barrier homeostasis in skin and restore esophageal muscle
tone in experimental esophagitis.

The pharmacological promise of phospholipase A_2 as drug targets remains to be
realized. Until recently the major difficulty in the characterization of the interfacial
enzymes included setting up suitable assays and screens for inhibitors without false
positives. In the recent years, with the evidence for multiple proteins with similar
activities in the same tissue, selection of the target enzyme has become a challenge.
For example, in addition to the sPLA2s, which do not exhibit significant substrate
specificity for the structure of the acyl chain, there are several serine hydrolases that
also cleave the *sn*-2-acyl chain of phospholipids with a pronounced specificity. These
include type IV (87 kDa) and type VI (17 kDa) PLA2, acidic calcium-independent
aiPLA2 with a peroxidase activity (28 kDa), PAF hydrolase with a preference for
truncated *sn*-2-chain (45 kDa) and numerous bacterial phospholipid hydrolases and
acyltransferases.

1.6 Secreted PLA2 (sPLA2) Prototype for Interfacial Enzymology

Pioneering studies of DeHaas and coworkers with pancreatic PLA2 set the
phenomenological foundation of interfacial enzymology (Verger and de Haas,
1976; Verheij *et al.*, 1981, 1995). The intellectual drive for the study of interfacial
enzymes has come from the challenge of understanding the effect of the 'quality
of the interface' on the observed kinetics (de Haas *et al.*, 1971; Dennis, 1973; Jain
and Cordes, 1973a, 1973b). As a result, sPLA2s have emerged as the prototype
for interfacial enzymology in general, for much the same reasons as trypsin and
lysozyme have come to represent the solution enzymology. In addition to their
availability from venom and pancreas, sPLA2s are small, abundant, stable, water-
soluble proteins that are purified with relative ease. Although limitations of the
assays used in the earlier studies have been a source of controversy and debate
(Jain and Berg, 1989), recent progress in the design of suitable assays and protocols
for the interfacial kinetic analysis (Gelb *et al.*, 1995; Jain *et al.*, 1986a-d, 1995; Berg
et al., 2001) has bridged much of the conceptual gap between the soluble versus
interfacial enzymes.

1.6.1 Molecular characteristics of the sPLA2 family

sPLA2s appear to be ubiquitous in virtually all higher organisms (Valentin and
Lambeau, 2000). For example, 10 sPLA2s have been cloned from the human gene
library (Table 1.4). As a class, the evolutionarily divergent sPLA2s contain about
120 to 170 amino acid residues in a single peptide chain (except IX) with 20 to 50 %
sequence identity. They have His–Asp as the catalytic residues and require calcium
for catalytic activity. The Ca–His–Asp architecture is retained through a rigid
three-dimensional structure stabilized by partially overlapping five to eight disulfide bridges. sPLA2s differ in patterns of chain deletions in the 60 to 70 loop region

and the insertions at the C-terminus. Crystal structures of scores of sPLA2s in the Protein Data Bank reveal that the gross features of the active site architecture do not vary significantly, although there are indications of marked flexibility in certain segments (Thunnissen *et al.*, 1990; Scott and Sigler, 1994; Scott *et al.*, 1994; Verheij, 1995; Yuan and Tsai, 1999; Berg *et al.*, 2001).

1.6.2 The catalytic site versus the interface binding region (i-face) of sPLA2

Operationally, two structural features of PLA2 are of functional interest in the context of the paradigm of Figure 1.8:

(a) The catalytic active site with His–Asp–Ca mediates the catalytic event in the active site of E*S, and the substrate specificity for the interfacial catalytic turnover cycle is the property of the active site. A conserved His–Asp pair and a calcium ion bound through a loop are the key features of the catalytic site. Although not required for the binding of the enzyme to the interface, calcium bound through a separate loop is an obligatory cofactor for the binding of the substrate to the active site of the enzyme and also for the chemical step.

(b) The i-face, interface binding region, determines the interface preference for the E to E* equilibrium. A remarkably flat surface of 1700 Å^2 has been assigned as the i-face of pancreatic PLA2 (Ramirez and Jain, 1991; Pan *et al.*, 2001). The active site is accessible from a hydrophobic slot in the middle of the i-face. The hydrophobic and hydrogen-bonding residues with five distinct anion binding sites are also a part of the i-face.

The functional significance of such features is developed in the subsequent chapters. Analytical methods and site-directed mutagenesis results support the structural and functional distinction between the i-face and the active site. Note that although the i-face and the catalytic active site are structurally and functionally distinct, such a dissection does not necessarily rule out a functional coupling between the interface binding and the catalytic cycle events.

1.7 Summary and Outlook: Towards the Paradigm for Interfacial Enzymology

About half of the cellular protein is membrane-associated to mediate metabolism of a large variety of hydrophobic nonpolar and amphiphilic solutes that participate in virtually all cellular functions. Thus interfacial enzymes have evolved to deal with the biophysical realities of the environment where their substrates are to be found and accessible. The interfacial enzymology paradigm has come of age in recent years with sPLA2 as the prototype (Verheij *et al.*, 1981; Berg *et al.*, 2001). However, not even a cursory coverage is accorded to the interfacial enzymes in standard biochemistry textbooks. As introduced in this chapter, the challenge of functional characterization of interfacial enzymes lies in evaluating the variables for the ensemble environment of the steady state catalytic turnover environment.

A kinetic path for the catalytic turnover must be unequivocally identified for evaluating the observed catalytic behavior in terms of the primary kinetic parameters for the elemental events with structural significance. As a condition for ensemble averaging of all the events on the dispersed particles, the path must explicitly consider all the rate and equilibrium processes that influence the substrate replenishment in the microenvironment of each and every enzyme in the ensemble during the course of the reaction progress. Towards realizing this goal, in the following chapters we build further on the two-step paradigm (Figure 1.8) that explicitly distinguishes the binding of the enzyme to the interface mediated by the i-face from the turnover cycle events mediated by the catalytic active site. The catalytic turnover occurs through a productive enzyme–substrate Michaelis complex at the interface, and the turnover cycle is formulated by including all the distinguishable steps of the kinetic path. The explicit dissection of the binding and turnover events makes it possible to take into account the interfacial processivity of the turnover cycles, which in conjunction with a defined kinetic path is absolutely necessary for determination of the interfacial catalytic turnover parameters.

Virtually all the evolutionary experience built into a gene product is expressed through the primary rate and equilibrium parameters. The characteristic constants and parameters for each of the elementary steps relate the observed kinetic behavior to the underlying variables, including the concentration, composition and the structural changes in the enzyme, substrate, cofactor and inhibitor. Analytical relations grounded in a well-defined reaction path permit insights into the structure function correlation for inhibitor or substrate specificity, the catalytic efficiency and the catalytic and allosteric mechanism. Such information is ultimately integrated with site-directed mutagenesis studies to surmise the cellular functions of a gene product. This conceptual strategy continues to guide all aspects of enzymology, and it is also beginning to provide insights into the unique features of interfacial enzymes.

As a prelude to the discussion in the following chapters, it can hardly be overemphasized that there are several distinguishing features of the interfacial enzymology paradigm. For example, the way variables come into play is not obvious from the perspective of standard solution enzymology. The concentration variable in homogeneous solutions is the number density per volume of the solution as in moles per liter. In a somewhat modified form this description is relevant for the binding of the enzyme to the interface, where the variable for the E to E^* step is the area of the interface. It often changes linearly with increasing concentration of the dispersed amphiphile. However, for the formation of the E^*S complex one has to invoke the number density of the substrate per unit area of the enzyme-containing interface. In addition to defining the variables in the microenvironment where the turnover events occur, it is particularly important to define what the enzyme 'sees' during the course of the observed reaction progress, which is interpreted as the ensemble average of the sum total of all the observed turnover events over a period of time. Thus in addition to the fraction of the total enzyme at the interface and the number density of the substrate that the bound enzyme 'sees', one must consider if and how the underlying variables change with time during the reaction progress.

Detailed interfacial kinetic analysis developed in the later chapters is useful for establishing a rigorous quantitative basis of virtually all functions of an enzyme.

Some of the basic concepts leading to an appreciation of the interplay of the interfacial and enzymatic processes are developed in the next two chapters. Besides the active site events and the catalytic mechanism, such considerations include the interfacial processivity of catalytic turnover cycles and the allosteric effects of the interface. Thus an understanding of the structural features of the enzyme, coupled with the organization and exchange properties of the substrate interface, is critical for evaluating the interplay of the events of interfacial catalytic turnover.

Finally, the primary functional parameters for an unequivocally defined kinetic path provide a rigorous quantitative basis for structure function analyses. When supported by suitable experimental controls, such fundamental information provides a quantitative basis for evaluating the behavior of an enzyme in the biological environment and for learning from the evolutionary rationale for the family of enzymes. sPLA2s continue to guide on a broad range of such objectives. Although the catalytic specificity of sPLA2s is similar, their interface preference, their ability to bind to cellular and extracellular receptors and their noncatalytic effects (antigenicity and toxicity) have made this small molecule the subject of intensive scrutiny.

Further Reading

For a critical review of the evidence (with bibliography) leading to interfacial kinetic paradigm see Berg *et al.*, 2001.

Berg, O.G., Tsai, M.D., Gelb, M.H. and Jain, M.K. (2001). Interfacial enzymology: the secreted phospholipase A_2 paradigm. *Chem. Rev.,* **101**.

Carey, M.C., Small, D.M. and Bliss, C.M. (1983). Lipid digestion and absorption. *Ann. Rev. Physiol.,* **45**, 651–77.

Gelb, M.H., Jain, M.K., Hanel, A.M. and Berg, O.G. (1995). Interfacial enzymology of glycerolipid hydrolases: lessons from phospholipase A_2. *Ann. Rev. Biochem.,* **64**, 653–688.

Homan, R. and Krause, B.R. (1997). Established and emerging strategies for inhibition of cholesterol absorption. *Curr. Pharm. Des.,* **3**, 29–44.

Jain, M.K. (1988). *Introduction to Biomembranes*, II edition, Wiley, New York, p. 423.

Smith, W.L. (1989). The eicosanoids and their biochemical mechanism of action. *Biochem. J.,* **259**, 315–24.

Tanford, C. (1980). *The Hydrophobic Effect: Formation of Micelles and Biological Membranes*, 2nd edition, Wiley, New York, 233 pp.

Verheij, H.M., Slotboom, A.J. and de Haas, G.H. (1981). Structure and function of phospholipase A2. *Rev. Physiol. Biochem. Pharmacol.,* **91**, 91–203.

Waite, M. (1987). *The Phospholipases*, Plenum, New York.

Woolley, P. and Petersen, S.B. (Eds.) (1994). *Lipases: Their Structure, Biochemistry and Application*, Cambridge University Press, Cambridge, p. 363.

2 Interface Phenomena: Accessibility and Exchange

The reason for this journey has not yet been revealed to me and may never really fully be revealed. As I continue I am transformed. I no longer feel the individual pounding of each heart-beat, the breathlessness. But rather I am in a trance like state – in a state of wonderment and amazement. Wondering how I will ever gain the strength to move my leg the next step. Wondering if I stop, will I ever be able to begin again. Wondering how will I find the mental energy to encourage me along. Amazed that indeed my leg extends forward to complete each step. Amazed that indeed after stopping I can begin again. Amazed that indeed there is an endless store of mental strength that seeps into my subconscious to propel me forward. So I continue on, ever so slowly, slower than I ever imagined possible.

Notes of **Nicole Reilly,** *On Trekking Himalayas*

Aggregates of amphiphile molecules dispersed in aqueous phase are of interest for a variety of reasons, including the need for suitable interfaces for functional studies with membrane-associated enzymes. As is apparent from Table 2.1, dispersed dissimilar phases make up objects of everyday life, such as ice-cream, soaps, aerosols, paints, smoke and smog. Dispersions of dissimilar phases in the biological milieu include milk, lung surfactant, serum lipoproteins, fat deposits of adipose tissues, lipidic granules, lamellar bodies, secretions and digestive juices. The key feature of all such dispersions is that solutes of opposing polarity are accommodated as a suspension of small discrete particles in an otherwise incompatible bulk phase. The molecular basis for the stability of the dispersed particles in a medium of opposite polarity lies in the organization of amphiphilic solutes at the interface to prevent particles from coalescing into larger particles or a separate phase. The rates of exchange of amphiphiles between the aggregate and the aqueous phase span over 10 orders of magnitudes with equilibration times of submicroseconds to weeks. Irrespective of such kinetic differences, we treat all dispersions as ensembles of discrete particles made up of a phase distinct from the surrounding aqueous solution of the monodisperse amphiphile.

Aggregated amphiphile concentration at the heart of the interface phenomena

Dispersed phases represent the underlying physics of insoluble and sparingly soluble solutes in a bulk phase. Beyond a certain total concentration, amphiphile

Table 2.1 Examples of the dispersed and organized phases

Dispersion	Phase	Products
Aerosol	Solid in gas	Smoke, smog, mist, graphite lubricant
	Liquid in gas	Mist, fog, sprays, teargas
Foam	Gas in liquid	Suds, lung surfactant, whipped cream
	Gas in solid	Styrofoam, cake
Emulsion	Liquid in liquid	Milk, digestive fluids, Napalm, bath oil
	Liquid in solid	Ice-cream, tar, butter
	Solid in liquid	Paints, ink, glue, floor wax
Sol	Solid in solid	Glass, ceramic
Monolayer	Air/water	Langmuir film
	Liquid/liquid	Scum
	Water/solid	Langmuir–Blodgett film
Micelle	Water	Soap and detergent solutions
Reverse micelle	Bulk organic phase	Dry-cleaning fluid
Vesicle	Bilayer	Biomembrane
Liquid crystal	Anisotropic phase	Electronic display

molecules dispersed in aqueous phase aggregate into particles with organized interfaces. The primary variable that drives the organization and controls the exchange dynamics of amphiphilic solutes at the interface is the hydrophobic effect due to disruption of the organization in liquid water. Such incompatibility drives the nonpolar groups and molecules away from the aqueous phase. To further minimize the unfavorable contacts with liquid water, the area of the interface between the dissimilar phases is also minimized. Such factors favor separation of bulk phases as oil–water mixtures. However, under certain conditions the separated phases can be dispersed into small particles. In addition, the packing constraints within an aggregate depend on the geometrical factors of the solute and the amphiphiles present in the particles. As developed in this chapter, the balance of such energetic and packing constraints ultimately determine the equilibrium and exchange properties of the species present in the ensemble.

The total amphiphile concentration

For an appreciation of the monomer-aggregate dynamics in a dispersion consider the significance of the total dispersed amphiphile concentration, $[A_T]$ (or $[S_T]$ if A is the substrate), expressed as moles of amphiphile per liter of the aqueous phase. In the context of the interfacial kinetic paradigm in Figure 1.8, substrate is the amphiphile at the interface. In such situations certainly $[S_T]$ is not a measure of the concentration that is accessible to the active site of the bound enzyme. On the other hand, for the E to E* equilibrium the primary variable is the concentration of the interface accessible to the enzyme in the aqueous phase. The interface concentration of interest here is related to but is not necessarily equal to $[A_T]$. Obviously, one must subtract the amount of the amphiphile that remains monodispersed in the aqueous phase and also discount for the fraction of the amphiphile that is not in direct contact with the aqueous phase containing E. At this stage the substrate, S, and other active site-directed amphiphiles are conceptually distinguished from those that do not bind to the active site. The binding equilibrium for the interface is an apparent value with contributions from E to E* and the E* + S* to E*S equilibrium.

Different consequences of the commonly used total concentration variable, $[S_T]$, in the individual steps emphasize the need for understanding and evaluating the microenvironment where such events occur. Thus, in agreement with the kinetic paradigm in Figure 1.8, it is necessary to distinguish the interface-forming amphiphiles on the basis of their affinity for the active site of the bound enzyme. As a prelude to the functional analysis of membrane-associated enzymes in general, and of interfacial enzymes in particular, it is also necessary to evaluate fully the organization and the exchange properties of the substrate interface. Such factors ultimately define the steady state condition and local variables for the catalytic turnover in the microscopically heterogeneous environment.

2.1 Aggregates and Dispersions

Nonpolar and amphiphilic solutes with low solubility in water are ubiquitous in the biological milieu. Above the solubility limit, *excess* solute readily separates from water to overcome the hydrophobic effect. Such separations lead to precipitation and de-mixing of bulk phases, or to emulsification as liquid or crystals of solid as microdispersed phases (Figure 2.1), or to the formation of organized interfaces (Figure 2.2), or to partitioning of the solute into a hydrophobic phase in contact with the aqueous phase, the S to S* equilibrium in Figure 1.2.

2.1.1 Stability of dispersions

The properties of stable organized aggregates formed spontaneously in aqueous dispersions of amphiphiles are often distinctly different from those of a dispersed insoluble phase that is stabilized by suitable additives. Typically, the organized

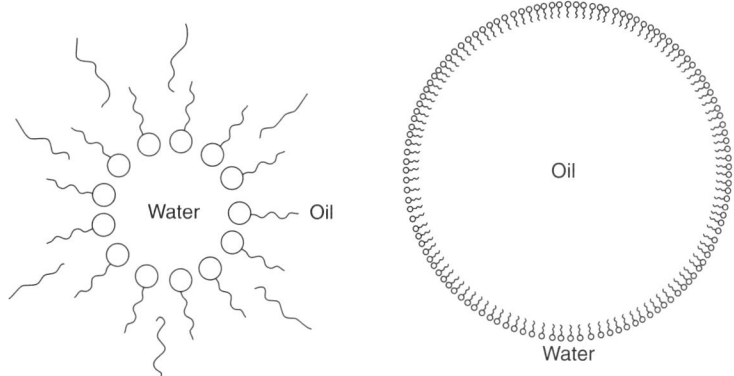

Figure 2.1 Microdispersed unorganized bulk phases of dissimilar polarity separated by organized interfaces into (left) water-in-oil or (right) oil-in-water emulsion droplet particles. Similar dispersed phases (foam or aerosol) are also formed with nonpolar air trapped between interfaces of organized amphiphiles dispersed in the aqueous phase

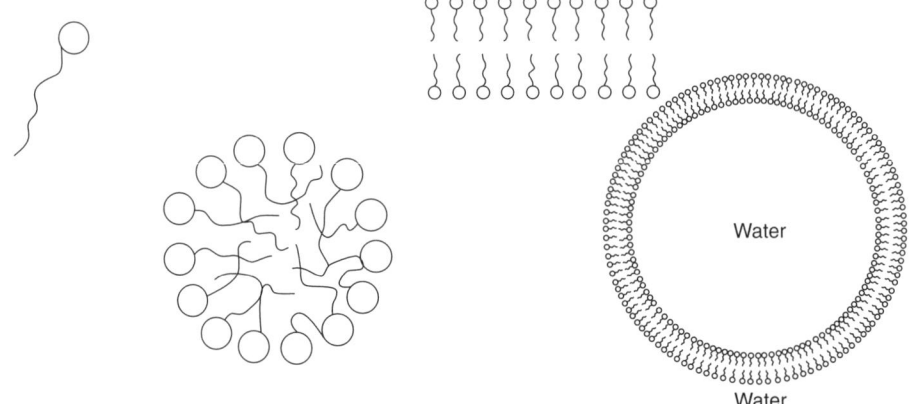

Figure 2.2 (From left) Monodispersed amphiphile monomer, aggregated amphiphile molecules in the micellar or bilayer organization, and a bilayer-enclosed vesicle with separate aqueous compartments on both sides

Table 2.2 Properties of the organized aggregates of amphiphiles

Property	Bilayer	Micelle	Monolayer
Aggregate shape	Vesicular	Spherical	Planar sheet
Accessible $[A^*] =$	$[A_T]/2$	$[A_T] - \mathrm{CMC}$	
Monomer $\mathrm{DC}_n\mathrm{PC}$	$n > 12$	$n = 5\text{--}9$	$n > 8$
Shape $(H/T)^a$	Cylindrical ($= 1$)	Cone (> 1)	0.7–1.5
Residence time	hours	milliseconds	<minutes

$^a H/T$ is the ratio of the effective cross-section of the head group to that of the tail (alkyl chain). As discussed later, the residence times are related to the particle-to-particle exchange times.

interfaces are thermodynamically and kinetically stable, whereas the emulsions may be kinetically stable against the bulk phase separation by additives that reduce the tendency of the particle to coalesce into larger particles to reduce surface curvature. Emulsions and froth, containing a volume of an unorganized nonpolar phase bounded by an organized interface separating immiscible phases of different polarity, have been used as substrates for interfacial enzymes. It is nearly impossible to define the microscopic variables at such interfaces. Therefore, they are not suitable interfaces for kinetic analysis. Our focus in this book is largely on the aggregates containing only the spontaneously organized interfaces (Table 2.2). Vesicles are of particular interest because all the amphiphile molecules are organized at the bilayer interface. In addition, phospholipid amphiphiles in vesicles do not exchange on the time scale of hours or more, and therefore their properties (Figure 2.3) are more easily controlled and defined.

2.2 Organized Micellar Aggregates in Aqueous Dispersions

Based on the substrate accessibility considerations, aggregates of interest for the kinetic characterization of interfacial enzymes are bilayers, micelles and monolayers

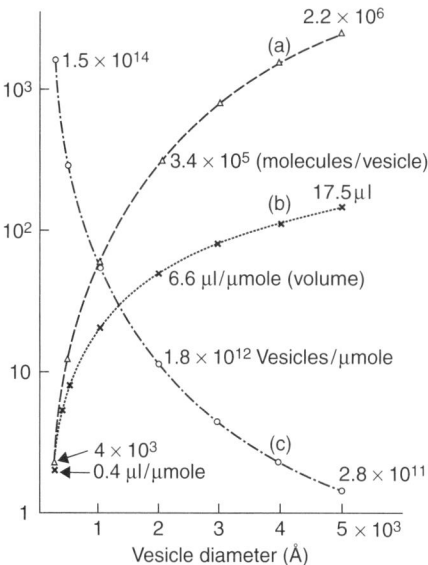

Figure 2.3 Properties (on a normalized logarithmic ordinate scale) of egg phospholipid (molecular cross-section about 70 Å^2) vesicle diameter to (a) the number of phospholipid molecules per vesicle, (b) the entrapped volume in liters per mole of phospholipid and (c) the number of vesicles per mole of phospholipid. Abbreviations in the text related to the vesicles: suv, small unilamellar vesicles of about 200–250 Å diameter are formed by extended sonication of phospholipid dispersions; luv, large unilamellar vesicles, typically with 500–5000 Å diameter, are formed by extrusion of the larger mlv dispersions under pressure through filters of well-defined pore size; mlv, multilamellar vesicles of up to 500 000 Å diameter with 7–15 concentric bilayer lamellae, each enclosing an aqueous compartment, are spontaneously formed from hydrated phospholipid films dispersed by mechanical shaking. Provided by Professor Demitri Papahadjopoulos

(Table 2.2). Only in these three types of dispersions does the interfacial organization permit evaluation of the total interface that is directly accessible in the aqueous phase for binding the enzyme. In all such aggregates the microscopic organization of the amphiphiles (Figures 2.2 and 2.4) is such that the nonpolar chains are sequestered with each other without significant unfavorable contacts with the aqueous phase or the polar parts of the amphiphile. The key difference between the micellar, monolayer and bilayer aggregates is in more than a 10^{10}-fold difference in the residence time of the amphiphile at the interface, which depends on the CMC. As a measure of the S to S* equilibrium (cf. Figure 1.2), the CMC is the concentration of the monodisperse amphiphile in the aqueous phase in equilibrium with the amphiphile in the aggregate. This exchange equilibrium determines the relationship between [S$_T$] and [S*], and its rate constants determine the residence time of the amphiphile at the interface and the exchange time between the interfaces of the ensemble of coexisting particles.

2.2.1 Onset of micelle formation

Ensembles of simple micellar aggregates, typically with less than 100 amphiphile molecules per micelle, are formed in aqueous solutions of short chain amphiphiles

Table 2.3 The CMC and aggregation number for micelle-forming amphiphiles (20 °C and pH 8)

Amphiphile	Charge	CMC(mM)	Aggregation number
Octylglucoside	Nonionic	25	27
CHAPS	Zwitterionic	8	10
Cholic acid	Anionic	13 $(11)^c$	5
Deoxycholic acid	Anionic	10 $(3)^c$	50
Dodecylsulfate	Anionic	8.1 $(1.4)^c$	58$(91)^c$
Dodecyl-N(CH$_3$)$_3$Br	Cationic	14.8	43
Triton X-100^a	Nonionic	0.24	140
Tween 80^a	Nonionic	0.012	60
Dipentanoyl-GPC^b	Zwitterionic	50	27
Dihexanoyl-GPC	Zwitterionic	15	34
Diheptanyl-GPC	Zwitterionic	1.5	40–155
Dioctanoyl-GPC	Zwitterionic	0.24	600–4500^d
Dinonanyl-GPC	Zwitterionic	0.0029	
1-Dodecanoyl-GPC	Zwitterionic	0.7	45
1-Myristoyl-GPC	Zwitterionic	0.09	60
1-Palmitoyl-GPC	Zwitterionic	0.01	110
1-Myristoyl-GPG^b	Anionic	0.16	50
1-Palmitoyl-GPG	Anionic	0.018	80
Dipalmitoyl-GPC	Zwitterionic	$<10^{-10}$ M	Infinite

^{a}Sold as polydisperse mixtures and the properties change from batch to batch.
^{b}GPC, -glycero-*sn*-3-phosphocholine; GPG, -glycero-*sn*-3-phosphoglycerol.
^{c}The value in parenthesis is in the presence of 0.1 M NaCl.
^{d}The micelle size changes with the bulk concentration and the dioctanoyl-GPC phase separates above 2 mM.

with a polar charged head group (Table 2.3). The solution behavior of an amphiphile leading to the onset of micellization is complex but informative (Mukerjee, 1967). Consider the exchange equilibrium between the monodisperse amphiphile and the two aggregated states shown in Figure 2.4. At very low amphiphile concentrations, properties of the surface of the solution exhibit characteristic monotonic changes due to the formation of a monolayer at the air–water interface (Tausk *et al.*, 1974; Bian and Roberts, 1991; also Chapter 10). Long chain amphiphiles form a virtually *insoluble monolayer*, whereas a significant concentration of monodisperse amphiphile remains in the aqueous phase in equilibrium with a *soluble monolayer*. Additional changes in the properties of the aqueous phase are seen at somewhat higher concentrations when the monolayer is saturated and additional amphiphile cannot be accommodated. Only above a critical total concentration of the amphiphile, when the excess amphiphile can neither be accommodated in the monolayer nor remain monodisperse in the aqueous phase, do the excess amphiphile molecules form micellar aggregates.

2.2.2 The CMC, critical micellization concentration

Equilibrium relationships between the three states of an amphiphile as a function of concentration is best shown by the results in Figure 2.5 for the short chain

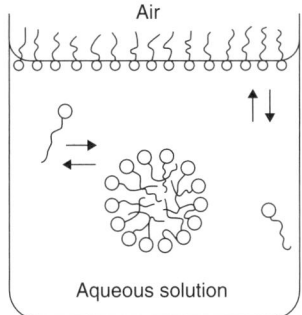

Figure 2.4 Monodisperse amphiphiles in equilibrium with the monomolecular (monolayer) layer at the air–water interface and the micelle in the bulk aqueous phase

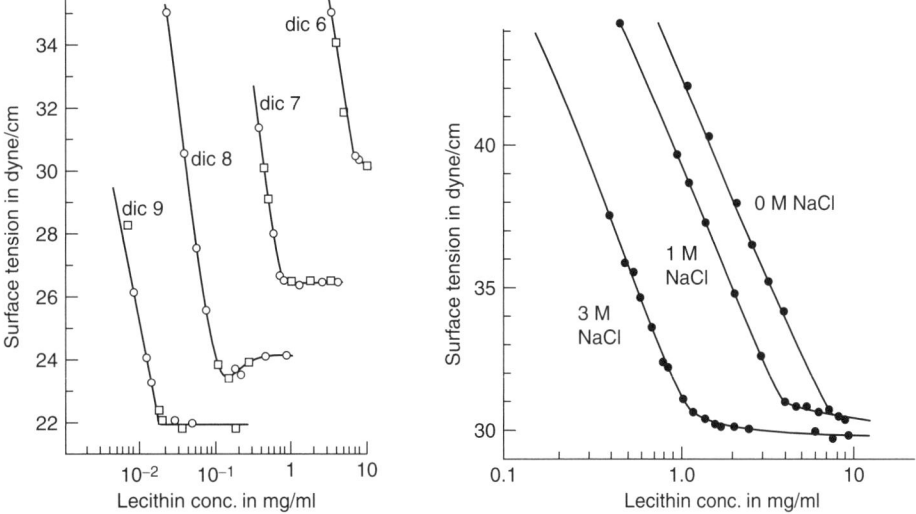

Figure 2.5 (Left) The change in the surface tension value of 10 mM phosphate buffer at pH 6.9 in water with increasing concentration of homologous short chain (6–9) diacylphosphatidylcholines. (Right) The change in the surface tension of dihexanoylphosphatidylcholine at 0, 1 and 3 M NaCl. Note that the *x* axis is in a logarithmic scale. Adapted with permission from Tausk *et al.* (1974). Copyright (1974), with permission from Elsevier Science

diacylphosphatidylcholines. Initially, the surface tension of the aqueous phase decreases as the amphiphile is added because the acyl chains of the amphiphile molecules at the air–water interface preferentially orient to make contact with air, a medium of very low polarity with a dielectric constant near zero. In this monolayer arrangement at the air–water interface, the glycerophosphocholine polar head group of the amphiphile remains in contact with the bulk aqueous phase of high polarity with a dielectric constant of 80. The chain-length dependence of the surface tension suggests that the short chain phosphatidylcholines form soluble monolayers with a significant proportion of the amphiphile remaining in the aqueous phase. As is apparent for all the curves in Figure 2.5, the surface tension of the aqueous solution decreases until the total amphiphile concentration reaches a value characteristic for each amphiphile. With the onset of micellization beyond this 'critical' concentration the surface tension remains relatively constant (Figure 2.5) as additional amphiphiles cannot be partitioned in the monolayer. Excess amphiphile molecules above the CMC form the micellar aggregates in which molecules are in equilibrium with the monodisperse amphiphile at the concentration that remains essentially at the CMC (Figure 2.6) corrected for the number of amphiphile molecules partitioned in the monolayer.

2.2.3   Significance of the CMC

The maximum concentration of an amphiphile that remains monodispersed in solution, but in equilibrium with the amount dispersed as micelles, depends on the structure of the amphiphile and the solution environment (Table 2.3). The CMC is one measure of the self-aggregation tendency of an amphiphile driven by the hydrophobic effect, i.e. minimization of unfavorable contact of the nonpolar chains with water. Above the CMC, amphiphiles spontaneously aggregate to form stable micelles of characteristic aggregation numbers, shapes and organization. In dilute solution, micelles are usually spherical with average aggregation numbers in the range of 50–100 with moderate dispersity. Micellized amphiphiles are oriented such that a hydrophobic core sequesters the nonpolar chains away from the bulk aqueous phase (Figure 2.2).

Amphiphiles in a micelle are in exchange equilibrium with those monodispersed in the aqueous phase. In short, as a first-order approximation, micellization is treated as the monomer–micellar aggregate equilibrium where [A] and [A*] change with [A$_T$] in a characteristic abrupt fashion (Figure 2.6). The monodisperse amphiphile concentration in the aqueous phase, [A], or the *interface density* of the amphiphile on a micelle, does not change with the total amphiphile concentration above the CMC. Thus the concentration of the micellar amphiphile is [A*] = [A$_T$] − CMC (Figure 2.6). Only the number of the micellar (or aggregated) particles increases with [A*], which is also proportional to the total area of the micellar interface. As a result, abrupt changes occur at the CMC in the concentration-dependent

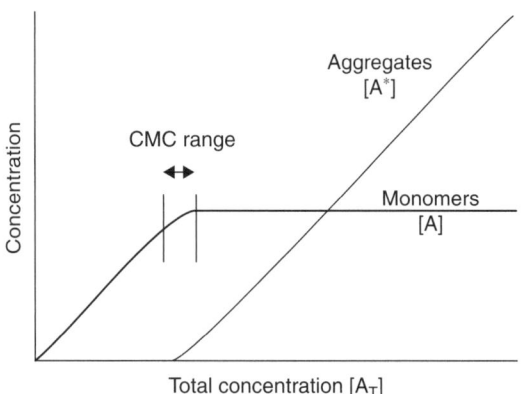

Figure 2.6 An idealized diagram of the dependence of the change in the monomer [A] and the number of micellar aggregate or the interface concentration [A*] as a function of the total amphiphile concentration in the solution [A$_T$]

properties that depend on [A] or [A*]. Such concentration-dependent anomalies result from the fact that only the monodisperse amphiphile concentration increases below the CMC and then only the number of micellar aggregates increases above the CMC. For example, the intensity of the signal from scattering, viscosity or dye binding increases above the CMC because these properties depend on the number of micelles. On the other hand, the osmotic pressure, surface tension and conductivity that depend on [A] reach a maximum at the CMC and then remain constant. Such a difference is also seen in the observed rates of soluble and matrix enzymes that sample the amphiphile monomers (Chapter 3) versus the interfacial enzymes that sample the micellar substrate (Chapter 8).

2.2.4 Nonidealities in micellization

As is apparent from the results in Figure 2.5, often the transition from the monomer- to micelle-dominated signal occurs over a finite concentration range. In addition to the problems associated with the measurements in a narrow but finite CMC region, there are other effects that also depend on the nature of the amphiphile and the environment. Below the CMC, amphiphiles in the bulk aqueous phase usually exist as monomers, although dimers and oligomers sometimes form as $[A_T]$ approaches the CMC. In most cases the aggregation number and therefore the micelle size also increases somewhat at higher amphiphile concentrations. In some cases, micelles also change shape from spherical to rod and discoidal with a concomitant increase in aggregation number at high amphiphile concentrations. Similarly, the monomer concentration for nonionic amphiphiles remains apparently constant above the CMC; however, the monomer concentration of some ionic amphiphiles actually decreases at bulk concentrations well above the CMC. Apparently, such non-idealities are related to changes in the micellar organization and dispersity, which may also have contributions from micelle–micelle interactions.

2.2.5 Structural requirements for micellization

The CMC is a measure of the balance between the hydrophobic and hydrophilic character of an amphiphile. The major drive for aggregation comes from the hydrophobic effect that sequesters the nonpolar tails away from the aqueous phase. Thus CMC is in the millimolar range for decyl derivatives and lower for the amphiphiles with longer chains (Table 2.3). The CMC of membrane phospholipids with two long alkyl chains is estimated to be in the subnanomolar range of concentration. CMC values also depend to a lesser extent on amphiphile structure as well as the solution conditions. The balance of forces underlying micellization shifts with the structure of the polar and nonpolar groups that may control the inter-amphiphile contacts in the organized interface. For example, repulsive forces of a strong hydrophilic and charged group can lead to complete solution, i.e. to a large upward shift in the CMC.

Synthetic amphiphiles have been developed for specific purposes. Such amphiphiles contain ionic (sulfate, sulfonate, phosphate, ammonium, carboxylate) or bulky oxygen-containing uncharged (amine oxide, polyoxyethylene, phosphine oxide, sugar residue) polar head groups capable of extensive hydrogen bonding and dipolar interactions with water. Note that weakly polar alkyl alcohols, -amines, -amides and unionized carboxylic acids do not micellize and they phase separate above their solubility limit. Also, molecular association through specific head group interactions results in the dimeric and oligomeric species of alcohols, amines, acids and salts in aqueous solutions.

2.2.6 Factors affecting the monomer–micelle equilibrium

The free energy (ΔG) for the monomer–micelle equilibrium is related to the enthalpy (ΔH) and entropy (ΔS) of micellization:

$$\Delta G = \Delta H - T\Delta S \tag{2.1}$$

The enthalpy of micellization is usually small because there is little change in specific bonding interactions. The negative free energy change associated with micellization is dominated by a significant increase in entropy. As discussed in Chapter 1, this results from the release of the forced ordering of the water molecules around a nonpolar group (Figure 1.5).

In addition to the dominant hydrophobic effect, the stability of micelles is also influenced by factors related to packing, head group and counterion interactions, and solution effects (Mukerjee, 1967). The A to A* equilibrium for homologous amphiphiles, as characterized by the CMC, changes with the alkyl chain length. With n methylene groups (Figure 2.5) the CMC usually follows the relationship:

$$-\Delta G/(2.3RT) = -\Delta G/1.36 = \log \text{CMC} = a - bn \tag{2.2}$$

Typically $b = 0.2$ to 0.5, which corresponds to an incremental change in the free energy of -0.35 to -0.85 kcal/mole CH_2. The value of a ranges between 1 to 3.5, depending on the structure of the head group. Electrostatic repulsion between adjacent head groups in an aggregate increases the CMC largely due to an increase in the value of a. The incremental contribution to the hydrophobic effect favoring micellization, $\delta\Delta G$, is -0.6 to -0.85 kcal/mole per methylene residue. Thus the CMC value for homologous amphiphiles decreases by a factor of 2 to 3 for each added methylene residue. From the structure of most naturally occurring phospholipids with >30 methylene residues in two acyl chains, the monodisperse concentration in the aqueous phase is estimated by extrapolation to be in the subpicomolar range.

The CMC (Figure 2.5) and the aggregation number also change in the presence of added salt. Several effects are discernible. Adding an indifferent electrolyte has a pronounced effect on the CMC of ionic amphiphiles. For zwitterionic or nonionic amphiphiles only a small decrease in the CMC comes from a change in the degree of hydration of the head group or specific ion binding. Thus salting-out of the monomer amphiphile thermodynamically favors micellization and lowers the CMC, and the converse applies for the salting-in of the monomer by certain solutes. A *salting-out effect* is observed from the enhancement of the hydrophobic effect caused by the salt concentration c_i in the aqueous medium. The salt effect is given by the empirical relation:

$$\log CMC = -k_s c_i + (\log CMC)_{c_i=0} \tag{2.3}$$

where k_s reflects the 'salting out' of a nonpolar solute. Salt effects on the surface tension of phosphatidylcholines are shown in Figure 2.5. These effects are observed at high salt concentrations. With $k_s = 0.22$ the CMC of zwitterionic diheptanoylphosphatidylcholine decreases from 1.5 mM in water to 0.7 mM in 1 M and to 0.2 mM in 4 M NaCl.

At high concentrations, the water structure makers (xylose, fructose) and breakers (urea, formamide) also have a significant effect on the CMC. The structure breakers raise the CMC, presumably due to an increase in the entropy of the bulk water, and thus reduce the hydrophobic effect. Certain organic impurities also lower the CMC.

Electrolyte counter ions affect the activity coefficients of a charged monodisperse amphiphile and its counter ions through a competition between the surfactant head group and the electrolyte. The Hoffmeister series for the salting-out effect of anions is in the order:

$$SO_4^{2-} > F^- > Cl^- > ClO_4^- > Br^- > NO_3^- > I^- > SCN^-$$

where the counterion valence has a strong effect on the CMC of ionic amphiphiles. For example, the CMC of sodium dodecylsulfate is 8 mM in the absence of added NaCl and decreases to 0.83 mM in the presence of 0.4 M NaCl. Such a common ion effect of the counter ion at a concentration c_i on the CMC of amphiphiles with a net charge is proportional to the degree of counter ion binding, α typically being in the 0.6 to 1.2 range, by the empirical relation:

$$\log CMC = a - \alpha \log c_i \tag{2.4}$$

In addition to partially screening the electrostatic repulsion between the head groups, a modest effect on the solvation and specific ion interactions may also appear at higher counter ion concentrations. Additional effects on the CMC and the solubility of the amphiphile come into play as a function of temperature. For example, raising the temperature has quite different effects on the dispersion of the ionic and nonionic amphiphiles. Below a characteristic temperature (*the Krafft point*) the solubility of ionic amphiphiles is quite low, presumably due to strong intermolecular interactions in the crystal lattice. At higher temperatures, increases in amphiphile solubility are accompanied by the formation of micelles. Nonionic amphiphiles disperse readily at low temperatures; however, above a critical temperature (the *cloud point*) large aggregates separate out into a distinct phase.

2.3　Amphiphile Organization and the Monomer–Micelle Equilibrium and Exchange

As a first approximation we treat micellar solutions as dispersion of discrete particles. Thus properties of micellar solutions are evaluated on the basis of the ensemble-averaged and time-averaged organization. Based on the conceptualization so far, micellization is a dynamic process where the organization, size, shape, aggregation number and the exchange properties of micelles depend on the temperature, concentration and the presence of other amphiphiles. In this dynamics, the effective amphiphile structure is determined by the effective size of the head group, chain length and the presence of double bonds or branching. The weak effects of the temperature and pressure on the CMC and other properties of micelles are a reflection of only modest changes in volume, solvation, bonding and heat capacity of micelles. The basis for such effects in relation to the organization of micelles is still an area of active discussion. The Hartley model of the micellar organization shown in Figure 2.2 is certainly not adequate. It is clear that the packing and organization of the acyl chains in the core of small micelles is possible only if some of the chains make contact with water at least some of the time. As the alkyl chains in a micelle are packed to minimize contact with water, the steric factors due to a large curvature of the micelle require a disordered arrangement of the chains.

2.3.1   Cooperativity of micellization

Micelle formation occurs over a narrow concentration range where virtually all the solution properties show an anomalous or abrupt change. Formation of a micelle is not a single-step kinetic event because the kinetic order of micellization is not high, i.e. not all the monomers that make up a micelle come together at the same time in a single event to form a micelle. Although smaller oligomeric aggregates may be present in the monomer range, theoretical considerations suggest that a high cooperativity of micellization, over a narrow concentration range at the CMC, depends on the number of monomers per micelle (aggregation number). Thus a series of discrete equilibria in which a monomer is added at each step proceed until a critical aggregation number is attained. Thus the micelle growth kinetics occurs with the addition of a single amphiphile, A, at a time to a growing aggregate as

$$A + A \rightleftharpoons \ldots A_{n-3}^* \rightleftharpoons A_{n-2}^* \rightleftharpoons A_{n-1}^* \rightleftharpoons A_n^* \quad . \qquad (2.5)$$

Shortcuts in this scheme could also occur as transient small aggregates coalesce to form larger ones. The cooperative effect for the position of the overall equilibrium is caused by increasing values of the association equilibrium constants for the later steps relative to the earlier ones. Of course, it is not possible to measure the individual equilibrium constants unless the individual aggregated species can be experimentally distinguished.

2.4   The Concentration Issues in the Monomer–Aggregate Equilibria

Consider an amphiphile A with a limited solubility and in equilibrium with its aggregated state A*. For a simplified picture of the monomer–aggregate concentration dynamics consider the equilibrium:

$$A + A^* \underset{k_d}{\overset{k_a}{\rightleftarrows}} A^* \tag{2.6}$$

between monodispersed amphiphile molecules in the aqueous phase at concentration [A] and those in the interface of the aggregates A^* corresponding to a bulk concentration $[A^*]$. This equilibrium is most easily envisaged as a dynamic equilibrium in an ensemble under a given set of conditions such that the flux from the forward reaction is equal to that of the reverse reaction. The rate, J_a, of monomers A joining aggregates must be proportional to the concentration [A] of molecules that can join and is also proportional to the concentration of interface molecules to which they can join, $[A^*]$. Thus, $J_a = k_a[A][A^*]$, where the proportionality constant k_a is the bimolecular association rate constant (counted per monomer in aggregates). For most interfaces k_a appears to be approximately diffusion limited, about 10^8 to 10^6 M^{-1} s^{-1} (Theory Box 2.1), on the basis of the interaction of the monomer with the aggregate particle. Similarly, the rate of leaving an aggregate must be proportional to the bulk concentration of interface molecules, $J_d = k_d[A^*]$. At equilibrium, $0 = d[A]/dt = J_d - J_a$, these fluxes are equal, $k_a[A][A^*] = k_d[A^*]$, so that the free concentration of amphiphiles is a constant:

$$[A] = \frac{k_d}{k_a} = K'_A = CMC_A \tag{2.7}$$

The monomer concentration at this 'solubility' or micellization limit equals the dissociation constant, $K'_A = k_d/k_a$, for the amphiphiles from the aggregates. Since it corresponds to the critical micelle concentration for A, the concentration dispersed as aggregates equals the total concentration, $[A_T]$, minus the free monodispersed concentration:

$$[A^*] = [A_T] - K'_A \tag{2.8}$$

For much of our purpose in the following chapters we will not go beyond this simple picture described above. Note that, as described later, near the CMC, as well as at the higher amphiphile concentration or in the presence of other amphiphiles, other kinds of aggregate phases with different sizes and dissociation constants can form and the behavior can become much more complex than these basic equations indicate.

It can be noted that this picture corresponds to the description of a liquid–gas equilibrium where the equilibrium constant is determined by the heat and entropy effects in moving single molecules between the two phases. For the micelles, this constant would correspond to the last step(s) in Equation (2.5), where removal or addition of individual molecules presumably has a small effect on the properties of the micelle. In this way, the micellization equilibrium constant, CMC, carries no information about the cooperativity of the micellization process.

2.4.1 The monomer residence time in micelles

As a reflection of the monomer–micelle equilibrium in micelles with large CMC virtually all molecular motions are rapid (Theory Box 2.1). In a simple micelle of a short chain amphiphile, with the CMC in the micromolar range, the residence time of an amphiphile in the micelle will be in the micro- to millisecond range. Thus the rate of intermicellar exchange of a short chain amphiphile with the CMC around 1 mM occurs on a time scale approaching a few microseconds (Aniansson *et al.*, 1976; Soltys and Roberts, 1994). As discussed in the next section, the monomer amphiphile concentration in equilibrium with virtually all other forms of aggregates is considerably lower, which influences the stability and exchange characteristics of such interfaces.

THEORY BOX 2.1 Diffusion-limited rates of exchange between aggregates

Consider a solution with a concentration [V] of spherical aggregates, each with radius R and N_T amphiphiles in the interface layer accessible from the bulk aqueous phase. Then the concentration of interfacial amphiphiles is $[A^*] = N_T[V]$. If each amphiphile contributes the surface area a_A with associated radius b, $a_A = \pi b^2$, to the aggregate, the interface area per aggregate is $4\pi R^2 = N_T a_A = N_T \pi b^2$. Thus the radius of an aggregate can be expressed as

$$R = \sqrt{\frac{N_T a_A}{4\pi}} = \frac{1}{2} b \sqrt{N_T} \qquad (1)$$

The *diffusion-limited association rate* constant for an amphiphile in solution to a spherical aggregate is determined by the Smoluchowski expression as

$$k_{assoc} = 4\pi D R N_{Av} = N_{Av} 2\pi b D \sqrt{N_T} \qquad (2)$$

(see also Theory Box 11.1, Equation (9)), where $k_{assoc}[A]$ corresponds to the association flux of monomers per aggregate per second. Here, D is the diffusion constant for a free amphiphile in solution and N_{Av} is Avogadro's number. Thus the diffusion-limited association rate to the whole aggregate is proportional to the square root of its size. In contrast, when the association is not diffusion-limited, the rate is directly proportional to the size N_T. (Note that the product DR has the units volume per second and that this volume must be in liters (dm^3) to give the association rate constant in the units M^{-1} s^{-1}.) The association rate constant counted per monomer in the aggregates is

$$k_a = k_{assoc}/N_T = N_{Av} 2\pi b D / \sqrt{N_T} \qquad (3)$$

Thus, the diffusion-limited association rate per monomer in the aggregate surface is reduced by a factor $1/(2\sqrt{N_T})$ compared to a free (spherical) monomer of radius b.

An amphiphile molecule in the interface will leave the interface with the *desorption rate* constant k_d. At equilibrium, the concentration of free amphiphiles equals the CMC, $[A] = K'_A$, and the total rate of desorption must equal the total rate of association:

$$k_d[A^*] = k_{assoc}[V]K'_A = k_a[A^*]K'_A \qquad (4)$$

Note that this corresponds to the net flux of molecules between aggregates and therefore the *diffusion-limited exchange rate* constant is

$$k_{exch} = k_d = k_a K'_A \approx 5 \times 10^8 K'_A / \sqrt{N_T} \quad s^{-1} \tag{5}$$

The last equality in Equation (5) is from $b = 4$ Å and $D = 3 \times 10^{-6} cm^2/s$. The result is in s^{-1} if K'_A is given in M. The exchange rate decreases with increasing aggregate size, reflecting the increased probability that a dissociated molecule encounters and returns directly to the same aggregate and therefore does not contribute to desorption or exchange.

The same expressions will be valid for the interaggregate exchange of enzymes, if the enzyme-interface binding is diffusion-limited. Thus if the diffusion rate of the enzyme is $D = 10^{-6} cm^2/s$ (the expected diffusion rate of a small spherical protein of ca. 18 kD), the rate of desorption or exchange of interface-bound enzyme is

$$k_{off} = k_{exch}^E \approx 1.5 \times 10^8 K_d / \sqrt{N_T} \quad s^{-1} \tag{6}$$

where K_d (M) is the dissociation constant for enzyme-interface binding.

By definition, the exchange rate corresponds to the rate by which one molecule leaves an aggregate and another one comes in. In contrast, the time it takes for a *particular* bound molecule to exchange will equal the sum of the time it takes to desorb and the time it takes to rebind another aggregate, i.e. $\tau_{exch} = 1/k_d + 1/k_a[A^*]$.

2.5 Exchange of Amphiphiles between Organized Interfaces

Thus the lifetime of micelles, as well as the residence time and the exchange time of a single short chain amphiphile, can be as small as a microsecond. The CMC, a measure of the propensity of an amphiphile to form an organized interface, is also related to the exchange dynamics between aggregates via the monomer amphiphiles in the aqueous phase (Theory Box 2.1). The CMC and monomer exchange rates for simple micelles are more than 10^{10}-fold larger compared to those for bilayer vesicles (Nichols, 1985, 1988). Also, as shown in Figure 2.7, other mechanisms that facilitate the exchange and transfer of amphiphiles between aggregates are known. For example, the rate of amphiphile exchange or transfer facilitated by a specific binding or exchange protein depends on the rate of the complex formation or decomposition, or on the rate of diffusion of the complex across the intervening aqueous phase. Virtually all other mechanisms for the transfer or exchange of amphiphiles involve mixing due to direct interface-to-interface contacts either through formation of mixed micelles, or the (hemi)-fusion of the outer monolayer of the vesicles, or a direct protein mediated contact between the interfaces without (hemi)-fusion.

The kinetics of the amphiphile exchange and transfer processes involving direct contact between interfaces (Figure 2.7) is complex. Such rate processes depend on direct aggregate-to-aggregate interactions that occur on the time scale of seconds to days. For example, under optimized conditions fusion and fission of mixed micelles occurs on the time scale of seconds to minutes. In contrast, fusion of vesicles occurs

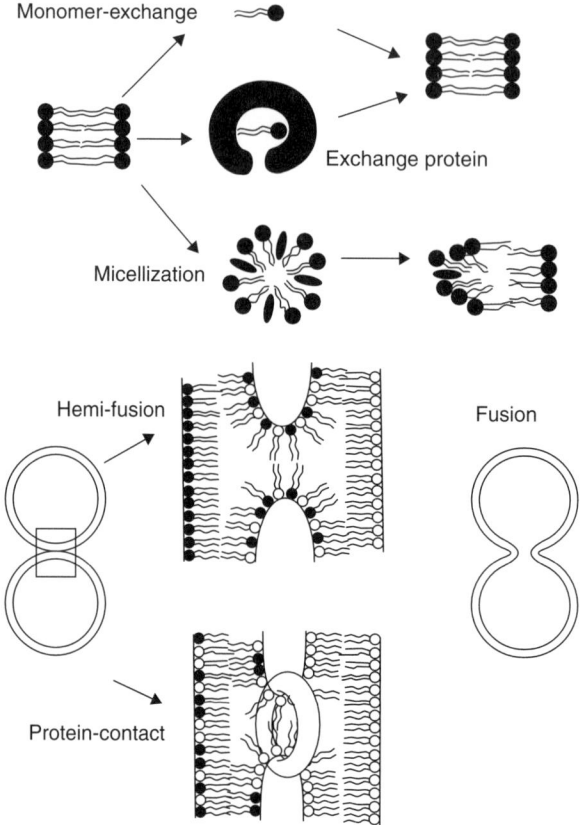

Figure 2.7 Possible modes of phospholipid exchange and transfer between interfaces. (From top) Monomer exchange predominates for amphiphiles with high CMC values. The exchange proteins 'ferry' the stoichiometric complex with the amphiphile monomer across the aqueous phase. 'Solubilization' or micellization of bilayer by detergents results in mixed micelles, where the mixing of components takes place by fusion and fission of mixed micelles. Hemifusion of vesicles results in mixing of the lipid molecules in the fused outer layers. Spontaneous nonleaky fusion involves mixing of both the layers of the bilayer and mixing of the aqueous compartments of the vesicles. Certain proteins and peptides mediate specific contacts between outer layers of vesicles and induce a rapid, selective and regulated exchange of phospholipids between the layers in contact. Specific examples are given in later chapters

on the time scale of days unless mediated by solutes and proteins. Spontaneous unaided fission of vesicles into smaller vesicles is usually not observed unless there is a dramatic change in the environmental conditions. A quantitative description of such multistep physical processes is not trivial and the primary kinetic events have not been identified. In addition, contributions from excluded volume effects, followed by the solvation, electrostatic, steric and van der Waals interactions would come into play successively as solutes, particles and the interacting interfaces are brought together.

Significant features of phospholipid exchange and transfer in bilayers, monolayers and mixed micelles are outlined below to provide an appreciation of their suitability as the substrate for the study of interfacial enzymes.

Table 2.4 Molecular, energetic and motional properties of phospholipids in bilayer

Property	Value
Extended length per CH_2	1.4 Å
Cross-sectional area	47 Å^2 (cylinder), 70 Å^2 (micelle)
Volume per CH_2	27 Å^3
Volume per CH_3	54 Å^3
Trans-gauche C—C bond rotation	
Free energy	0.5 kcal/mole
Activation energy	2.5 kcal/mole
Rotation time	10^{-12} s
Chain rotation along the long axis	10^{-11}–10^{-10} s
Segmental motion	10^{-9}–10^{-7} s
Lateral diffusion coefficient	10^{-7}–10^{-9} cm^2/s
Transbilayer movement (flip-flop)[a]	>100 min
Vesicle-to-vesicle exchange[a]	>10 000 min
C6-NBD-phosphatidylcholine	0.8 min
C12-NBD-phosphatidylcholine	150 min
Trisialoganglioside	30 min
Monosialoganglioside	120 min
C12-NBD-diacylglycerol	>2000 min
16,18:1-Phosphatidylcholine	>5000 min
Sphingomyelin	>10 000 min
Pyrene cerebroside	>10 000 min
Half-times for fusion–fission of mixed micelles[b]	0.2–40 s

[a]The half-times are for the naturally occurring phospholipids in bilayer vesicles (Nichols and Pagano, 1982).
[b]From Fullington *et al.*, 1990.

2.5.1 No exchange between the bilayer-enclosed vesicles

Cylindrical amphiphiles with long chains, including the membrane phospholipids (Figures 1.3 and 1.6), spontaneously form bilayer-enclosed vesicles (Figure 2.2). The monomer concentration in equilibrium with the vesicles is exceedingly low, estimated to be well below 10^{-12} M for most naturally occurring phospholipids. Therefore, the spontaneous monomer-mediated exchange of phospholipids between vesicle populations occurs on the time scale of days. Thus stable vesicle dispersions of definite size can be prepared by suitable methods (Figure 2.3) and their aqueous compartments remain intact without mixing of the phospholipid at the two interfaces. Several types of molecular motions occur (Table 2.4) within the constraints of the bilayer organization such that the polar or nonpolar regions of the amphiphile molecules in transition do not make unfavorable contacts with the aqueous phase or the hydrophobic interior. The half-time for the spontaneous vesicle-to-vesicle exchange and for the transbilayer exchange of natural phospholipids is of the order of several hours to days. Thus the bilayer-enclosed vesicles are stable on the time scale of hours and days during which the contents of the aqueous compartment do not leak. In an ensemble the phospholipid molecules do not exchange from vesicle to vesicle. The major challenge to stability and integrity of vesicles comes under conditions where vesicles fuse or are disrupted by other mechanical means. Freezing and thawing disrupt vesicles and induce fusion–fission. Fusogens include certain solvents as well as amphiphiles with a net charge, multivalent ions, peptides and proteins. Sharp pH shifts may fuse charged vesicles. Vesicles are also disrupted by addition of detergents at concentrations larger than the 0.2 mole fraction and larger than the inter-micellization concentration (Chapters 3 and 11) for the mixed micelles of detergents and phospholipids.

2.5.2   Trough geometry determines the exchange in monolayers at the air–water interface

Virtually all nonpolar solutes and amphiphiles added to the aqueous phase partition into the air–water interface to form a monomolecular layer with the polar groups in contact with the aqueous phase and the chains extend into the air medium of a virtually zero dielectric constant (Figure 2.4). Transfer of amphiphiles to the monolayer is monitored as a change in the surface tension, often expressed as the surface pressure of the amphiphiles at the interface equal to the decrease in the surface tension of water in the presence of the amphiphile. A maximally close-packed monolayer, just below a characteristic collapse pressure, is formed at a characteristic bulk aqueous phase concentration of an amphiphile (Figure 2.5). Essentially 'insoluble' monolayers are formed with medium and long chain phospholipids, such that the monomer amphiphile concentration in the bulk aqueous phase is virtually negligible. Distribution and equilibration of an amphiphile in an organic solvent spread on the surface of the aqueous phase takes several minutes. Similarly, partitioning of a solute added to the aqueous phase takes several minutes for the equilibration into the air phase to form a monolayer. As developed in Chapter 10, the exchange dynamics at the monolayer interface is rather complex because diffusion through the unstirred aqueous layer in contact with the monolayer interface dominates the rate of transfer or exchange. The monolayer technique is also not suited for equilibrium binding and partitioning studies because the large hydrophobic surface of the trough is often a nonspecific and uncharacterizable sink for enzymes and hydrophobic solutes.

2.5.3   Mixing by fusion–fission of mixed micelles

The vesicle organization collapses in the presence of detergents. Mixed micelles of phospholipids with detergents often exhibit a rather complex phase behavior suggestive of a transition between the various organized aggregates as a function of the phospholipid-to-detergent ratio. Even at a constant ratio, the exchange behavior of the two amphiphiles between the coexisting mixed micelles is rather complex. The spontaneous exchange rate of the detergent monomer is rapid because its intermicellar concentration is significant but generally lower than its CMC. The monomer phospholipid concentration in the aqueous phase also decreases in equilibrium with the mixed micelles.

Therefore, mixing of phospholipids between mixed micelles (Figure 2.8) is mediated by fusion–fission. In this process two micelles fuse during a collision-limited encounter and the components mix in the fused micelle which breaks (fission) again into two micelles. With the protocol in Figure 2.8, depending on the nature of the

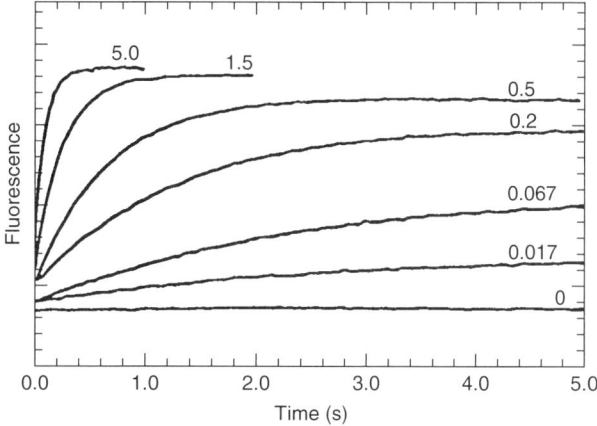

Figure 2.8 Effect of the acceptor micelle concentration (in mM as indicated on the traces) on the rate of transfer of N-NBD-phosphatidylethanolamine fluorescent probe between DOPC-chenode-oxycholate (1:2) mixed micelles. The donor micelle concentration is 0.033 mM and the acceptor micelle concentration (mM) is given with the trace. The calculated monomer dissociation rate is 29 s^{-1} and the bimolecular collision rate constant is 143 M^{-1} s^{-1}. Reprinted with permission from Fullington *et al.* (1990). Copyright (1990) American Chemical Society

bile salt in the mixed micelle, the half-times for the mixing of N-NBD-PE with excess mixed micelles without the probe ranges from 0.19 to 43 s (Fullington *et al.*, 1990). As a second-order collisional process, fusion–fission is micelle concentration dependent. Typically, as developed further in Chapter 11, under the normally used assay conditions with millimolar concentrations of the mixed micelles, substrate replenishment by fusion–fission can support the turnover rates below 10 s^{-1}; i.e. the effective rate of mixing of the contents of mixed micelles by fusion–fission is often too slow for the catalytic turnover rates, which are several-fold faster.

2.6 Dispersed Phases

2.6.1 Stable emulsions

Dispersions containing microdroplets of immiscible unorganized bulk phases are of considerable biological and industrial interest because they offer a range of polarity gradients to 'dissolve' and disperse solutes of wide-ranging polarities and sizes. Depending on the organizational and exchange characteristics, such dispersions are called emulsions, colloids and mesomorphic phases. A common feature of water-in-oil or oil-in-water emulsions (Figure 2.1) stabilized by amphiphiles is that only a fraction of the total solute in the droplet, and possibly also the amphiphile, is accessible at the interface. Nonpolar solutes can often be dispersed as kinetically stable microdroplets in an otherwise immiscible medium. Molecular associations, dominated by noncovalent interactions between all the components in the droplet, control the size and the organization of the aggregates. The lower size limit for dispersed droplets is also around 1 nm, and smaller droplets ultimately become indistinguishable from true solutions.

The free energy associated with creating and maintaining the interface of a dispersed medium is relatively large. The size of the dispersed particles depends on the surface tension of the interface, as well as on the kinetic factors that promote coalescence of the dispersed droplet. Oil-in-water and water-in-oil emulsions of hydrophobic solutes are often stabilized by suitable amphiphiles, which organize at the interface to effectively lower the interfacial tension and reduce the propensity of the enclosed droplets to coalesce. Droplet size in most emulsions is >100 nm; however, thermodynamically stable microemulsions of 5 to 50 nm diameter are also formed spontaneously. In a narrow range of the variables of the phase diagram, often protected as industrial secrets, even the thermodynamically unstable emulsions do not 'break' on the time scale of months.

The composition and the exchange dynamics of the nonpolar solute or the amphiphile at the surface of an emulsion droplet are difficult to ascertain because of the size dispersity, accessibility and polymorphism in the phase diagram. Together the uncertainties in such properties make it impossible to determine the local variables from the bulk-averaged values. In addition to the consideration of the size of the droplet of the dispersed phase, the range of morphologies of the dispersed phases becomes even larger if one considers the effects of composition, organization, phase separation, exchange dynamics and hysteresis. Since the interface concentration or the solute concentration cannot be easily defined in such dispersions, they are not suited for the characterization of interfacial enzymes.

2.6.2   Serum lipoproteins

As a potential target interface for the enzymes of lipid metabolism, these particles in blood serum represent an interesting range of organizational features. Typically their surface has an organized monolayer of phospholipid embedded with characteristic amphipathic proteins. The smallest of these particles, the high-density lipoprotein (HDL), essentially consists of a bilayer disk of phospholipid and cholesterol with the edges sealed with several types of amphiphilic protein molecules to protect the hydrophobic core from contact with water. Somewhat larger low-density lipoprotein (LDL) particles contain a microdroplet as the spherical nonpolar core of cholesterol ester and triglyceride bounded by a monolayer of phospholipid, cholesterol and proteins. The enclosed droplet core in the very-low-density lipoproteins (VLDL) and chylomicrons have even more triglycerides and cholesterol esters on a per weight basis. In addition to stabilizing the particles, the proteins also serve as anchors for the initial recognition step for the receptor-mediated endocytosis. As is the case with bilayers and biological membranes, the exchange of phospholipids between the lipoprotein particles is slow, and is possibly mediated *in vivo* by transfer or exchange proteins.

2.7   Choice of Interface for Study of an Interfacial Enzyme

The morphology and organization of the aggregates determine the accessibility of the interface for the binding of an interfacial enzyme, and the exchange properties between the aggregate particles determine the substrate replenishment for the steady state catalytic turnover. An appreciation of the considerations and criteria for the assay of an interfacial enzyme is critical because the underlying variables in a microscopically heterogeneous environment are far more complex than those encountered by a soluble enzyme in a homogeneous solution. The generic assay criteria summarized in Table 2.5 are not complete or rigorous, and as such they are useful for pruning potential problems. As developed in the following chapters, a far more significant consideration for the fidelity of an assay comes from the fact that the residence time of the enzyme at the interface determines the interfacial processivity of the successive turnover cycles in the interface.

In designing an assay for a quantitative steady state kinetic analysis of an enzyme-catalyzed reaction, the most critical consideration is the choice of the interface where the microscopic variables can be quantified and systematically varied without introducing unwanted changes in the microenvironment of the bound enzyme. Such conditions are readily satisfied with bilayer vesicles (Chapters 4 to 7) and to a limited extent with simple micelles of short chain phospholipids (Chapter 8). On the other hand, while it is possible to change the surface density of substrate in a monolayer, virtually all other variables required for kinetic analysis remain uncontrolled in such a system (Chapter 10). Microemulsions and dispersions offer sufficient area for the interfacial reaction with the enzyme in the aqueous phase. However, use of emulsions for the study of the equilibrium and kinetic processes at interfaces is limited because the local variables for the steady state kinetic path cannot be adequately defined. The problem is further compounded by the eutectics and boundaries in the phase diagram of mixed micelles and emulsions. For example, often in such systems the number of particles does not necessarily change with the total lipid concentration. Also, the fusion–fission rate, related to the substrate replenishment rate, changes in a nonlinear fashion with the concentration and mole ratio of the dispersed components. Far more serious quantitative problems emerge as one considers the number density of the substrate and other variables for the

Table 2.5 Critical criteria for the assay of interfacial enzymes

Ability to monitor the signal in real time (continuous assay)

Substrate interface with well-defined area and accessibility from the aqueous phase

Stability of the interface through the measured reaction progress

Well-defined and identical microenvironment for each enzyme molecule

Eliminate contributions of the extraneous surfaces

Eliminate effects of unwanted shifts in the E to E* equilibrium to identify the kinetic path for the steady state kinetics turnover cycle events in the microenvironment of the enzyme at the interface. Together, such factors also preclude the use of water droplets in organic solvents, emulsion particles, reverse and mixed micelles, aerosols, supported films, soap bubbles, froth and vessel walls as the interface for interfacial kinetic analysis.

Of course, considering the diversity of the nonpolar solutes and amphiphiles of interest, there are no general solutions to such problems. The suitability of the substrate interface, along with the changes introduced during the reaction progress by an interfacial enzyme, emphasizes the need to develop a biophysical understanding of the interface. It is fair to start with the assumption that a single interface will not meet all the strict criteria for the kinetic analysis, and therefore often it is necessary to build on the unique windows offered by suitable assays. As a criterion for suitability, it is needless to say that an assay in which the substrate replenishment due to exchange across the interface is rate-limiting cannot provide interpretable kinetic results. In short, assay conditions that do not give an absolute measure of the reaction rate or assays in which the rate-limiting step lies outside the events of the catalytic turnover cycle of the kinetic path do not give reliable quantitative readouts of the primary enzymatic processes.

2.8 Summary: Interface Determines the Accessibility and Replenishment Rate

Significant progress has been made in our understanding of the organization and exchange dynamics of amphiphiles in aqueous dispersions. The assay mixtures for interfacial enzymes contain ensembles of such aggregates. Two key considerations for the choice of substrate interface are the accessibility and the replenishment rate by the substrate exchange between the particles. Most, if not all, assay criteria are satisfied with interfaces of vesicles and micelles of suitably chosen phospholipids. Properties of micellar and bilayer aggregates in phospholipid dispersions are reasonably well characterized. As developed in subsequent chapters, with suitable choices, virtually all the primary rate and equilibrium parameters for the binding and the interfacial catalytic turnover can be obtained.

The accessibility of amphiphile molecules can be fully evaluated only in a few types of organized aggregates. The stable vesiculated bilayer organization assumed by natural phospho-, glyco- and sphingolipids provides a remarkably versatile system for the characterization of the interface properties without exchange of amphiphiles. At the other extreme, simple micelles also offer an interface where the catalytic turnover events remain rate-limiting. Under these fast-exchange conditions, the underlying concentration variables relate to the thermodynamic properties expressed in terms of the primary constant K_S' described in terms of the association and dissociation rates of the amphiphile from the aggregate. Thus K_S' relates to the parameters such as CMC and the partition coefficient of a solute (Chapter 3), as well as to the rates of exchange and transfer of a substrate between the coexisting interfaces in an assay mixture.

In short, the degrees of freedom of a solute change at interfaces. Such changes relate to the concentration or the number density of a solute per unit area, which is a basic property of the system with a direct thermodynamic significance. In this chapter we have identified the interfaces of the various organized aggregates with more than a 10^{10}-fold range of the exchange rate of amphiphile between the coexisting interfaces.

Further Reading

Cevc, G. and Marsh, D. (1987). *Phospholipid Bilayers: Physical Principles and Models*, Wiley, New York, pp. 442.

Helenius, A. and Simons, K. (1975). Solubilization of membranes by detergents. *Biochim. Biophys. Acta*, **415**, 29–79.

Hofman, A.F. and Mysels, K.J. (1988). Bile salts as biological surfactants. *Colloids Surf. Sci.*, **30**, 134–73.

Hunter, R.J. (Ed.) (1989). *Foundations of Colloid Science*, Vols. I and II, Clarendon Press, 1089 pp.

Jain, M.K. (1988). *Introduction to Biomembranes*, II edition, Wiley, New York, p. 423.

3 To Be or Not To Be: Dilemma for the Substrate in Solution

If Shakespeare had Hamlet say 'I think, therefore I am', would that prove to us that Hamlet exists?

Robert Nozick

The general kinetic paradigm for enzymology in Figure 3.1 provides a recurring theme and conceptual focus for this book. It has a kinetic path for the turnover in the aqueous phase and a separate path is also introduced for the turnover at the interface via the species marked with an asterisk:

(a) For the limiting case of solution enzymology, in accord with the classical Michaelis–Menten paradigm, the catalytic *turnover in the aqueous phase* is described only in terms of the upper branch where the enzyme, E, reacts with the substrate, S, to form the ES complex; occasionally the substrate in the complex undergoes a chemical change and the product may be released from EP to regenerate E.

(b) In the other limit, the *interfacial turnover steps* (in the oval) are mediated only by the interface localized species marked with an asterisk, such as S^*, E^*, E^*S, E^*P and P^*.

Each of the elementary steps for the turnover cycle in the aqueous phase or the interface is characterized by a separate rate constant, e.g. k_2 or k_2^*, for the chemical steps. The equilibrium dissociation constants for the binding of an active-site-directed ligand, L ($=$ S, P, I or an imperfect interface diluent referred to as ND), to the enzyme in the aqueous phase (K_L) or the interface (K_L^*) are also different. L may be the substrate such that E^*S undergoes a catalytic change. For a reversible reaction E^*P may also undergo a catalytic change; however, the dead-end complex with an inhibitor, EI or E^*I, does not undergo such a change.

In addition, Scheme I explicitly takes into consideration the exchange relationships between the coexisting species in the microscopically heterogeneous environment of the interface. Thus the issues related to accessibility and replenishment of the substrate are accommodated in terms of the equilibrium parameters that determine the partitioning of the ligand (K_L') and the dissociation of E^* (K_d) or E^*L (K_d^L) from the

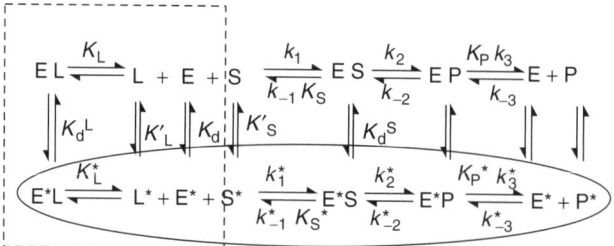

Figure 3.1 General kinetic paradigm (Scheme I) for enzymology in a microscopically hetero-geneous environment. In this adaptation of the Michaelis–Menten formalism, several kinetic paths are possible. Without the interface it simplifies to the classical kinetic path for soluble enzymes (upper branch with the species without an asterisk). The interfacial path is constrained to the steps shown in the oval. Other kinetic paths come into play if one or more of the species partition between the two phases. As developed in Chapter 6, the detailed balance condition for the equilibria in the square box provides insights into the effect of the dimensionality change on the interaction of an active-site-directed ligand (L = I as implied in this scheme, but also P or S with suitable assumptions). The reacting species in the interface, as well as the corresponding constants and parameters in the interface, are marked with an asterisk. The equilibrium dissocia-tion constants are given in upper case and the rate constants in the indicated directions are given in lower case. Note that sometimes the binding of an active-site-directed ligand may require a cofactor; for example, the binding of L to secreted PLA2 requires calcium (Chapter 6). In such a case, the scheme is valid when enzyme is saturated by cofactor

interface. Under certain conditions the underlying rate constants may also influence the observed kinetics.

Need to constrain the variables

The E to E* step, as an extension to the classical Michaelis–Menten formalism, was first explicitly introduced as a pre-steady state step by Verger and de Haas (1976) to account for the catalytic turnover by phospholipase A$_2$ (PLA2) at the substrate interface under a wide variety of conditions (de Haas *et al.*, 1971; Dennis, 1973; Jain, 1973; Jain and Cordes, 1973a, 1973b; Verger *et al.*, 1973). Over the years, the idea of the enzyme binding to the interface as a prerequisite for the interfacial catalytic turnover garnered considerable circumstantial support and the pancreatic PLA2 emerged as the prototype for interfacial enzymology (Verheij *et al.*, 1981; Dennis, 1983). Although introduction of the E to E* step permitted rationalization of the *quality of interface* effects on the observed kinetics, as such the observed kinetics could not be analyzed in terms of the underlying elementary events.

The kinetic path necessary for the analysis of the steady state kinetics could not be defined without a full appreciation of the interfacial processivity for the successive turnover cycles (Upreti and Jain, 1980) and the substrate replenishment rates in the microenvironment of the bound enzyme (Jain and Berg, 1989). Throughout this book we outline the key arguments that led to the resolution of the kinetic contribution of interfacial processivity (Jain *et al.*, 1986a–1986d) and permitted determination of the primary rate and equilibrium constants for defined kinetic paths under several sets of specified conditions (Berg *et al.*, 1991, 1997, 1998, 2001).

Versatility of Scheme I due to multiple kinetic paths

It is useful to appreciate numerous kinetic paths that are possible within the paradigm in Scheme I. In its various versions the paradigm has been useful for rationalizing the experimental results and for building a consensus for the catalytic turnover by the enzyme localized at the interface. For example, a reaction progress with virtually any shape can be generated, including those in Figure 3.2 and others reported in the literature under a variety of conditions.

In accord with the dictum that 'with six constants one can fit an elephant', it is obvious that Scheme I cannot be analyzed in its entirety. To be useful, Scheme I must be truncated with explicitly specified constraints to define valid boundary conditions. Two obvious extremes are when the steady state turnover occurs only in the aqueous phase or only in the interface. Such limits reduce the number of parameters, especially by discounting the contributions of the steps that relate to the transfer of the reacting species between the two phases. These steps control the substrate accessibility and replenishment (Chapter 2) and the underlying processes are grounded in the biophysical reality of the interface. In other words, the critical first step is the choice of a suitable assay system where the kinetic path can be unequivocally defined. For analytical rigor it is also preferable to minimize the number of steps in a path. If not, the number of parameters increases rapidly with the number of steps, and more so with the introduction of the branches. In order to develop an appreciation of the underlying assumptions and constraints necessary to evaluate the interfacial turnover cycle, we start by explicitly identifying the physical

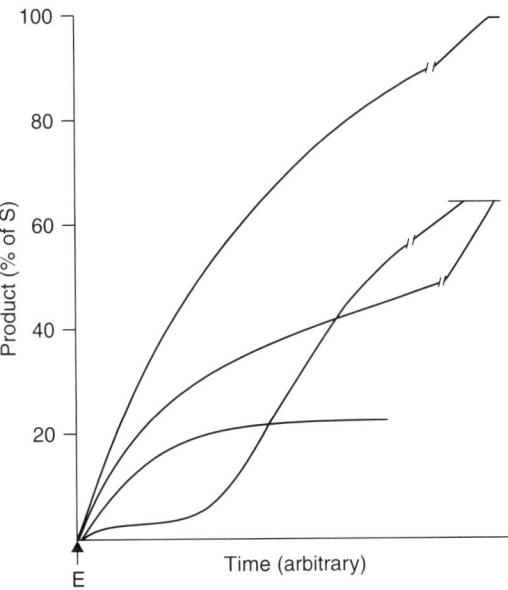

Figure 3.2 Some of the experimentally observed reaction progress curves for PLA2 with the same substrate at different interfaces. All such families of diverse reaction progress curves are quantitatively generated from different reaction paths intrinsic in Scheme I. Reprinted from Jain and Berg (1989). Copyright (1989) with permission from Elsevier Science

basis and the assumptions that go into ascertaining the kinetic path for a soluble enzyme.

3.1 Constrained Reaction Path for the Turnover in the Aqueous Phase

The observed reaction progress is the sum total of all the changes that occur in an ensemble of reactants. A kinetic path is the explicit and complete statement of the kinetic and equilibrium relationships between all the 'players in the game'. First and foremost, for this purpose not only must we define the preferred kinetic path for the turnover but also rule out all other possible paths, i.e. to establish that all players play by the same rules. For the appreciation of the consequences of not doing so, imagine the kinetic effects of the adsorption of the enzyme on the vessel walls, or consider the effect of sparingly soluble or aggregated substrate on the concentration variable (see below).

The challenge of kinetic analysis is to describe the entire observed behavior of the system in terms of the primary constants for the steps in the kinetic path. It is axiomatic that the larger the number of steps, the more independent observations will be required for analysis. Analytical relations are derived from the rules of the game grounded in the physical reality. For describing the behavior of an enzyme, as a first step one defines the events that make up a single catalytic cycle which typically takes less than a few milliseconds for all the steps of the turnover path. Under identical conditions, each molecule behaves identically within the Boltzmann distribution of the kinetic and potential energies. Under most conditions it is nearly impossible to measure a single turnover cycle, let alone any one step of the cycle. Therefore the task of kinetic analysis is to account for the ensemble behavior of all the reactants over the period of the observation window. Individual turnover events are *summated* to obtain the *average behavior of the ensemble* included in the data set. Thus the variables must be constrained so that the kinetic path remains invariant for the observed data set.

3.1.1 Assumptions, valid constraints and boundary conditions

As a paradigm for an enzyme-catalyzed reaction, a turnover cycle consists of three sequential steps: the initial binding of the substrate to the active site of the enzyme, the chemical step during which the bound substrate changes to the product and the final step where the enzyme is regenerated as the bound product leaves. Reversibility of each of these three steps gives a total of six primary rate constants for the six elemental events:

$$E + S \underset{k_{-1}}{\overset{k_1}{\rightleftharpoons}} ES \underset{k_{-2}}{\overset{k_2}{\rightleftharpoons}} EP \underset{k_{-3}}{\overset{k_3}{\rightleftharpoons}} E + P$$

The need to constrain the steps of the kinetic path for the turnover in the aqueous phase is already obvious at this stage. To determine six primary constants we need at least six independent data sets. On the other hand, as we will see below, the steady state substrate concentration dependence of an enzyme operating according to this path can give information about only four parameters at most, the effective k_{cat} and K_M values in each direction, i.e. with $[S] \gg [P]$ and with $[P] \gg [S]$.

The number of parameters can be reduced if the back reaction is eliminated, as would be the case if the product concentration is insignificant. The kinetic path is further simplified if the chemical step is essentially irreversible ($k_2 \gg k_{-2}$). Further simplification is achieved if the rate of product release is rapid and the chemical step is rate-limiting ($k_2 \ll k_3$). Within these valid constraints for an enzyme, with $[S] \gg [P]$ for the steady state assumption, the turnover cycles during the initial rate is essentially dominated only by three rate constants, k_1, k_{-1} and k_2.

Now consider the *implicit constraints*. For the turnover per enzyme it is assumed that all the enzyme molecules are active and not absorbed on the interface, or some such complication. Also we must assure that all the substrate molecules are equally accessible to all the enzyme molecules in the form considered for the binding to the enzyme. This would not be the case if the substrate is not uniformly dispersed, or is sparingly soluble or aggregated. Such assumptions about modeling the turnover cycle are generally valid within the constraints of *Brownian motion* in a perfectly mixed *homogeneous* solution. It ensures local equilibration of all the species in the aqueous phase at any given point in time and also at all points along the reaction progress. Such considerations assure that all E and all S molecules have an *equal probability of encounter and collision* to form the ES complex. The diffusion-limited collision rate of an enzyme molecule with the substrate, driven by the thermal motions at room temperature in a homogeneous one molar solution of the substrate, is in the range of 10^8 to 10^{10} s^{-1}. Although some enzymes have evolved to the perfection of a diffusion-limited turnover, for most enzymes one or more of the other steps in the kinetic path for the turnover cycle still remain rate-limiting.

3.1.2 The steady state condition for flux

Valid constraints not only define a unique kinetic path, but also help satisfy the requirements for the ensemble averaging. Consider the steady state 'initial' rate, where the change is monitored over a period of time as the sum total of all the changes mediated by a large number of enzyme molecules. For an appreciation of what assumptions go into the 'summation' of events consider an observed noise-free signal corresponding to the formation of 0.1 nmole of product in one second by one picomole (6.2×10^{11} molecules) of enzyme added to an *excess* of the substrate in the assay mixture. This *flux* per second amounts to 6.2×10^{13} product molecules, each being produced during an individual turnover cycle mediated by one of the 6.2×10^{11} enzyme molecules present in the reaction mixture. From this ensemble behavior we obtain information about the individual turnover cycle only if it can be assumed that all enzyme molecules experience an identical environment over this time interval. Such a large ensemble assures that *all* events reported by the signal conform to the Boltzmann distribution of probabilities and are not corrupted by fluctuations from the individual events.

As the basis for the steady state condition for the initial rate, the excess of the substrate ensures that the substrate concentration does not change noticeably over the observation period. The situation is simplified further if the substrate concentration is so high that all the enzyme molecules are saturated with the substrate, i.e. the [ES] is the same as the total enzyme concentration, $[E_0]$. Within such constraints, for the example above the *maximum flux* or turnover per enzyme is 100 s^{-1}.

In short, analysis of a kinetic path is based on the assumptions that assure that all enzyme molecules go through the same turnover steps during the observed period. More precisely, the average over all enzymes in the ensemble is the same as a suitable time average over any individual enzyme. This constitutes the starting point for the analysis of the steps of the kinetic path based on the information about the observed fluxes under different conditions. An appreciation of the effect of the microenvironment in an ensemble during the reaction progress is developed in the next chapter.

3.2 Analysis of the Steady State Turnover Cycle in the Aqueous Phase

Dissection of the individual steps of the turnover cycle is based on the constraints that assure that all the observed turnover cycles occur through one and the same reaction path. For the purpose of deriving a closed analytical relationship for the ensemble-averaged kinetic behavior from the observed flux, the steady state is defined as the conditions under which there is no significant change in the number or concentrations of the various molecular species in the kinetic path. With the simplifications discussed above, based on the original proposal by Michaelis and Menten, the simplest 'bare-bones' kinetic path is

$$E + S \underset{k_{-1}}{\overset{k_1}{\rightleftharpoons}} ES \overset{k_2}{\longrightarrow} E + P$$

In this situation the flux of reaction across each of the two consecutive steps must be the same; otherwise the concentration [ES] would change with time. Therefore the net flux, say v, would be given by the balance of fluxes that determine the fraction of the enzyme in the ES form. Thus,

$$v = k_1[E][S] - k_{-1}[ES] = k_2[ES] \tag{3.1}$$

Assuming that the total enzyme concentration is $[E_0] = [E] + [ES]$, the steady state flux is solved from Equation (3.1) in the classical form of the *Michaelis–Menten equation*:

$$v = \frac{k_{cat}[E_0][S]}{K_M + [S]} \tag{3.2}$$

where

$$k_{cat} = k_2 \tag{3.3}$$

and

$$K_M = \frac{k_{-1} + k_2}{k_1} \tag{3.4}$$

K_M is the *Michaelis constant* which inversely relates to the *affinity* of the enzyme for the substrate. If the chemical change is much slower than the dissociation rate of S from ES, i.e. $k_2 \ll k_{-1}$, K_M is simply the equilibrium dissociation constant for ES defined as $K_M = k_{-1}/k_1 = K_S$ (for the first equilibrium step in a standard kinetic nomenclature). The units of v are M/s, i.e. the concentration change per unit time, and k_{cat} is the *turnover number*, i.e. the rate per enzyme when saturated by substrate. Considering instead the average flux per enzyme,

$$j = \frac{v}{[E_0]} = \frac{k_{cat}[S]}{K_M + [S]} \tag{3.5}$$

one gets the average number of products produced per enzyme per unit time with units of s^{-1}. In this text we will focus on the flux j per enzyme rather than v.

3.2.1 The substrate concentration dependence of the flux

During the reaction progress, the concentration of free substrate, [S], decreases with time. Therefore a steady state with constant [ES] cannot be upheld unless substrate is replenished in a separate process. Alternatively, a steady state can be upheld approximately for as long as depletion of substrate is insignificant, i.e. for times much smaller than

$$\tau_0 = \frac{[S_0]}{v} = \frac{1}{j_0} = \frac{K_M + [S_0]}{k_{cat}[E_0]} \tag{3.6}$$

where $[S_0]$ is the initial concentration of free substrate. In this limit the *initial rate*, j_0, of the enzyme is given by Equation (3.5) with [S] replaced by $[S_0]$.

One way to ensure the steady state and an initial rate that lasts for a longer time is to use a large substrate-to-enzyme ratio, as given by Equation 3.6 (also see below). This has the consequence that the concentration of free substrate, [S] or $[S_0]$, is roughly the same as the total concentration of substrate. This is the common situation in most experiments with soluble enzymes. As will be seen later, for an interfacial enzyme, it is the binding to the interface that depends on the total substrate concentration present as the interface.

3.2.2 Time-dependent change in the rate

If there is no substrate replenishment, the rate of change is

$$\frac{d[S]}{dt} = -v = -\frac{k_{cat}[E_0][S]}{K_M + [S]} \tag{3.7}$$

This differential equation has the solution

$$\frac{k_{cat}[E_0]t}{K_M} = \ln\left(\frac{[S_0]}{[S]}\right) + \frac{[S_0] - [S]}{K_M} \tag{3.8}$$

This is the *integrated Michaelis–Menten (MM) equation*, which depends explicitly on the time, t. Introducing the amount of product formed, $[P] = [S_0] - [S]$, Equation (3.8) can be written as

$$\frac{k_{cat}\,[E_0]\,t}{K_M} = -\ln\left(1 - \frac{[P]}{[S_0]}\right) + \frac{[P]}{K_M} \tag{3.8a}$$

The integrated MM equation is based on the results from a continuously declining substrate concentration and contains the same information as the initial rates from Equation (3.5) determined at a number of different and constant substrate concentrations.

Physical chemistry has been called the science of drawing straight lines. In that tradition, both the steady state (Figure 3.3) and the integrated MM equation

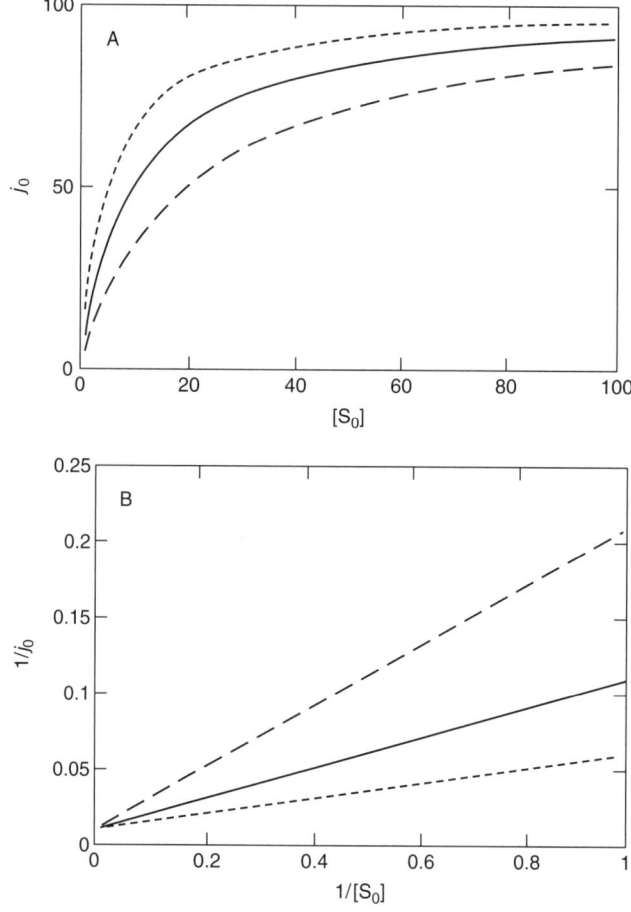

Figure 3.3 Panel A: the initial rate per enzyme as a function of the concentration of substrate (Equation 3.5). $k_{cat} = 100\ \text{s}^{-1}$, $K_M = 5$ (dot), 10 (continuous), 20 (dash) from upper to lower curve. Concentration units are arbitrary, but K_M is in the same units as $[S_0]$. Half of the maximum rate is reached when $[S_0] = K_M$. Panel B: Lineweaver–Burk plot (or inverse plot) of the same data as in panel A, where the inverse flux per enzyme, $1/j_0$, is plotted as a function of the inverse concentration, $1/[S_0]$, to generate a linear plot. The y intercept gives the inverse of the maximum rate, $1/k_{cat}$, and the slope is K_M/k_{cat}

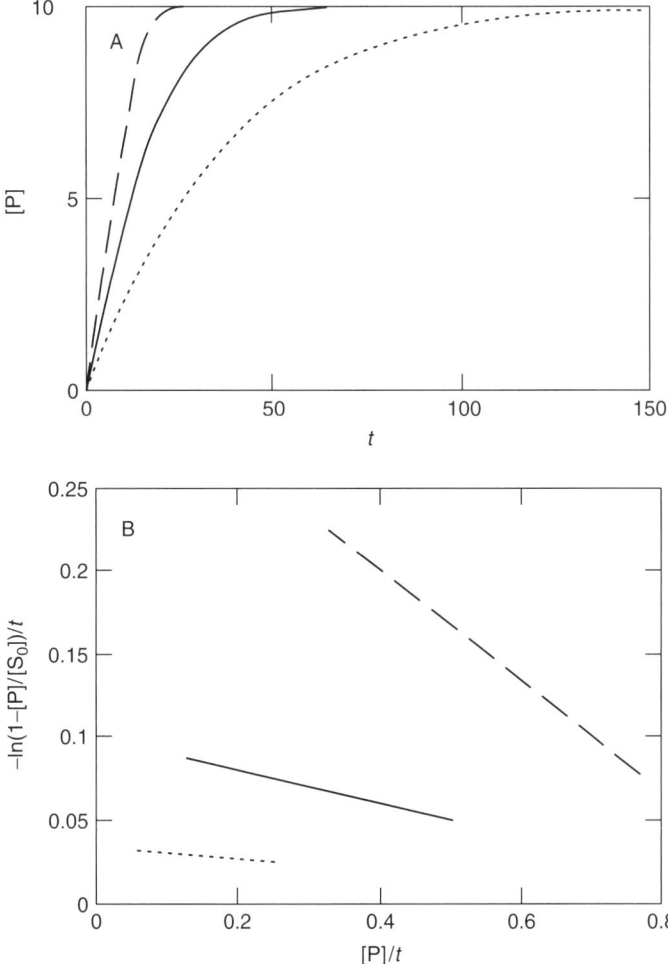

Figure 3.4 Panel A: concentration of product produced after time t from the integrated MM Equation (3.8a). $[S_0] = 10$ and $k_{cat}[E_0] = 1$ in all curves and $K_M = 3, 10, 30$ from upper to lower. Panel B: a linear plot of the same data as in the top panel: $-\ln(1 - [P]/[S_0])/t$ plotted versus $[P]/t$. The slope is $-1/K_M$, the extrapolated y intercept is $k_{cat}[E_0]/K_M$ and the extrapolated x intercept is $k_{cat}[E_0]$

(Figure 3.4) can be arranged to give linear plots. Although, for computational fits, it is often better to do nonlinear regression on the original data, such straight-line representations can be useful as diagnostics to investigate whether the basic relations are satisfied.

3.2.3 More assumptions

In addition to the constraints and the steady state assumption there are also a number of tacit assumptions behind this analysis, some of which may not be valid under certain conditions for interfacial turnover and therefore need to be spelt out:

(a) *Perfect mixing* of the components is required because the rate of formation of the enzyme–substrate complex is given by $k_1[E][S]$ as a product of the average concentrations. This requires that every free enzyme has the same probability of binding a substrate molecule, which in turn requires that all enzymes 'see' the same environment.

(b) There is *no back reaction* in the initial rate since the product concentration that the enzyme 'sees' is arranged to be vanishingly small initially.

(c) The *variables*, such as the initial concentrations of the various reactants as well as the environmental conditions, can be *manipulated independently* by the experimenter.

These conditions are usually satisfied in homogeneous aqueous solution. The substrate, product and enzyme molecules are dispersed as single molecular species (monodispersed) as explicitly considered for the description of the turnover cycle. Of course, if the substrate is sparingly soluble or if the substrate forms an aggregate, these relations break down because the $E + S$ to ES equilibrium depends on the concentrations of the monodisperse enzyme and the monodisperse substrate which interact to form monodisperse ES. As we will see in this and subsequent chapters, other factors that influence the observed rate for a soluble enzyme include not only the allosteric effects but also the formation of a microaggregate of the enzyme with excess substrate and the interactions at the extraneous interfaces under certain conditions. In short, the variables for the turnover cycle in the local environment of the enzyme must be explicitly defined.

3.2.4   Significance of the primary kinetic parameters

It can hardly be overemphasized that virtually all the experience of evolution of an enzyme is ultimately reflected in the functional parameters. The kinetic constants and derived parameters quantitatively describe the behavior of the kinetic path in relation to the variables that are included. Such information about an elementary step is related to the unique structural features including the conformational state. Therefore, it is useful for correlation of the structural features, with virtually all functions ranging from the substrate and inhibitor specificity to the catalytic mechanism and allostery. In short, the primary constants for the elementary steps of a kinetic path have fundamental kinetic and thermodynamic significance with specific structural and conformational origins. Note that a change in the kinetic path amounts to a change in the mechanism concomitant with a change in the significance of the parameter.

3.3   Ascertaining the Reaction Path in the Aqueous Phase

Consider a leading question of general interest: does an interfacial enzyme carry out the catalytic turnover through the solution path? Such information is useful not only to circumvent problems intrinsic in most assays of interfacial turnover but also for understanding the role of the interface by comparing the parameters for the solution and interfacial paths (Scheme I). The conceptual strategy and the protocols used to resolve issues of steady state turnover in the aqueous phase are outlined in this section. Use of the aqueous solution of a substrate, below its solubility limit or below the critical micellization concentration (CMC), offers the possibility of studying the catalytic turnover through the monodisperse ES complex. For the integrity of the turnover cycle, here the assumption would be that not only are the free substrate and the free enzyme monodisperse but the ES complex is also monodisperse as it undergoes the chemical change to EP. Not only must the critical assumptions be verified, but other possible reaction paths must also be eliminated.

The results in Figure 3.5 clearly show that the *Humicola* lipase catalyzed rate of hydrolysis of monodisperse PNPB (*p*-nitrophenylbutyrate) in the unstirred reaction mixture is virtually negligible. A burst of activity is observed during the mixing of the cuvette contents, and a sustained high unstirred rate is observed in the presence of POPG vesicles. This protocol is useful for evaluating the contribution of the reaction that occurs on extraneous surfaces in contact with the bulk solution phase. Note that the jump in the signal occurs only with shaking and stirring, and after that the slope is the same as before stirring. This is because the accumulated product of reaction at the walls during the unstirred period is brought into the beam path only during mixing.

Note that shaking has little effect on the background rate of hydrolysis of PNPB, an activated ester, seen with the catalytically inert S144A mutant of the

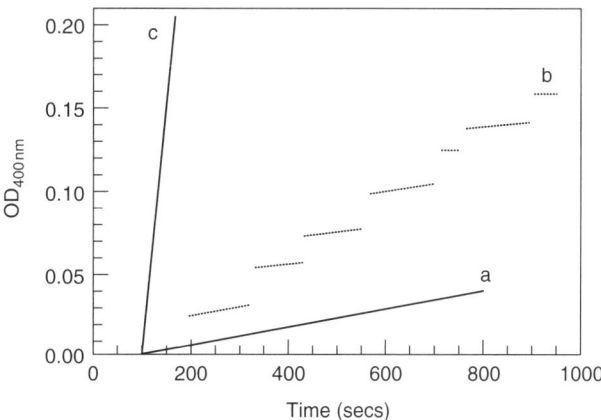

Figure 3.5 Effect of occasional shaking of the reaction mixture on the time course of the hydrolysis of unstirred solution (after the initial mixing of all the components) of monodisperse *p*-nitrophenyl butyrate by *Humicola* triglyceride lipase. The reaction progress monitored (a) continuously, (b) with intermittent shaking during the breaks and (c) continuously in the presence of POPG (1-palmitoyl-2-oleoyl-phosphatidylglycerol) vesicles. Reaction progress virtually identical to that seen in (a) was also obtained with the catalytically inert S144A mutant, where shaking or the presence of POPG had no noticeable effect (not shown). The increase in the optical density during the shaking is due to mixing of the products formed at the vessel walls as well as the enhanced rate of hydrolysis on the surfaces of air bubbles. As developed in Chapter 7, the rate enhancement with the POPG vesicles is due to the activation of the bound enzyme by the interfacial anionic charge. Reprinted with permission from Berg *et al.* (1998). Copyright (1998) American Chemical Society

lipase (Figure 3.5). Such controls show that shaking the reaction mixture brings the products from the vessel walls into the beam path, and also transiently creates extraneous interfaces of the air bubbles where the interfacial reaction of the lipase could occur. The contribution of the reaction on the two cuvette surfaces directly in the beam path, as well as that in the intervening bulk aqueous phase, is recorded at all times during the reaction progress under the unstirred conditions. However, the contributions from the other surfaces of the cuvette and the variable contribution from the reaction on the surfaces of air bubbles depend on the vigor and time of shaking. Such assay conditions are not reliable for the kinetic analysis because the variables at the extraneous surfaces cannot be quantified.

Another protocol for eliminating the contribution of the reaction on all the surfaces of a cuvette is shown in Figure 3.6 (Yu *et al.*, 1999a). Here the time course of hydrolysis of the monodisperse dihexanoylphosphatidylcholine by pancreatic phospholipase A2 (PLA2) is monitored as a change in the fluorescence of a suitable pH-indicator dye in the linear range of the signal. The difference between the rate without and with stirring clearly shows that virtually all the hydrolysis occurs on the cuvette walls. The upper limit estimate for the rate through monodisperse ES is less than 0.05 s^{-1}, compared to a rate of 300 s^{-1} or more at the interfaces (Chapters 5 and 8). Comparable differences between stirred and unstirred rates have also been found with several other lipolytic enzymes (Yu *et al.*, 1999a, 2000).

The anomalous effects of mixing, stirring and shaking of the reaction cocktail raise a cautionary flag. Contributions of extraneous surfaces are not entirely surprising for the interfacial enzymes. The problem from the reaction at the extraneous surfaces is likely to be more significant under the conditions where the observed activity in the bulk aqueous phase is low. Also the extraneous surface contribution would be

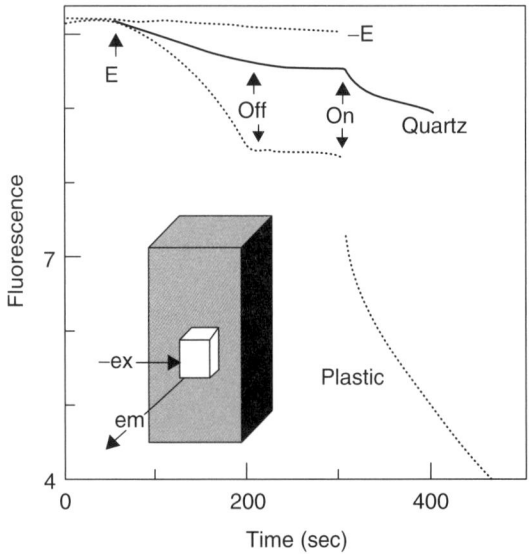

Figure 3.6 Effect of stirring on the reaction progress for the hydrolysis of dihexanoylphosphatidylcholine by pancreatic PLA2 in quartz or plastic fluorescence cuvette. Reprinted with permission from Yu *et al.* (1999a). Copyright (1999) American Chemical Society

relatively more prominent under the conditions where the turnover at the interface is low due to a less favorable E to E^* equilibrium. Such tell-tale signs of uncontrolled variables include unpredictable changes in the observed rate with a change in the reaction conditions. Even under apparently constant or comparable conditions, such uncontrolled variables from the extraneous surfaces introduce noise, as well as operator- and instrument-dependent variability.

3.3.1   Premicellar aggregates

The problem of designing suitable controls for measuring the monomer rate and the interpretation is not straightforward, even when significant activity is observed with the monodisperse substrate in the unstirred reaction mixture. Here the reaction may occur through the monodisperse ES complex, or the E^*S complex may be present as a premicellar aggregate of the enzyme with substrate amphiphiles. For example, the unstirred rate of hydrolysis of dihexanoyl-PC by PLA2 from bee venom PLA2 is significant and only modestly lower than the stirred rate (Figure 3.7). Also note that the pig PLA2 catalyzed rate of hydrolysis of monodisperse anionic dioctanoylphos-phatidylmethanol is significant and it does not change with stirring. Such differences observed under the monomer assay conditions are not necessarily due to an intrinsic difference in the substrate specificity of pig PLA2. In fact, virtually all secreted PLA2 form premicellar aggregates with monodisperse anionic amphiphiles, and certain PLA2s also form such aggregates with short chain phosphatidylcholines (Van Eijk *et al.*, 1983; Van Oort *et al.*, 1985a, 1985b). The assertion that there is a kinetic path through the premicellar aggregates formed under the standard 'monomer' kinetic conditions is also supported by the inhibition kinetics, which suggests formation of E^*I complex in the presence of monodisperse zwitterionic phosphatidylcholines

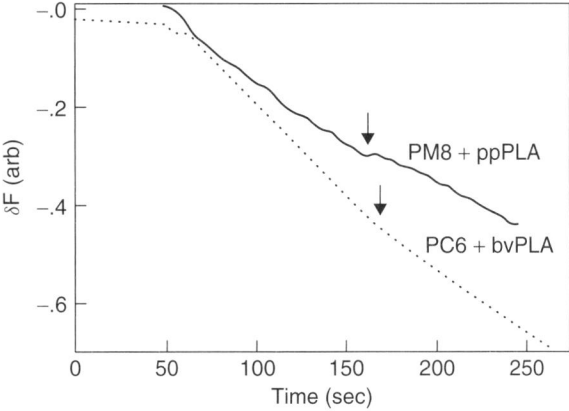

Figure 3.7   The reaction progress curves for the hydrolysis of 0.048 mM dihexanoylphosphatidyl-choline (PC6) with bee venom PLA2 (dots) or for the hydrolysis of dioctanoylphosphatidyl-methanol (PM8) by pig pancreatic PLA2 (line). Under the conditions of Figure 3.6, the reaction was started in a plastic cuvette by adding the enzyme at 50 s, and stirring was turned off at about 170 seconds, as indicated by the arrows. Based on the results in Yu *et al.* (1999a)

as the substrate (Rogers *et al.*, 1992). Together, observations with monodisperse substrate show that the ES form of at least some of the interfacial enzymes is catalytically inert and that premicellar aggregate formation needs to be carefully considered in other cases.

3.4 PAF-Acetylhydrolase Hydrolyzes the Monodisperse Substrate

In this section we summarize the kinetic consequences of the suggestion that matrix enzymes access their substrate from the aqueous phase (Min *et al.*, 1999). For numerous membrane-associated enzymes acting on polar water-soluble substrates, such as the sodium + potassium activated ATPase, it is assumed that the protein directly accesses the substrate from the aqueous phase. The assumption may be reasonable for polar solutes. However, evidence for such an assertion cannot be developed from the steady state kinetic analysis alone. Consider PAF-hydrolase which cleaves the *sn*-2 substituent of monodisperse PAF analogs, 1-alkyl- or 1-acyl-2-acetyl-phosphatidylcholines (Min *et al.*, 1999, 2000). Results in Figure 3.8 show that the rate increases with increasing substrate concentration. The observed rate is constant above the CMC, reflecting the constant concentration of monodisperse substrate. The observed rate per enzyme reaches a maximum of about $40\,s^{-1}$ for all three substrates, implying a lack of k_{cat} specificity for the sn-1-chain. Independent checks, analogous to those discussed in the preceding section, showed that the reaction with monodisperse substrate does not occur at the vessel walls. This is consistent with the solution reaction path either through ES or through E*S in premicellar aggregates. There is no direct evidence for the formation of premicellar aggregates, and unlike the case in Figure 3.8 the concentration dependence for the formation of the premicellar aggregates is likely to be sigmoidal, reflecting cooperativity of the aggregation process. However, the evidence that PAF-hydrolase binds to the interface raises several interesting possibilities discussed below.

Note that the solution mechanism is not incompatible with the matrix mechanism. However, the results in Figure 3.8 suggest that the intrinsic rates for the free or bound forms are indistinguishable; i.e. there is no interfacial activation. For the solution reaction path, the monodisperse substrate concentration dependence is given by

$$E + S \rightleftharpoons ES \rightarrow \text{(to products)}$$

However, in accord with the suggestion in Figure 1.2 there are two other paths that are possible above the CMC for the enzyme bound to the interface. In the *matrix mechanism* the path will be

$$E \longrightarrow E^* + S \rightleftharpoons E^*S \longrightarrow E^* + \text{products}$$

where the interface-bound enzyme accesses the substrate directly from aqueous solution. On the other hand, in the *interfacial mechanism* the path will be

$$E + S \rightleftharpoons E^* + S^* \rightleftharpoons E^*S \longrightarrow E^* + \text{products}$$

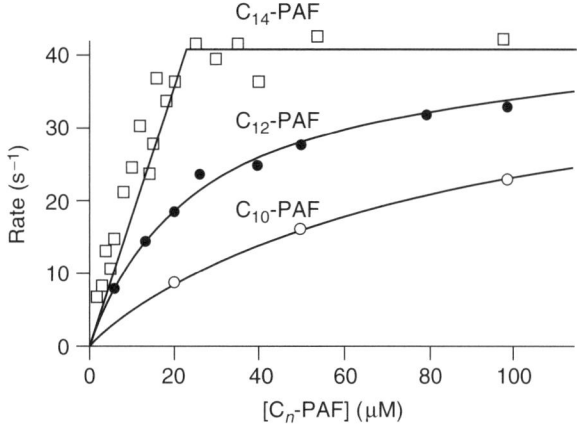

Figure 3.8 The rate of hydrolysis of 1-alkyl-2-acetyl-*sn*-glycero-3-phosphocholines (C_n-PAF) by PAF-acetylhydrolase as a function of the substrate concentration. The CMC for the three homologs are about 20, 110 and 1000 μM. Reprinted with permission from Min *et al.* (1999). Copyright (1999) American Chemical Society

where the substrate is accessed directly from the interface. Above the CMC, the interface-bound enzyme will also 'see' a constant concentration ($X_S^* = 1$) of substrate in the interface, and the rate would be constant even though the total substrate concentration, $[S_T]$, in the mixture increases. The interfacial mechanism could also be consistent with the substrate concentration dependence of the rate below the CMC, if the enzyme triggers the formation of premicellar aggregates.

3.4.1 Effects of the interface in a matrix mechanism

The dilemma posed by the matrix versus interfacial mechanisms outlined above can be formulated to define the problem. If the substrate amphiphiles are the only interface-forming molecules in the system, the concentration of the accessible substrate in solution, $[S]$, will be

$$[S] = [S_T] \qquad \text{for } [S_T] < K_S'$$

and

$$[S] = K_S' = CMC \qquad \text{for } [S_T] > K_S'$$

Therefore, the Michaelis–Menten equation for the reaction in solution, Equation (3.5), will assume the form

$$j = \frac{k_{cat}[S]}{K_M + [S]} = \begin{cases} \dfrac{k_{cat}[S_T]}{K_M + [S_T]} & \text{for } [S_T] < K_S' \\[3mm] \dfrac{k_{cat}K_S'}{K_M + K_S'} = j_{max} & \text{for } [S_T] > K_S' \end{cases} \qquad (3.9)$$

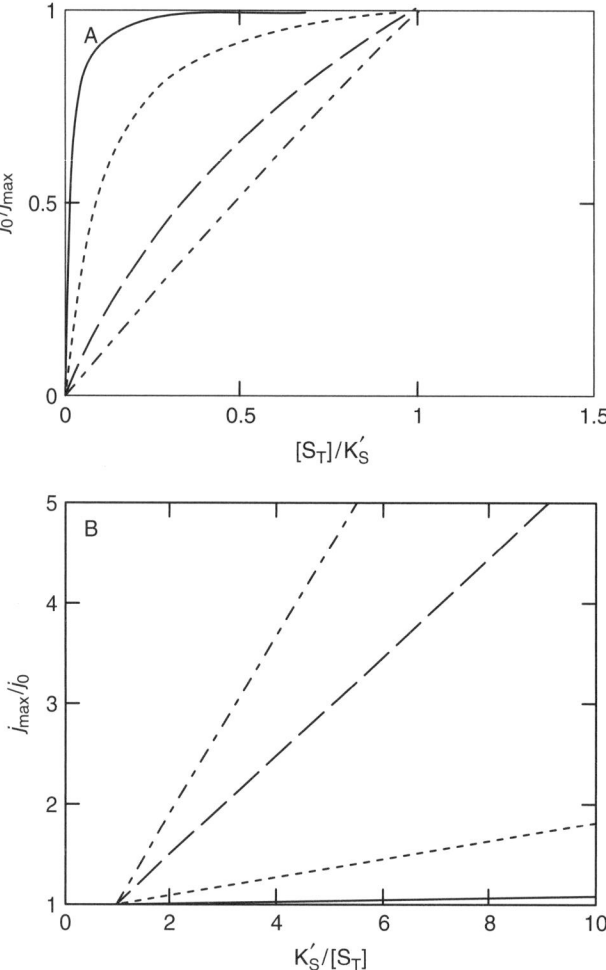

Figure 3.9 Panel A: predicted substrate concentration dependence for the matrix mechanism as described by Equation 3.9. j_0/j_{max} versus $[S_T]/K'_S$ for $K_M/K'_S = 0.01, 0.1, 1, 10$ (from upper to lower curves). Panel B: Lineweaver–Burk plot, j_{max}/j_0 versus $K'_S/[S_T]$ below the CMC for the results in the panel above. From upper to lower, the curves are for $K_M/K'_S = 10, 1, 0.1, 0.01$. From Equation (3.9), this is a linear plot where the slope equals $K_M/(K_M + K'_S)$

With increasing $[S_T]$, the flux increases to a maximum at $[S_T] \geqslant K'_S$. The results for a range of different values of K_M/K'_S are plotted in Figure 3.9, which are analogous to the experimental results in Figure 3.8. Thus, k_{cat} and K_M for the matrix mechanism can be found by fitting the rate to the MM equation at changing substrate concentrations below the CMC, assuming that the kinetic path is indeed that of a matrix enzyme.

Equation (3.9) describes the flux from an enzyme, solution or matrix, which accesses its substrate from solution. The expression assumes that the interface has no effect on the activity of the enzyme whether it is bound or not. Thus one expects the activity to follow a Michaelis–Menten relation below the CMC and be constant above. For an interfacial enzyme, on the other hand, no activity is expected below the CMC, unless premicellar aggregates are formed. It is also possible that a matrix enzyme is active only when bound in the interface, in which case no activity is expected below the CMC either.

3.4.2 Partitioning of the monodisperse substrate

The results in Figure 3.10 show that the rate of hydrolysis of two homologous PAFs decreases with added concentration of phospholipid vesicles. The decrease is much larger with the longer chain substrate. As a first approximation this is consistent with the matrix mechanism because the monodisperse concentration of the longer chain C_{16}-PAF would decrease more rapidly than that of C_{10}-PAF. However, as developed below, these results do not necessarily rule out the interfacial mechanism because increasing concentration of the vesicle interface not only increases the fraction of the bound enzyme as related to K_d but increasing the surface area also dilutes the interfacial mole fraction of the partitioned substrate.

3.4.3 Nonequilibrium effects

From the discussion of the equilibrium relations above and those developed in the following section, it is clear that without independent evidence steady state kinetics cannot distinguish between the cases when the enzyme and/or substrate are in solution or at the interface for the catalytic turnover. As long as the enzyme accesses the substrate and the substrate partitioning is at equilibrium, it does not matter for steady state kinetics where the substrate or the enzyme is localized. Thus both the effect of the CMC (Equation 3.9 and Figure 3.8) and the equilibrium partitioning of the substrate (Figure 3.10 and Equation 3.10 in Section 3.5.2 below) can be rationalized in terms of the interfacial as well as the matrix mechanism.

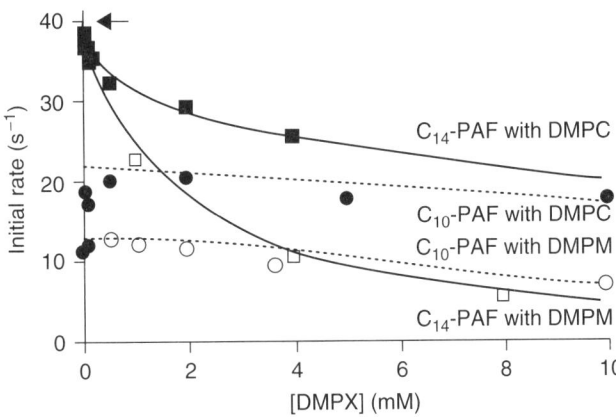

Figure 3.10 The rate of hydrolysis of $50\,\mu M$ 1-alkyl-2-acetyl-*sn*-glycero-3-phosphocholines (C_n-PAF) by PAF-acetylhydrolase in the presence of added DMPM or DMPC vesicles. Reprinted with permission from Min *et al.* (1999). Copyright (1999) American Chemical Society

Conditions under which all the interacting species are not equilibrated offer a window to manipulate the substrate concentration in the microenvironment of the enzyme. If the enzyme binds tightly to vesicles, the substrate concentration in the aqueous phase can be manipulated during the steady state turnover by addition of a fresh aliquot of vesicles. The outcome of suitably designed order-of-addition experiments under such conditions was tested under conditions where the substrate and PAF-hydrolases are localized on different vesicles (Min *et al.*, 1999). Imagine two sets of vesicles, V containing only DMPM and V_{nbd} containing 5 mole % of C_{12}-NBD-PC in DMPM, subjected to the hydrolysis by PAF-hydrolase under the following sets of conditions:

(a) The observed rate of hydrolysis of V alone was $0.012 \, s^{-1}$, and the rate remains unchanged with equal amounts of V_{nbd} added after the onset of the reaction progress. This would be the case if the enzyme bound to V cannot access the substrate added as V_{nbd}, i.e. both NBD substrate and the enzyme are irreversibly bound to the vesicle population to which it is added first.

(b) The rate of hydrolysis of V_{nbd} alone was $0.024 \, s^{-1}$, and it decreased to $0.01 \, s^{-1}$ after the addition of an equal amount of V. Soon after the addition of V, the enzyme bound to V_{nbd} 'sees' a lower concentration of NBD substrate in the aqueous phase.

(c) The difference due to the order of addition of vesicles is not seen with vesicles containing C_6-NBD-PC, which equilibrates rapidly between vesicles because its concentration in the aqueous phase is significantly higher.

These order-of-addition experiments are designed to take advantage of the fact that the exchange rate for the long-chain substrate between the vesicles is exceedingly slow so that the reequilibration time for substrate over the vesicles is very long. The time to reach a new equilibrium concentration, [S], in solution, however, is expected to be very short (Theory Box 3.1). Qualitatively result (a) suggests that the bound enzyme does not see the substrate added as V_{nbd}. This is consistent with an irreversibly bound interfacial enzyme, or with a matrix enzyme if the substrate concentration in solution is too small to increase the enzyme rate above the background. The main result is in experiment (b), which shows that the enzyme cannot access the substrate that is available in the same vesicles where it is bound once the added substrate-free vesicles have been added. The added vesicles quickly soak up most of the free substrate in solution and the enzyme rate decreases to background level. This is consistent with a matrix enzyme but not an interfacial one. However, quantitatively result (b) is not consistent, because with equal amounts of the two kinds of vesicles the aqueous phase concentration can only be halved, and not reduced to its background level. Therefore the anticipated rate would be $0.017 \, s^{-1}$ assuming a background rate of $0.01 \, s^{-1}$ and the initial rate of $0.024 \, s^{-1}$.

THEORY BOX 3.1 Desorption of an amphiphile from the interface at the steady state

Consider the situation where the enzyme accesses the free substrate from solution at concentration [S] while a large amount is bound in the interface at concentration [S*]. If [S] is small, the enzyme is not saturated and substrate will disappear with the rate $k_{cat}[E_0][S]/K_M$ (Equation 3.2). At the same time, free substrate can be replenished by desorption from the interface at rate $k_d[S^*]$ or bind to the interface with rate $k_a[S][M^*]$. Thus the rate of change of [S] is given by

$$\frac{d[S]}{dt} = k_d\left[S^*\right] - k_a[S]\left[M^*\right] - \frac{k_{cat}\left[E_0\right]}{K_M}[S] \qquad (1)$$

At the steady state, this is zero and we find that

$$[S] = \frac{k_d\left[S^*\right]}{k_a\left[M^*\right] + k_{cat}\left[E_0\right]/K_M} \qquad (2)$$

If the first term dominates in the denominator, i.e. if $k_a[M^*] \gg k_{cat}[E_0]/K_M$, we recover the equilibrium partitioning (cf. Equation 3.10)

$$[S] = \frac{k_d\left[S^*\right]}{k_a\left[M^*\right]} = K_S'X_S^*$$

When the second term dominates in the denominator, we get instead

$$[S] = \frac{k_d\left[S^*\right]K_M}{k_{cat}\left[E_0\right]}$$

In this limit, the enzyme rate becomes

$$v = \frac{k_{cat}\left[E_0\right][S]}{K_M} = k_d\left[S^*\right] \qquad (3)$$

This is the desorption-limited rate. The distinguishing property of this limit is that v is independent of enzyme concentration [E_0], in contrast to the usual MM kinetics (Equation 3.2). Furthermore, it requires a free substrate to bind and be processed by the enzymes much faster than it can bind to the interface; this is not likely to hold except in very special cases. More often, binding to the interface is diffusion-limited, or nearly so. Since the interface constitutes a much larger target than the enzymes, the rate of binding the interface is expected to be much larger than the rate of binding the enzymes. Thus, even if one finds, as for PAF-acetylhydrolase, that the steady state enzyme rate is proportional to the desorption rate of substrate from the interface, this does not imply a desorption-limited rate. It is more likely to be a reflection of the equilibrium partitioning of substrate where the free concentration is proportional to k_d,

$$[S] = \frac{k_d\left[S^*\right]}{k_a\left[M^*\right]} = K_S'X_S^*$$

as discussed above.

From Equation (1) one finds that the time required to reach a new steady state or equilibrium distribution of free S is

$$\tau = \frac{1}{k_d + k_a\,[M^*] + k_{cat}\,[E_0]/K_M} \approx \frac{1}{k_a\,[M^*]} \tag{4}$$

In the diffusion-limited case (Theory Box 2.1) and with $[M^*] = 0.5$ mM, the equilibration time is 0.01 s. It should be noted that this is independent of the desorption rate (if $k_d \ll k_a[M^*]$), so that the equilibration time does not depend on the desorption rate even if all substrates are in the interface initially. This somewhat counterintuitive result can be rationalized since the amount that needs to go into solution is given by K_S', which in turn is proportional to k_d. Thus, as the rate of going into solution is also k_d, the equilibration time is independent of k_d. This is the time to equilibrate the solution concentration. The time required to redistribute solutes between different vesicles, on the other hand, is the inverse of the exchange rate, i.e. $1/k_d$ (see Theory Box 2.1), which could be very large.

3.4.4 Teleology

The belief in the validity of the matrix mechanism for PAF-hydrolase comes from other circumstantial evidence. PAF-hydrolase, a water-soluble protein, is found in human serum bound to lipoprotein particles. Members of this family are also found in other eukaryotes including fungi. The substrate specificity of the human enzyme is broad and not limited to phospholipids (Min *et al.*, 2000). It shows a preference for a smaller *sn*-2-acyl substituent in phospholipids and also for the smaller acyl substituent, as in nitrophenylbutyrate or tributylglycerol. However, the reaction rate decreases as the length of the *sn*-2 chain of PAF is increased from acetyl up to pentanoyl. Yet addition of a polar substituent, COOH or CHO, in place of the methyl at the end of the pentanoyl chain gives a substrate that is hydrolyzed nearly as efficiently as PAF. Although PAF hydrolase binds with high affinity ($K_d < 10$ µM) to vesicles and micelles of phospholipids, the rate of hydrolysis seen with such dispersions is generally lower than 0.1 s^{-1}. It appears that the natural substrate of PAF-hydrolase includes oxidized phospholipids bearing −COOH or CHO group at the end of a truncated *sn*-2-acyl chain.

3.5 Equilibrium Partitioning and Binding to Aggregates

In the context of the enzyme-catalyzed reaction at the interface an amphiphile can play two roles. It may form the interface and it may act as an active-site-directed ligand, L, that interacts with the enzyme. A combination of these two properties raises the challenge of distinguishing the various roles, including that of a neutral diluent amphiphile, which essentially serves only as a two-dimensional interface solvent without affinity for the active site of the bound enzyme.

3.5.1 Neutral diluent

In order to grasp the concept of an interface solvent, consider the second type of amphiphile, L, that can 'join in' or partition into the existing aggregates of the amphiphile (A) molecules. For example, in Figure 3.5, the POPG interface provides the two-dimensional solvent for the catalytic turnover. We define these as *neutral diluent* amphiphiles, which do not have the active site affinity. In this case the substrate by itself does not form the interface (developed further in Chapter 7). Similarly, in Figure 3.8, L is the PAF substrate (L) and its micelles also provide the interface above the CMC. On the other hand, in Figure 3.10, DMPM and DMPC provide the interface into which PAF is partitioned. As conceptualized in Figure 3.11, depending on the experimental conditions, a neutral diluent not only permits a control over the free and the bound forms of an amphiphile (L and L*) as well the E, E* and E*L forms of the enzyme. As developed in Chapter 6, this strategy is useful for determining the various equilibrium constants.

3.5.2 Defining equilibrium partitioning as the dissociation constant, K'_L

Consider the equilibrium partitioning of L into the A* aggregates:

$$L + A^* \underset{K'_L}{\rightleftharpoons} L^* + A^*$$

The rate of binding of L to the aggregates (forward reaction) must be proportional to the concentration of free L molecules, [L]. For the partitioning the total concentration of the molecules at the interface is $[M^*] = [A^* + L^*]$. Similarly, the rate of dissociation (reverse reaction) is proportional to [L*]. Thus at equilibrium one finds the concentration of the free ligand in the aqueous phase

$$[L] = K'_L \frac{[L^*]}{[M^*]} = K'_L X^*_L \tag{3.10}$$

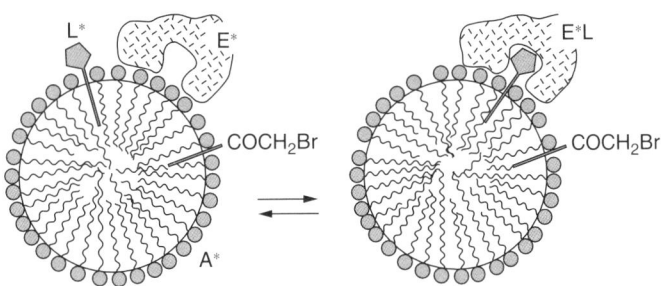

Figure 3.11 Conceptualization of a neutral diluent amphiphile (A), which forms the interface (in this case micellar), where A* has little affinity for the active site of E*, the enzyme bound to the diluent interface. On the other hand, an active-site-directed ligand, L*, partitioned in the interface of A* binds to the active site in relation to $X^*_L = [L^*]/([A^*] + [L^*])$. Thus A* permits a dissection of the binding of the enzyme to the interface along the i-face, versus the occupancy of the active site. As developed in Chapter 6, a reagent (such as α-bromoketone) modifies an active site residue in E* but not E*L

where X_L^* is the mole fraction of L in the interface. K_L' is the *dissociation constant* for an L molecule from an aggregate. It has the units of concentration per mole fraction, i.e. the $[M^*]$ concentration at which half of the total L is partitioned in the interface, or $[L] = [L^*]$. However, writing the relationship as $K_L' = [L][M^*]/[L^*]$ shows that K_L' can also be considered as a conventional dissociation constant with units M. Also note that when there is no A present, i.e. $X_L^* = 1$ in Equation (3.10), it would correspond to the CMC for L (Equation 2.7).

Traditionally the partition or distribution coefficient is defined as the ratio of the solute concentrations between the immiscible solvents. This definition is not easily extrapolated to the partitioning of solutes in the interface. In addition to the effect of the partitioned solute on the $[M^*]$ term, it is not trivial to define the volume of the membrane phase. Since the density of the bilayer is approximately 1, often the membrane volume is approximated from the lipid concentration. The problem of defining the accessible volume in the membrane or the interface is discussed further in Chapter 12. Thus, $1/K_L'$ corresponds to a partition coefficient that depends on the concentration units employed in the two phases, mole fraction and M.

3.5.3   Nonidealities

It is most likely that K_L' will depend on the amount of L present in the interface. For small X_L^*, the L molecule would be partitioning into a region where it would be interacting mostly with A molecules. For large X_L^*, on the other hand, each L molecule would be interacting mostly with other L molecules in the aggregates. Since these interactions may be different, the dissociation constant will also be different in this limit. Together such interactions can give rise to intractable nonideal behavior (Chapter 11).

If the presence of L molecules changes the partitioning of L molecules, the reverse must also be true, so that the addition of L molecules must change the aggregation of the A molecules. When L interacts differently with A than with other L, it must also be true that A interacts differently with L than with other A. This reciprocal dependence on the partitioning of L and A can be quantitated by application of the Gibbs–Duhem relation in the interface. Consider A^* as the interface solvent with mole fraction $X_A^* = 1 - X_L^*$ and L^* is the interface solute at mole fraction (concentration) X_L^*. This can be treated as a nonideal solution with activity coefficients that are different from 1 (Theory Box 11.3).

3.5.4   Ideal partitioning of a solute in micelles

Neglecting these nonideal contributions for the moment, we can consider the relations for the coupled equilibria for the partitioning of L and A:

$$[L] = K_L' X_L^* \tag{3.11}$$

$$[A] = K_A' \left(1 - X_L^*\right) \tag{3.12}$$

Combined with the relations for the total concentrations,

$$[A_T] = [A] + [A^*] = \left(1 - X_L^*\right)\left([M^*] + K_A'\right) \tag{3.13}$$

$$[L_T] = [L] + [L^*] = X_L^* \left([M^*] + K_L'\right) \tag{3.14}$$

$[M^*] = [A^*] + [L^*]$ from Equations (3.13) and (3.14) can be eliminated to give X_L^* as the root of a quadratic equation:

$$X_L^* = \frac{[A_T] + [L_T] + K_L' - K_A'}{2\left(K_L' - K_A'\right)} \left\{ 1 \pm \sqrt{1 - \frac{4[L_T]\left(K_L' - K_A'\right)}{\left([A_T] + [L_T] + K_L' - K_A'\right)^2}} \right\} \tag{3.15}$$

The sign must be chosen such that $0 \leqslant X_L^* \leqslant 1$. Furthermore, one finds that the concentration of interface molecules, $[M^*]$, can be expressed as

$$[M^*] = [A^*] + [L^*] = \frac{[A_T]}{1 - X_L^*} - K_A' \tag{3.16}$$

$[A_T] - K_A'$ is the expected concentration of interface molecules in the absence of L (Equation 3.11). Addition of L will drive both A and L molecules into the interface by increasing $[M^*]$. Thus the concentration of interface molecules depends on the amount of added L in a complicated way. When L is present at mole fraction X_L^*, the concentration of A* has increased by $X_L^* K_A'$ (cf. Equation 3.13).

From Equations (3.13) and (3.14) it can be seen that $[M^*] > 0$ only when $1 - [A_T]/K_A' < X_L^* < [L_T]/K_L'$ so that no interface forms when $[A_T]/K_A' + [L_T]/K_L' < 1$. Thus the combined concentrations of A and L that satisfy

$$[A_T]/K_A' + [L_T]/K_L' = 1 \tag{3.17}$$

corresponds to the CMC for A plus L if L* and A* form an ideal mixture. When the concentrations are below this limit, X_L^* is undefined and $[L] = [L_T]$, $[A] = [A_T]$ hold rather than Equations (3.11) to (3.16). The intermicellar concentrations (IMC) of L and A are determined by $[L]$ and $[A]$ from Equations (3.11) and (3.12) using the X_L^* value from Equation (3.15). If the mixture of L* and A* is not ideal, the relations above hold only within a narrow range of X_L^* values; alternatively, the constants K_A' and K_L' would have to be considered effectively dependent on X_L^* (see Theory Box 3.2).

THEORY BOX 3.2 Choice of concentration units in the interface

As in solution, the choice of concentration units in the interface is dictated both by thermodynamics and convenience. There are two basic choices: the concentration of L in the interface can be expressed either as the mole fraction:

$$X_L^* = [L^*] / \left([A^*] + [L^*]\right) \tag{1}$$

or as the surface density (molecules per unit surface area):

$$\rho_L = \frac{[L^*]}{a_A [A^*] + a_L [L^*]} = \frac{X_L^*}{a_A \left[1 - X_L^* \left(1 - a_L/a_A\right)\right]} = \frac{X_L^*}{a_A \left(1 + \varphi X_L^*\right)} \tag{2}$$

Here a_A or a_L is the surface area contributed by each molecule of A or L, respectively. In the last equality, the *area correction factor*,

$$\varphi = a_L/a_A - 1 \tag{3}$$

has been introduced. The equilibrium for the partitioning can be expressed either through the mole fraction or through the surface density

$$[L] = K'_L X^*_L = K''_L \rho_L \tag{4}$$

For a (strictly) ideal mixture in the interface, there would be no change in enthalpy or surface area when one A molecule in the interface is removed and replaced by an L molecule. In this case, $a_L = a_A$ and $\varphi = 0$, so that the relationship between the equilibrium constants is simply a multiplicative factor

$$K''_L = a_A K'_L \tag{5}$$

converting between the different concentration units. The same result (Equation 5) holds in the *dilute ideal* limit when $\varphi X^*_L \ll 1$. When this ideal limit is not satisfied, the conversion factor becomes concentration dependent:

$$K'_L = \frac{K''_L}{a_A \left(1 + \varphi X^*_L\right)} \tag{6}$$

and the equilibrium constants, K''_L and K'_L, cannot both be considered fundamental constants. It can be argued that ρ_L and therefore also K''_L are thermodynamically more fundamental (Theory Box 11.3), in which case K'_L becomes concentration-dependent in the general case.

Thus for molecules L and A that are similar in size and interactions, mole fraction is not only adequate but also the most convenient choice of concentration units. However, for dissimilar molecules, nonideal effects may become important and corrections for size and interactions may be needed, as discussed further in Chapters 7 and 11. The factor $1/(1 + \varphi X^*_L)$ can be considered as an activity coefficient accounting for nonideal effects to first order in X^*_L, and the partitioning equilibrium (Equation 3.11) becomes

$$[L] = \frac{K'_L X^*_L}{1 + \varphi X^*_L} \tag{7a}$$

or, equivalently,

$$X^*_L = \frac{[L]}{K'_S - \varphi[L]} \tag{7b}$$

Thus the total concentration of L molecules in the system is

$$[L_T] = [L] + [L^*] = \frac{K'_L X^*_L}{1 + \varphi X^*_L} + \frac{X^*_L}{1 - X^*_L} [A_T] \tag{8}$$

assuming for simplicity that all of the diluent A is in the interface. Equation (8) gives a quadratic relation from which X^*_L can be solved as a function of $[L_T]$ and $[A_T]$:

$$X_L^* = \frac{1}{2}\frac{[A_T]+[L_T](1-\varphi)+K_L'}{K_L'-\varphi([A_T]+[L_T])}\left\{1\pm\sqrt{1-\frac{4[L_T]\left[K_L'-\varphi([A_T]+[L_T])\right]}{\left[[A_T]+[L_T](1-\varphi)+K_L'\right]^2}}\right\}$$

(9)

This replaces Equation (3.15) to first order in a nonideal partitioning if $K_A' = 0$.

With three or more components the equilibrium relations quickly become intractable except in certain limits (Theory Box 9.1). Such generally intractable situations prevail in assays involving mixed micelles and other interfaces formed by mixing high mole ratios of dissimilar solutes such as detergent, phospholipids and the products of hydrolysis (Chapters 2 and 11).

3.5.5 Effect of equilibrium partitioning of the substrate on the steady state rate

To describe the activity of a matrix enzyme that accesses a substrate or inhibitor primarily from solution, it is important to know how these molecules are partitioned. Thus, the mole fraction X_L^*, which is a measure of the concentration of a ligand that the enzyme 'sees' in the interface, can be determined from Equation (3.15) as a function of the total concentrations $[A_T]$ and $[L_T]$ if K_A' and K_L' are known (Figure 3.12). The concentration of ligand that a matrix enzyme can access from

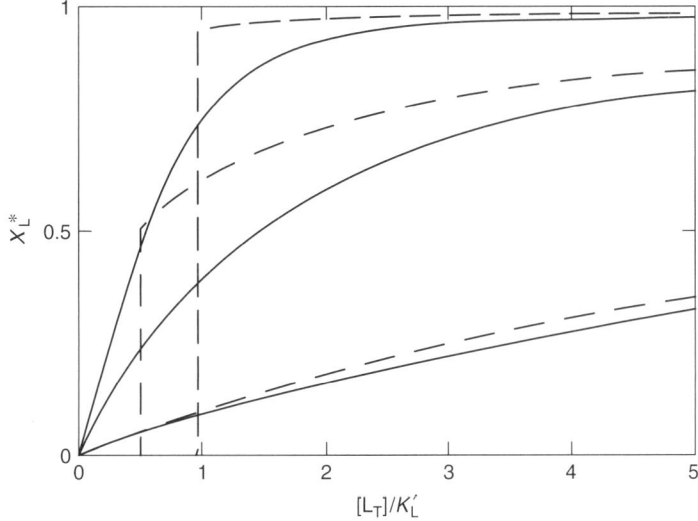

Figure 3.12 Mole fraction of L in the interface as a function of the total concentration of L (Equation 3.15). Solid curves are for $K_A' = 0.1K_L'$ and dashed curves are for $K_A' = 2K_L'$. In each set, the upper curve is with $[A_T] = 0.1K_L'$, the middle one for $[A_T] = K_L'$ and the lower one for $[A_T] = 10K_L'$. The break points correspond to the effective CMC of the combined A plus L from Equation (3.17)

solution is also determined by X_L^* through Equation (3.11). Similarly, the concentration of interface molecules to which the enzyme can bind is determined from Equation (3.16). The simplest case is, of course, when K_A' and K_L' are very small so that all molecules are in the interface. This does not change the relations derived above, but makes it a lot easier to calculate and account for the variations in X_L^* and $[M^*]$.

From the equilibrium relations developed above one can predict the effect on the enzyme-catalyzed rate of adding a second (inert) component. If L in Equations (3.11) to (3.16) above is the substrate S and A is a neutral diluent, the steady state reaction rate for the matrix enzyme is determined by

$$j = \frac{k_{cat}[S]}{K_M + [S]} \qquad (3.18)$$

where, from Equations (3.11) and (3.15),

$$[S] = K_S' X_S^*$$

$$= \frac{K_S'([A_T] + [S_T] + K_S' - K_A')}{2(K_S' - K_A')} \left\{ 1 \pm \sqrt{1 - \frac{4[S_T](K_S' - K_A')}{([A_T] + [S_T] + K_S' - K_A')^2}} \right\}$$

$$\text{for } \frac{[A_T]}{K_A'} + \frac{[S_T]}{K_S'} > 1 \qquad (3.19)$$

in the presence of an interface. At high dilution in the interface, i.e. for $X_S^* \ll 1$, the expression simplifies to

$$[S] \approx \frac{[S_T]K_S'}{[A_T] + [S_T] + K_S' - K_A'} \qquad \text{for } \frac{[A_T]}{K_A'} + \frac{[S_T]}{K_S'} > 1 \qquad (3.20a)$$

When there is no interface, all substrate is in solution

$$[S] = [S_T] \qquad \text{for } \frac{[A_T]}{K_A'} + \frac{[S_T]}{K_S'} < 1 \qquad (3.20b)$$

It is obvious from Equation (3.11) that the concentration of free substrate in solution, [S], will change in exactly the same way as its concentration in the interface, X_S^*, except of course when there is no interface present. Thus, as discussed also in Chapter 2, there will be little difference in the concentration dependence of the activity for an enzyme that accesses its substrate from the interface or from solution. The results are shown in Figure 3.13 as a function of $[S_T]$ and in Figure 3.14 as a function of $[A_T]$. The apparently linear Lineweaver–Burk plot in Figure 3.13, panel B, also for large values of $[S_T]$, shows that the partitioning relation (Equation 3.19) can lead to an apparent MM relation where $j_0 = k_{cat}^{app} [S_T] / (K_M^{app} + [S_T])$, also above the CMC.

These results are based on a steady state initial rate analysis and requires that accessible substrate in solution is not depleted during the course of the experiment. Under these conditions, the partitioning can be described by the equilibrium relations used above. As outlined in Theory Box 3.1, this is expected to hold under most situations even if desorption rates are very small.

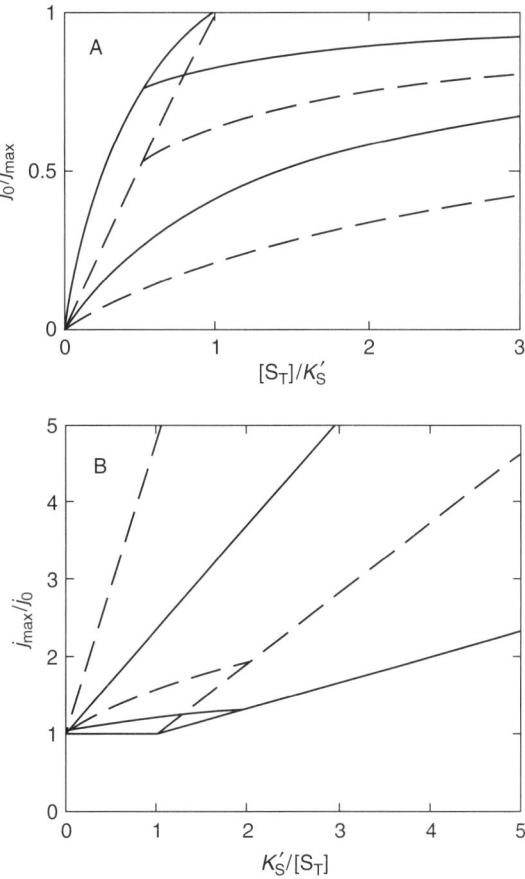

Figure 3.13 Panel A: the rate of a matrix enzyme in the presence of a neutral diluent A with $K'_A = 2K'_S$ relative to the maximum rate $(= k_{cat}/(1 + K_M/K'_S))$ in the absence of A. Solid curves are for $K_M/K'_S = 0.5$ and dashed curves are for $K_M/K'_S = 10$. In each set, the upper curve is for $[A_T] = 0$, the middle one for $[A_T] = K'_S$ and the lower one for $[A_T] = 5K'_S$. Panel B: Lineweaver–Burk plot of j_{max}/j_0 versus $K'_S/[S_T]$ with the same parameters as in panel A. The curves are approximately linear also below the break point, i.e. at large $[S_T]$, where S is partitioned into the neutral diluent interface

3.6 Summary: The Equilibrium Dilemma

In this chapter we considered the kinetic consequences of the equilibrium partitioning of the substrate, the S to S* equilibrium. Continuing on the theme of substrate accessibility and replenishment, the results selected for discussion in this chapter show that for a variety of reasons it is not trivial to ascertain the catalytic turnover path for an enzyme that acts on nonpolar and amphiphilic substrates:

(a) Results with PLA2 and triglyceride lipase emphasize the importance of alternative reaction paths that are possible under certain assay conditions, including the extraneous surfaces of air bubbles and the cuvette wall. These results provide an upper limit estimate for the reaction rate through the bulk aqueous

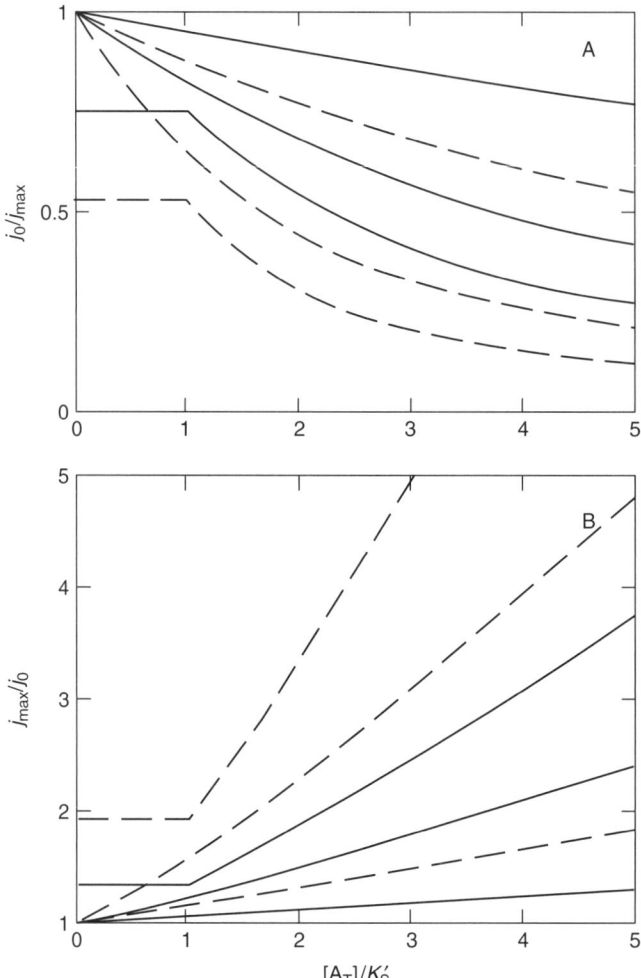

Figure 3.14 Panel A: the initial rate of a matrix enzyme as a function of the concentration of a neutral diluent A with $K'_A = 2K'_S$ relative to the maximum rate in the absence of A. Solid curves are for $K_M/K'_S = 0.5$ and dashed curves are for $K_M/K'_S = 10$. In each set, the upper curve is for $[S_T] = 5K'_S$, the middle one for $[S_T] = K'_S$, and the lower one for $[S_T] = 0.5K'_S$. Panel B: the inverse flux j_{max}/j_0 plotted as a function of $[A_T]/K'_S$. The curves are approximately linear with slopes that approach $K_M K'_S / ((K_M + K'_S)[S_T])$ at large concentrations $[A_T]$. The same parameters as in panel A were used

phase. However, the situation is further complicated by the possibility of the reaction path through premicellar aggregates of the enzyme with the substrate amphiphile to form E*S.

(b)   The analysis in this chapter clearly illustrates the difficulties in dealing with the equilibrium steps coupled to a catalytic turnover cycle. With such coupled equilibria the problem of defining the substrate concentration as a variable for steady state kinetics poses a very real dilemma for distinguishing a matrix enzyme that accesses the substrate from the aqueous phase from an interfacial enzyme that accesses the substrate from the interface. This dilemma emanates from the equilibrium relation (Equation 3.11), which shows that the interface and solution concentrations of the substrate are directly proportional to each other and will change in parallel when total concentrations of S or A are changed.

(c) The role of the interface for interfacial enzymes is several-fold. As a two-dimensional solvent in contact with the aqueous phase, the interface binds the enzyme and partitions the ligands. The interface also provides the medium for the two-dimensional equilibrium between an enzyme and the ligand in the interface. Such a distinction is emphasized by the choice of suitable units. The conceptual dissection is experimentally facilitated by interfaces of a suitable neutral diluent. The organized dispersions provide the phases of molecular dimensions, and it is clear that nonidealities are built into the definitions of the variables with which one has to deal in microscopically heterogeneous environments.

Without an understanding of such limitations, it is evident that the power of Scheme I to rationalize the observed reaction progress becomes a liability for the steady state kinetic analysis. As developed in the following chapters, within certain constraints it is possible to unequivocally define the variables for the reaction progress under certain well-defined conditions. Under such conditions with the knowledge of what the enzyme 'sees', rather than what is present in the reaction mixture, a rigorous kinetic analysis of the interfacial turnover cycle becomes possible.

Further Reading

Segel I.H. (1975). *Enzyme Kinetics: Behavior and Analysis of Rapid Equilibrium and Steady-State Enzyme Systems*, Wiley, New York, 957 pp.

4 Interfacial Processivity: Ensemble Behavior in the Scooting Mode

A new scientific truth does not triumph by convincing its opponents and making them see the light, but rather because its opponents eventually die, and a new generation grows up that is familiar with it.

Max Planck

The interfacial kinetic paradigm in Scheme I (Figure 3.1) is amenable to analysis only in parts for which a well-defined kinetic path can be ascertained under the assay conditions. This requires eliminating or constraining some or all of the exchange processes between the coexisting interfaces in the ensemble. For example, the interfacial catalytic turnover path for the steady state turnover on vesicles emerges under the conditions where the bound enzyme, substrate and products do not exchange or leave the interface during or between the successive turnover cycles. As developed in this and the next chapter, the resulting *scooting mode* reaction progress occurs with virtually *infinite interfacial processivity* for the turnover cycles at the interface of the vesicles to which the enzyme binds initially. The reaction progress in the scooting mode provides unequivocal direct evidence for the catalytic turnover at the interface.

The reaction progress in the scooting mode for the action of pig pancreatic PLA2 on vesicles of 1,2-dimyristoylphosphatidylmethanol (DMPM) is shown in Figure 4.1. The key result is that the enzyme bound to the anionic vesicles does not leave the interface and therefore the reaction ceases after only a fraction of the total substrate vesicles in the reaction mixture is hydrolyzed. This is a consequence of the ensemble behavior of the initial binding and random distribution of the enzyme over the vesicle population. Since the bound enzyme does not leave the interface, the kinetic path is described by the steps of the interfacial catalytic turnover cycle. Analysis of the extent of hydrolysis as a function of the enzyme vesicle ratio helps in setting up experimental conditions under which there is only one enzyme molecule on each enzyme-containing vesicle. As developed in subsequent chapters, analysis of the scooting mode reaction progress permits unequivocal characterization of the catalytic behavior in terms of the interfacial turnover parameters.

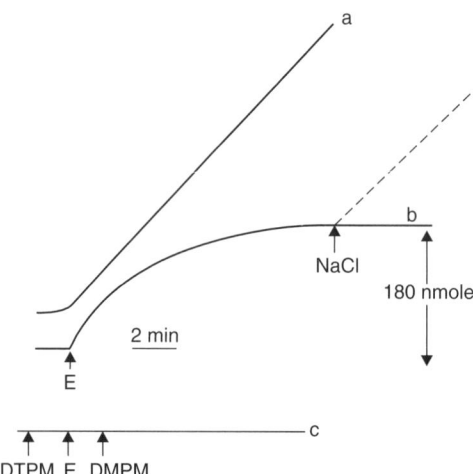

Figure 4.1   Reaction progress for the hydrolysis of small sonicated vesicles (with $N_S^0 = 4400$, the number of phospholipid molecules in the outer layer) of 1.3 micromole of 1,2-dimyristoylphosphatidylmethanol (DMPM) in (a) the presence of 0.4 M NaCl or (b) the absence of NaCl (or as indicated in a separate experiment NaCl was added near the end of the reaction progress) or (c) enzyme added to 1,2-ditetradecylphosphatidylmethanol (DTPM) vesicles followed by the DMPM substrate vesicles. Adapted from Jain *et al.* (1986a to 1986d). Copyright (1986) with permission from Elsevier Science

4.1   Conceptualizing Interfacial Processivity

For the interpretation of the apparently first-order scooting mode reaction progress in Figure 4.1, consider the idealized situation in Figure 4.2 for the reaction in an ensemble of one enzyme with two vesicles. The initial binding of the enzyme to the interface is a pre-steady state event, after which the bound enzyme does not leave the interface. Not only do the substrate and products remain in the vesicles, the vesicle-to-vesicle exchange or transfer of the substrate and products is also negligible on the time scale of the whole reaction progress. The bound enzyme does not leave the interface during, in between or after the turnover cycle. Therefore, *all* the successive catalytic turnover cycles of the observed reaction progress occur processively during the residence of the enzyme at the vesicle interface to which the enzyme binds initially. Two of the consequences of such high affinity binding of the enzyme to vesicles on the extent of hydrolysis at the end of the reaction progress are shown in Figure 4.3. The key result is that the extent of hydrolysis depends on the enzyme-to-vesicle ratio. Such an unusual dependence of the extent of hydrolysis results from the random distribution of the enzyme on vesicles, which depends on the enzyme-to-vesicle ratio.

4.1.1   Characteristics of the reaction progress in the scooting mode

Compared to the turnover in solution or the one with low processivity, the scooting mode reaction progress with infinite processivity (Theory Box 4.1) has several distinguishing features. For example, only the substrate present in the outer

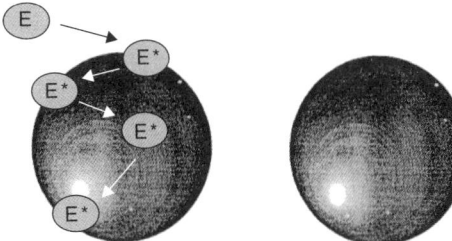

Figure 4.2 A cartoon to emphasize the virtually infinite processivity of the successive interfacial catalytic turnover cycles in the scooting mode by an enzyme bound to a vesicle. Under such conditions the bound enzyme does not exchange between vesicles, and thus excess vesicles, for example the second vesicle in this ensemble of one enzyme and two vesicles, are not hydrolyzed. Since E* 'sees' only the species present at the interface of the vesicle to which it binds initially, the substrate present in excess vesicles is not hydrolyzed. In the scooting mode, the vesicle-to-vesicle exchange rate of the substrate and products is also negligible on the time scale of the reaction progress. On the other hand, *hopping* of the enzyme, or the exchange of the substrate, would ultimately lead to the hydrolysis of all the accessible substrate present as excess vesicles. Adapted from Jain *et al.* (1986c) and Berg *et al.* (1991)

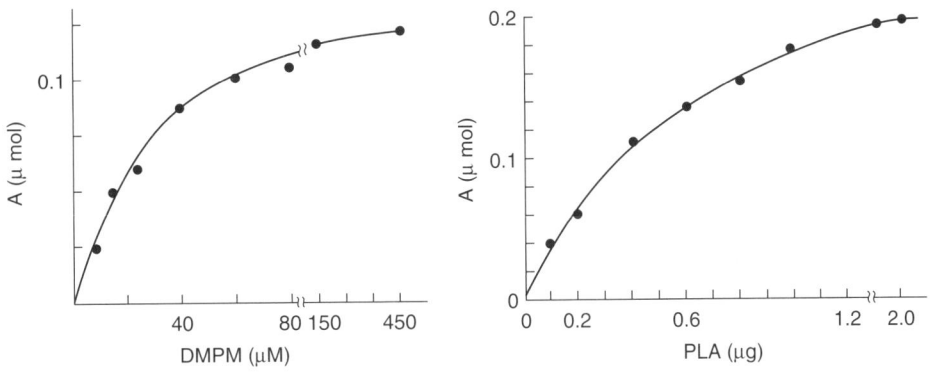

Figure 4.3 Effect of the two variables on the extent of hydrolysis (ordinate) of dimyristoylphosphatidylmethanol vesicles ($N_S^0 = 4400$) by pig pancreatic PLA2 at the end of the reaction progress in the scooting mode. These results were obtained under the conditions of Figure 4.1 (curve a). (Left) Effect of the total DMPM concentration (in 4 ml) with 30 pmole of pig pancreatic PLA2. (Right) Effect of the PLA2 concentration with 0.32 micromole of DMPM vesicles in a 4 ml reaction mixture. Reprinted from Jain *et al.* (1986c). Copyright (1986) with permission from Elsevier Science

monolayer of the enzyme-containing vesicles is hydrolyzed during the reaction progress. The substrate present in the inner layer of the vesicle is not hydrolyzed and neither is the substrate in excess vesicles, present initially or added to the reaction mixture during the reaction progress. Based on the model intrinsic in the large ensemble version of Figure 4.2 with many more vesicles and enzyme molecules, analysis of the scooting mode reaction progress is possible within the *constraints* that ensure that the microscopic conditions necessary for several types of *ensemble averaging* are satisfied. As developed later in Theory Box 4.2, the initial binding of the enzyme added to a vesicle population is governed by the *Poisson law for random distribution*. Therefore the extent of hydrolysis at the end of the reaction progress is determined by the fraction of vesicles that contain bound enzyme. As developed in the next chapter, a rigorous analysis of the integrated reaction progress in the scooting mode is rather straightforward because ensemble-averaging is possible with at most one enzyme per enzyme-containing vesicle. Only under such conditions can it be assumed that all enzyme molecules 'see' identical environments at all time points during the reaction progress.

4.1.2 Consequences of undefined processivity

Reaction progress in the scooting mode provides the near ideal conditions for the characterization and analysis of the events of the interfacial turnover cycle. The conditions for infinite processivity in the scooting mode are rather stringent. However, these are the only conditions in which the kinetic path for the steady state turnover is unequivocally defined only by the steps of the interfacial catalytic turnover cycle. Under assay conditions of unknown or low processivity, it is not possible to rule out the kinetic contributions of the hopping of the enzyme or of the rate-limiting exchange of the substrate and product from aggregate to aggregate. Such exchange processes introduce the possibility that the steps for the binding (k_{on}) and desorption (k_{off}) of the enzyme to the interface also make an unknowable kinetic contribution to the steps for the steady state turnover (Figure 1.8). In principle, such contributions can be adequately accounted for also in the extreme situation of the *hopping mode*, where only a small fraction of the substrates in a vesicle are hydrolyzed every time an enzyme binds the interface. In this case of very low processivity, all vesicles will be hydrolyzed to roughly the same extent throughout the reaction progress. The major problem is when the processivity is intermediate between very high and very low or changes during the reaction progress. For such reasons the reaction progress under the conditions of undefined processivity cannot be adequately analyzed to obtain insight into the primary events of interfacial turnover cycle.

THEORY BOX 4.1 Interfacial processivity

The processivity of the enzyme could be defined from the number of turnover cycles it performs during one binding event at the interface. In these terms, virtually infinite processivity would ensue when the enzyme is strongly bound. For the quantitative analysis it is more convenient to describe a processivity parameter, P, as the probability that the enzyme will perform one more catalytic cycle when one has just concluded. Thus, consider the reaction scheme

$$\uparrow k_{off} \quad E^* \underset{k^*_{-1}}{\overset{k^*_1}{\rightleftarrows}} E^*S \overset{k^*_{cat}}{\longrightarrow} \tag{1}$$

After one catalytic cycle is concluded at the interface, the enzyme starts in state E^*. There are only two possible paths: either the enzyme dissociates with rate constant k_{off} or it concludes another cycle by taking the full path to the right. This path to the right may include a number of back-and-forth steps between E* and E*S, but only one catalytic event is allowed. It is straightforward to calculate the probability that it takes the full path to the right rather than dissociating, and one finds

$$P = \frac{k_{cat}^* X_S^* / K_M^*}{k_{off} + k_{cat}^* X_S^* / K_M^*} \tag{2}$$

where $K_M^* = (k_{cat}^* + k_{-1}^*)/k_1^*$ is the interfacial Michaelis constant (cf. Equation 3.4). In this way, the processivity will depend on the concentration, mole fraction X_S^*, of substrate in the interface, since a large X_S^* will tend to drive the process in Equation (1) to the right. The calculation relies on the assumption that E*S does not dissociate. It is easily shown that the same expression (Equation 2) results from a more complete reaction involving also the state E*P as a separate species if k_{cat}^* and K_M^* are identified appropriately (Theory Box 5.1). In fact, this is the meaning of k_{cat}^*/K_M^* as the apparent second-order rate constant of catalysis for $E^* + S^*$. If X_S^* remains fairly constant during a binding event, the probability that a binding event involves exactly n catalytic cycles is

$$p_n = P^n(1 - P) \tag{3}$$

This is a geometric distribution with average $P/(1 - P)$. This gives the connection between the processivity parameter P (defined above as the probability of carrying out one more round of catalysis) and the processivity considered as the average number of catalytic cycles per binding event; when P approaches one, the processivity $P/(1 - P)$ approaches infinity.

4.2 The Fourfold Meaning of the Substrate Concentration

Concentration is a thermodynamic term related to the statistical probability of random encounter of the reactant molecules in thermal motion. In a heterogeneous system the local significance of such encounters will depend on the local environment. This raises the possibility of many microscopic state functions each with a unique significance for the microheterogeneity encountered by the reactants at the interface. Consider the meaning of the concentration variables that are based on the total substrate present in the reaction mixture. In a heterogeneous environment such a probability of encounter is not identical at all points in the reaction volume but differs with the local environment for the event. The significance of the concentration term would depend on the constraints that determine the encounter probability in the microenvironment of the species that participate in the event of interest. For our analyses, without consideration of the microscopic states, the *total solute concentration*, $[A_T]$ or $[S_T]$, is the average number density of A or S molecules per volume of aqueous phase. It is related by the Avogadro number to the commonly expressed units of moles of amphiphile per liter of the solution. In the limit of dilute solution, it is the same as the moles per liter of solvent.

Considering what an enzyme 'sees' microscopically in a reaction mixture within the context of Scheme I under different reaction paths and for different steps of the turnover cycle, at least four 'local' primary variables are derived from the total amphiphile concentration $[S_T]$:

(a)   The monomer substrate concentration, $[S]$, in the aqueous phase in equilibrium with the S^* species in the interface is commonly expressed in units of moles of monodisperse amphiphile per liter. The CMC and the $[S]$ in the partitioning equilibrium are related to K_S' (Chapter 2). For vesicles of long chain phospholipids, the CMC is exceedingly low, which also makes the spontaneous transfer or exchange of phospholipids between vesicles through the monomer path (Table 2.4) negligible on the time scale of hours, if not days.

(b)   As a first approximation, the interface concentration, $[S^*]$, of an amphiphile that controls the E to E^* equilibrium increases linearly with $[S_T] -$ CMC. This term and K_d are expressed as moles of amphiphile per liter of aqueous phase. However, the actual interface accessible to the enzyme in the aqueous phase may range from 100 % for micelles, about 70 % for the highly curved smallest unilamellar vesicles to about 50 % for the large unilamellar vesicles (Figure 2.3). The fraction of phospholipid molecules in the outermost monolayer of hand-shaken multilamellar vesicles, with about 10 to 15 concentric bilayer lamellae, is typically 10 to 6 %.

(c)   The Poissonian distribution of the enzyme population on the vesicle population is predicted by the enzyme over vesicle (E/V) mole ratio (Theory Box 4.2); E/V is unit-less. The vesicle concentration is equal to $[S_T] - [S]$ divided by the number of amphiphiles in both layers of a vesicle.

(d)   The two-dimensional substrate concentration that E^* 'sees' for the formation of the E^*S complex for the catalytic turnover at the interface is the number of substrate molecules per unit area or the number density of the substrate molecules in the interface in contact with the bound enzyme. At the contact point of the enzyme at the interface, the number density and the substrate conformation may not necessarily be the same as in the rest of the interface along the i-face; i.e. what matters for the formation of E^*S is the microenvironment in the contact region of the enzyme with the interface. As a first approximation within the limit of the equilibration of all such species present in an interface, the 'substrate concentration' variable for the interfacial turnover kinetic path is most conveniently approximated as X_S^* expressed in units of the mole fraction of the substrate in the interface (Theory Box 3.2). In a single component system, X_S^* has the maximum possible value of 1, and the value would be lower with codispersed amphiphiles. With X_L^* as the variable, the units of K_L^* (L = P, I, S), as well as K_M^*, are also mole fraction (not mole ratio). The local concentration can also be defined from the total volume that an interface-bound substrate molecule can access (Chapter 12).

4.2.1   Variables change with the quality of the interface

It is obvious that the concentration variables are ultimately related to the organization of the dispersed amphiphile. Although $[S_T]$ may not change for the various organized aggregates, the relative and absolute values of the microscopic concentration variables could change dramatically. The properties that control the accessibility and replenishment rates, and thus change one or more of the concentration variables, include the composition, morphology, shape, size, dispersity and the phase behavior. Unless the reacting species are completely equilibrated on the time scale of the single turnover time, the variables in the microenvironment of the bound enzyme are unlikely to be the same as the variables obtained as the average over the total environment. Such quality-of-the-interface effects have been the source of considerable confusion and debate. As a first step towards resolving such difficulties, a detailed knowledge of the concentration variables in a heterogeneous environment is necessary. As developed below, it requires a critical evaluation of the properties of dispersions under the microscopic conditions relevant for the events of a kinetic path.

4.3 Constraints for Defining the Variables for the Reaction Progress in the Scooting Mode

For interfacial kinetic analysis it is necessary to define the variables for the events that make up the kinetic path. The observed reaction progress is the sum total of the changes in an ensemble. Such changes are measured as the changes in the number of the substrate or product molecules in the observed space as a function of time. For the mechanistic interpretation of such changes by ensemble averaging we need to be sure that at any given point in time each and every enzyme molecule 'sees' and 'operates' in identical microenvironments. For the analysis of a PLA2-catalyzed reaction in the scooting mode, the ensemble-averaging over a time window is based on the following considerations and criteria:

(a) Reaction progress in the scooting mode is monitored after the enzyme is added to a population of substrate vesicles. The number of enzyme molecules per enzyme-containing vesicle is determined by the enzyme-to-vesicle (E/V) ratio by the Poisson distribution (Theory Box 4.2). For example, more than 95 % of the enzyme-containing vesicles will have at most one enzyme per vesicle at $E/V < 0.1$. The reaction progress in the scooting mode is a measure of the reaction occurring on the enzyme-containing vesicles; therefore the extent of hydrolysis at the end of the reaction progress is the measure of the enzyme-containing vesicles. The excess vesicles remain enzyme-free through the course of the reaction progress if the bound enzymes do not exchange.

(b) As a rule, the reaction progress in the scooting mode is possible only with vesicles of long chain phospholipids whose concentration in the aqueous phase is virtually zero. Negligible CMC eliminates the possibilities of the turnover through a monodisperse ES complex and of the substrate replenishment from the excess vesicles.

(c) The *integrity* of the vesicles through the reaction progress is retained. Only the substrate molecules in the outer layer of the enzyme-containing vesicles are hydrolyzed (Wilshut *et al.*, 1978, 1979; Jain and de Haas, 1983; Yu *et al.*, 1997a; Cajal and Jain, 1997; Cajal *et al.*, 1995, 1996a, 1996b). The integrity of a vesicle cannot be maintained if a significant amount of the product, lysophospholipid and fatty acid, is removed or introduced. Here the key point is that even though the products may be able to exchange with the products in other vesicles, such an exchange does not occur. What matters for maintaining the integrity of vesicles during the reaction progress in the scooting mode is that there is neither a net transfer of the products nor an exchange with the substrate present in excess vesicles.

(d) The microenvironment for each and every enzyme molecule at the interface during the course of the reaction progress is ultimately determined by the E/V ratio as well as by the vesicle size. Imagine a *population* of vesicles in which each vesicle is of nearly identical size, say with 5000 ($= N_S^0$) phospholipid molecules in the outer layer. With at most one enzyme per enzyme-containing vesicle the microenvironment of each vesicle changes with time in the same fashion. However, if the turnover number is of the order of 100 s^{-1} per enzyme, the substrate depletion in the enzyme-containing vesicle is rapid. Thus the microscopic steady state condition for the initial rate, say $X_S^* = 1$, is not maintained for more than a few seconds even though only a small fraction of the total substrate is depleted. The substrate depletion on the enzyme-containing vesicles in the ensemble is predicted by the integrated reaction progress in the scooting mode (see Equation 5.8). Also, as discussed later in this chapter, the initial steady state zero-order reaction progress can be extended either by using larger vesicles or by inducing rapid substrate replenishment from the excess vesicles.

(e) The *dispersity* of the vesicle size also contributes towards the assumption about the homogeneity of the microenvironment on the enzyme-containing vesicle (Theory Box 4.2). Imagine a reaction mixture containing a bimodal population of vesicles with 5000 and 500 000 substrate molecules in the outer layer accessible to the bound enzyme. If the turnover rate at $X_S^* = 1$ is 100 s^{-1}, after 10 seconds on the vesicle containing one enzyme, X_P^* would be approximately 0.2 on the smaller vesicle compared to 0.002 on the larger vesicle. Thus, in order to ensure that each enzyme in the reaction mixture sees an identical environment, it is necessary that the reaction mixture contain a vesicle population with a narrow size dispersity. In short, for ensemble-averaging it is necessary that, under the conditions with one enzyme per enzyme-containing vesicle, all the enzyme molecules 'see' an identical environment through the whole course of the reaction progress.

(f) Note that the microheterogeneity concerns for the ensemble-averaging also come into play if there is more than one enzyme per vesicle (Theory Box 4.2). In principle, the results can be analyzed if the number of enzymes per vesicle is exactly identical on all enzyme-containing vesicles. This is practically impossible except in the limit of one enzyme per enzyme-containing vesicle where other vesicles in the population would have no enzyme. According to the Poisson law, if E/V = 2 (i.e. on average two enzymes per vesicle), then a significant number of vesicles will also have one or three enzyme molecules, and the number of vesicles with none as well as four, five or six enzymes would not be negligible.

THEORY BOX 4.2 Poissonian enzyme distribution over vesicles: ensemble-averaging

If enzymes are distributed at random over the vesicles, the probability that a vesicle carries n enzymes is given by the Poisson law as

$$P_n = \frac{(E/V)^n}{n!} e^{-E/V} \qquad (1)$$

where $E/V = [E]/[V]$ is the ratio of the overall concentrations of enzymes and vesicles. Similarly, the fraction q_n of enzymes that are on vesicles with a total of n enzymes is also Poissonian:

$$q_n = \frac{(E/V)^{n-1}}{(n-1)!} e^{-E/V} \qquad (2)$$

Thus the *probability* that an enzyme is alone at a vesicle is $q_1 = e^{-E/V}$. When $E/V < 0.1$, $q_1 > 0.905$ and less than 10 % of the enzymes will be on vesicles carrying more than one enzyme.

At time t, a vesicle with n enzymes will have produced a mole fraction $X_P^*(t;n)$ of product (e.g. as given in Equation 5.8 for $n_E = n$). Consequently, the total product concentration at time t will be determined by an average over the vesicle population:

$$[P(t)] = [V]N_S^0 \sum_n P_n X_P^*(t;n) = [S_0^*] \sum_n P_n X_P^*(t;n) \qquad (3)$$

Here, N_S^0 is the number of substrate molecules initially on the accessible surface of a vesicle, while $[S_0^*]$ is the total initial concentration of interfacial substrate that is accessible to enzyme. For simplicity, let us consider the case where the equilibrium is far to the product side so that $X_P^*(t = \infty) = 1$, except for the vesicles lacking enzymes ($n = 0$) where $X_P^* = 0$. Then, the final concentration of product will be

$$[P(\infty)] = [S_0^*] \left(1 - e^{-E/V}\right) \qquad (4)$$

Thus the Poissonian distribution and the lack of exchange can be tested by studying the final product concentration at varying $[E]/[V] = E/V$. With large amounts of enzyme, the maximal amount of product produced equals the initial amount of accessible substrate.

Equation (3) is the relation for *ensemble-averaging* over the vesicles with different extents of substrate depletion. When there is no exchange, each subsystem will develop independently of all others. The sum cannot be calculated in general. However, when E/V is small so that almost all enzymes appear singly on a vesicle, only one term, $n = 1$, contributes to the sum in Equation (3), and Equation (5.8) can be used with $n_E = 1$. The averaging in Equation (3) still assumes that all vesicles are of the same size, N_S^0. If there is a very heterogeneous size distribution, there should be another averaging over this distribution as well (Berg *et al.*, 1991). On the other hand, a very heterogeneous size distribution would probably also break the Poissonian distribution in Equations (1) and (2) since larger vesicles would be more likely to bind more enzymes, possibly in relation to their surface area. Thus, to study the enzyme kinetics beyond the initial rates in systems without exchange requires both that the vesicles are of a single size and that the enzyme-to-vesicle ratio is sufficiently small that there is at most one enzyme per vesicle (Figure 5.3).

4.4 Integrity of Vesicles during the Scooting Mode Reaction Progress

Ensemble-averaging is required for the analysis of the observed reaction progress obtained with an ensemble of vesicles and enzyme molecules. This condition is satisfied with phospholipid vesicles of narrow-size dispersity. These are suited for the reaction progress in the scooting mode because the vesicles retain their integrity through the course of the reaction progress and none of the interacting species exchange between vesicle population. As outlined in Chapter 2, there is a linear relationship between the accessible interface area and the phospholipid concentration. The vesicle morphology is stable over a broad concentration range, which makes it possible to determine the accessible area, surface density, size distribution, partitioning of solutes and exchange of the components between the coexisting interfaces. The motional and conformational properties of phospholipids in bilayer vesicles are outlined below.

4.4.1 Stability of bilayer vesicles

Phospholipids, with a strong polar group at one end of the cylindrical shape (Figure 4.4), spontaneously disperse in water as bilayer-enclosed vesicles. Only the number of vesicles and the overall surface area, but not the organization or morphology, change with the total concentration of dispersed phospholipid. Vesicles are formed spontaneously from the components without the help of a catalyst. The thermodynamics and kinetics of bilayer assembly and vesiculation of the hydrated lamellae during the hydration of phospholipids is very favorable. Driven by the hydrophobic effect, phospholipids are held together in a lamellar plane, generally without specific intermolecular interactions. The magnitude of the hydrophobic drive for the long acyl chains of natural phospholipids is about 25 kcal/mole of a phospholipid with about 30 to 35 methylene residues. This ensures that both the chains of no more than one phospholipid molecule in 10^{15} are in contact with bulk water. This lowers the CMC to the subpicomolar range and drives vesiculation by 'sealing the frayed ends' of a bilayer sheet.

4.4.2 Hydration of phospholipids

Phospholipids crystallize in organized arrays in which polar groups are in contact with each other and not in intermolecular contact with the hydrophobic chains (Figure 2.2). For example, different phospholipids crystallize in bilayer arrays

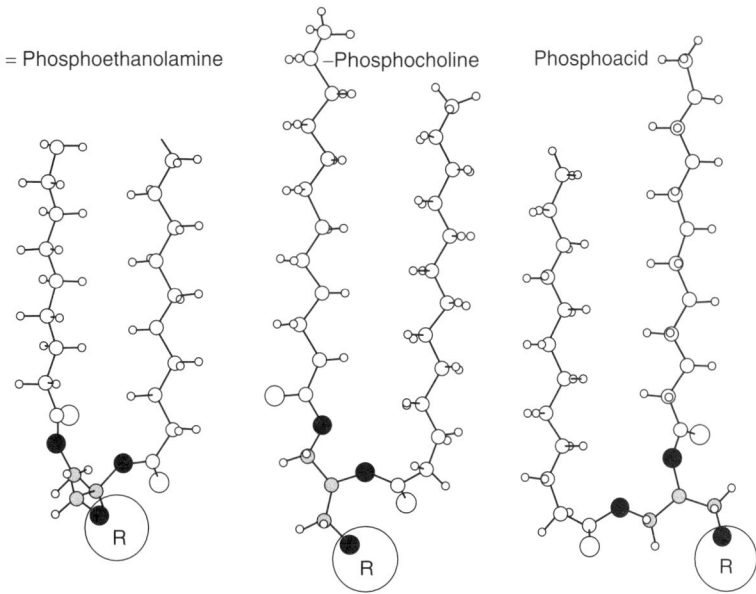

Figure 4.4 The preferred conformations of the diacylglycerol moiety in the crystallographic structures of phospholipids (Hauser *et al.*, 1981; Hauser, 1984; Pascher and Sundell, 1986) with different polar head groups with phosphoethanolamine (PE), phosphocholine (PC), and phosphate (PA) moieties (=R) in Figure 1.6. Note that the orientation of the glycerol backbone carbons (gray circles) is quite different in the three structures

with the glycerol carbons oriented at different angles from the lamellar plane (Figure 4.4). The *sn*-1-acyl chain continues the *trans* configuration; however, in all cases the *sn*-2-acyl chain initially projects at 90° from the glycerol and then bends another 90° at C-2 and then continues parallel to the *sn*-1 chain. Lateral packing of the head groups in the crystal structure depends on the detailed structure of the head group. Ionic and hydrogen-bonding intermolecular interactions are seen only under certain conditions. In phosphatidyl choline crystals, the phosphate to choline intermolecular distance is about 4 Å, which may also permit intermolecular hydrogen bonding with a neighboring amino or a phosphate group. Neighboring phosphates are bonded through two water molecules. Based on the NMR evidence, the bend in the *sn*-2 chain C-2 apparently persists in the vesicles. Hydration of the phospholipids at interfaces requires 10 to 20 water molecules for each phospholipid head group, and the excess water coexists as a separate phase in equilibrium with the water partitioned into the bilayer. Available evidence suggests that at least some of the conformational features seen in the crystal packing persist in the organized interfaces formed in dispersions of hydrated phospholipids in excess water.

4.4.3 Size and dispersity

The size and dispersity of vesicles depends largely on the method of vesicle preparation; however, the lipid composition and the environment (pH, temperature, salt) also make significant contributions. For example, the smallest vesicles of narrow size dispersity (20 to 25 nm diameter) with a single bilayer are formed by ultrasonic disruption of the multilamellar dispersions of hydrated phospholipids (Figure 2.3). Typically, the smallest unilamellar vesicles have about 4000 to 7000 molecules of saturated phospholipids with a cross-sectional area of about 50 Å^2 or about 3000 molecules of unsaturated phospholipids with a cross-sectional area of about 75 Å^2. Extruded unilamellar vesicles of narrow-size dispersity are obtained with filters of defined pore size of 20 to 2000 nm diameter. Such preparations contain less than 5000 to over 100 000 phospholipid molecules in each vesicle. About 65 % of the total phospholipid molecules in the smallest vesicles is exposed to the external aqueous phase, and of course the proportion changes with the vesicles size and approaches 50 % in vesicles of >100 nm diameter.

4.4.4   Conformational and motional constraints within the bilayer

As summarized in Table 2.4, the various intra- and intermolecular motional changes in bilayer occur on the time scale of a few picoseconds to weeks. Some of the motional properties of the phospholipid molecules are constrained by the bilayer organization. The segmental, rotational and lateral diffusion motions within the plane of the bilayer are rapid. In contrast, the vesicle-to-vesicle exchange rate is slow because it requires the chains to make contact with water. Similarly, the transbilayer exchange (flip-flop) is slow partially because the polar head group has to be transiently removed from water.

The bilayer organization, driven by the hydrophobic effect, has weak inter-chain interactions and sometimes a two-dimensional net of hydrogen bonds involving the polar group and the water molecules. The temperature and solute-induced changes in the ratio of the *trans–gauche* rotamers along the CH_2-CH_2 bonds in alkyl chains can generate a range of conformers of the acyl chains. Considering the low free-energy difference between the *trans – gauche* rotamers with a low activation energy barrier (Table 2.4), the population distribution of such conformers is probably dictated largely by the packing constraints related to the environmental variables and the statistical factors. Such changes within the confines of the bilayer organization are probably also responsible for the apparent cooperativity of the thermotropic or solute-induced gel-fluid (order–disorder) transition in the acyl chains of phospholipids. If both the chains are saturated and longer than 14 carbons the transition temperature exceeds room temperature. Solutes, including other amphiphiles, partitioned in the bilayer can induce isothermal phase changes. The transition temperature of the dispersions of natural phospho- and glycolipids are well below the physiological temperature. Considering the fact that the local organization of phospholipids is based on noncovalent interactions, it is fair to conclude that the organization in contact with an interfacial enzyme will not necessarily follow the rules for the phase propensities of phospholipids alone.

4.4.5   Organizational polymorphism and metastability

A full understanding of the organizational and the phase properties of the substrate dispersion is critical because such factors can result in a variety of anomalous

Table 4.1 Enzymes that show scooting mode reaction progress

Enzyme	Vesicle charge/substrate
Secreted 14 kDa PLA2	Anionic
Cytoplasmic 87 kDa PLA2	Anionic
Acyltransferase, bacterial	Anionic
Sphingomyelinase, bacterial	Sphingomyelin
PI-3′-hydroxykinase	Phosphatidyl inositol
Phosphatidic acid-PLA1	Anionic

kinetic effects (Chapters 10 and 11). Only general guidelines follow. Certain phospholipids do assume a non-bilayer organization depending on their structures, as well as on the composition and environment. Occasionally, metastability and complex kinetics of the changes in the dispersed phases show hysteresis on the time scale of minutes and hours. In fact, an understanding of the equilibrium phase behavior may not be entirely relevant. Stability and lifetime of the organizational phases and polymorphs formed with the products of hydrolysis is the key consideration during the processive reaction progress at the interface. As the substrate-to-product ratio changes, the bilayer organization could also change monotonically or abruptly. The origins of such effects are difficult to evaluate from the equilibrium measurements. Such concerns have to be addressed for each dispersed form of the substrate. As described below, anomalous changes in the bilayer organization do not occur during the scooting mode reaction progress with PLA2 and several other enzymes (Table 4.1). It is not unlikely that the integrity of target vesicles may be maintained under a broader set of conditions if only the outer layer lipids are attacked.

4.5 Tests for the Scooting Mode Reaction Progress by PLA2

The conditions for the kinetics of pancreatic PLA2 catalyzed hydrolysis of anionic phospholipid vesicles have been extensively examined to ascertain that the scooting mode reaction progress is due to processive turnovers at the interface. The analysis of the turnover cycle developed in the next chapter is based on the assumptions about the variables in the microenvironment of the reaction path for the steady state turnover, developed from the following constraints.

4.5.1 Extent of hydrolysis increases with the vesicle size

At high vesicle-to-enzyme ratios, only a fraction of the total substrate is hydrolyzed at the end of reaction progress. The extent of hydrolysis per enzyme in the presence of excess vesicles increases with the size of the vesicles (Figure 4.5, Theory Box 4.2). The extent of hydrolysis remains the same for PLA2 from different sources.

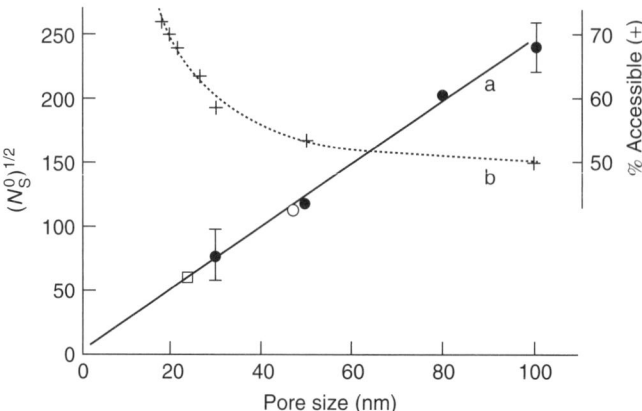

Figure 4.5   Dependence of (a) the square root of the number (= N_S^0) of phospholipids hydrolyzed per enzyme at E/V < 0.1 or (b) the percent of the total phospholipid molecules accessible to the excess of added PLA2 (at E/V > 10) on the POPG vesicle size expressed as the pore size of the filters through which the vesicles were extruded. Reprinted with permission from Berg *et al.* (1991). Copyright (1991) American Chemical Society

4.5.2   Product and substrate do not exchange

The reaction progress in the scooting mode occurs without the exchange of product with the substrate present as excess vesicles. Only the substrate present in the outer layer of the vesicles to which the enzyme is bound is hydrolyzed. For example, only 7 to 10 % of the substrate is hydrolyzed in multilamellar vesicles. As shown in Figure 4.5, the fraction of the substrate accessible for the total hydrolysis at the end of the reaction progress decreases from 68 % for the smallest vesicles to 50 % for the large unilamellar vesicles. Due to geometrical factors, the proportion of the total phospholipid that is in the outer layer increases for a smaller radius of curvature of the vesicles.

Additional evidence that the integrity of hydrolyzed vesicles is retained is shown in Figure 4.6, where only the probe present in the outer monolayer is quenched by dithionite. The NBD probe in the inner layer remains inaccessible unless vesicles are disrupted by detergent. Thus, during the scooting mode reaction progress by PLA2, only the fraction of the total substrate corresponding to that present in the outer layer is hydrolyzed. As pointed out before, even if the self-exchange of the products distributed in vesicles were to take place, the exchange with the substrate in the inner layer or in the excess vesicles is unlikely through the monomer exchange path. Of course net transfer of fatty acid and lysophospholipid out of the bilayer, as brought about in the presence of serum albumin in the aqueous phase, disrupts the PLA2 hydrolyzed vesicles.

4.5.3   Vesicle integrity is maintained

A key assumption for the analysis of the reaction progress in the scooting mode is that the number of substrate molecules on the enzyme-containing vesicles does

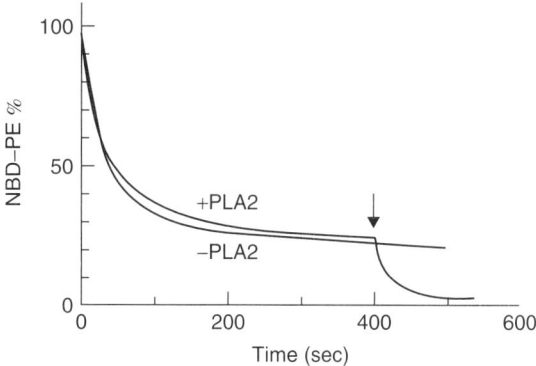

Figure 4.6 Time course of the reaction of dithionite with NBD-PE (0.6 mole %) in POPC/DMPM vesicles treated or not treated with PLA2. All the probe molecules become accessible when vesicles are treated with deoxycholate added at the arrow. Reprinted with permission from Yu *et al.* (1997a). Copyright (1997) American Chemical Society

not change except through the catalytic turnover. One of the simplest protocols to ascertain the integrity of hydrolyzed vesicles is shown in Figure 4.6 (Yu *et al.*, 1997a), which shows that neither the NBD-probe lipid from the inner layer nor the water-soluble dithionite reagent from the aqueous phase cross the bilayer.

4.5.4   Binding to the interface as a pre-steady state event

Results in Figures 4.5 and 4.6 show that the integrity of vesicles in the scooting mode is maintained after the binding of PLA2 and even when virtually all the available substrate in the outer monolayer is hydrolyzed by PLA2. Thus hydrolysis of DMPM vesicles takes place without exchange or transbilayer transfer of the enzyme, substrate or the product of hydrolysis on the time of the reaction progress. This ensemble behavior during the scooting mode reaction progress suggests that the affinity of PLA2 for the interface is very high. This conclusion is consistent with the direct binding results (Chapter 6) as well as the order-of-addition protocols under the kinetic conditions (Figure 4.1).

4.5.5   The order of addition and the distribution of the components

As shown in Figure 4.1, the enzyme added to the nonhydrolyzable DTPM vesicles does not hydrolyze the DMPM substrate vesicles added afterwards. On the other hand, DTPM vesicles added after initiating the reaction with DMPM do not change the course of the reaction progress. Inability of the enzyme, substrate and product molecules to exchange freely and rapidly creates a clear distinction between the pre-steady state binding and the conditions that develop during the steady state reaction progress. This difference is particularly dramatic in terms of the effect of the order of addition of the components because the microenvironments are not identical under such conditions even though the averaged global conditions are identical. For example, the substrate vesicles added after the initial addition of the enzyme to the reaction mixture do not react with the bound enzyme in the same way as the excess vesicles present at the start of the reaction.

4.5.6 Substrate replenishment conditions

Compelling evidence for the substrate depletion during the scooting mode reaction progress on the enzyme-containing vesicle also comes from two other sets of conditions. First, as described next, the extent of hydrolysis per enzyme (Figure 4.3) follows the Poissonian distribution of enzyme over vesicles. Second, as developed later in this and the next chapter, the reaction progress can be resumed by inducing the substrate replenishment on the enzyme-containing vesicles through the vesicle-to-vesicle transfer of phospholipids or by the salt-induced exchange of the bound enzyme (Jain *et al.*, 1991c). As expected, the substrate replenishment effects do not depend on the source of the enzyme.

4.6 Ensemble Behavior of the Binding of the Enzyme to Vesicles

The observations outlined in the preceding section leave little doubt that the DMPM vesicles exhibit a near-ideal ensemble behavior in which the extent of hydrolysis provides a direct measure of the number of enzyme-containing vesicles. This property is useful for designing protocols that provide otherwise inaccessible but useful information.

4.6.1 Counting the catalytically active enzyme species

Under suitable conditions, the measure of the extent of hydrolysis per enzyme molecule provides an independent check for the size or number of catalytically active units in a preparation of the enzyme or mutants. Since the vesicle size can be independently calibrated, from the weight of the enzyme in the reaction mixture it is possible to calculate the weight of the catalytically active species per vesicle in the presence of excess vesicles (see below).

4.6.2 Poisson distribution of enzyme over vesicles

The extent of hydrolysis at the end of the reaction progress in the scooting mode is due to the hydrolysis of the phospholipid in the outer monolayer of all the enzyme-containing vesicles. As shown in Figure 4.7, according to the Poisson law, the fraction of the vesicles that contain one or more enzyme molecules per vesicle is related to the enzyme-to-vesicle ratio in the reaction mixture. The slope of -1.2

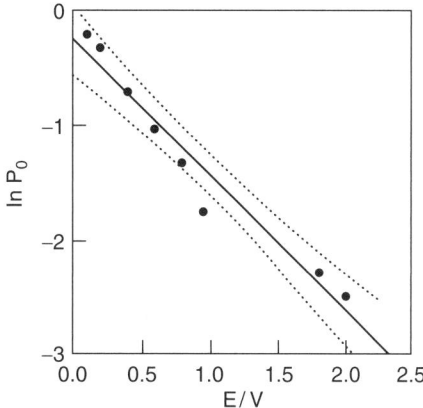

Figure 4.7 The Poisson distribution plot (Equation 1 in Theory Box 4.2) for the extent of hydrolysis of small sonicated DMPM vesicles by pig pancreatic PLA2 with increasing E/V. Data from Figure 4.3. The slope of -1.2 (line with 95% confidence limit) is consistent with the assertion that one molecule of the 14 kDa enzyme species is fully catalytically active in the processive scooting mode. $P_0 = (Y_{max} - Y)/Y_{max}$, where $Y_{max} = [S_0^*]$ is the total hydrolyzable substrate and $Y = [P(\infty)]$ is the total substrate hydrolyzed at a given E/V ratio (cf. Equation 4 in Theory Box 4.2)

in this plot shows that the catalytically active unit of the enzyme at the interface is monomeric, as also supported by independent evidence (Jain *et al.*, 1991b).

4.6.3 Substrate replenishment through polymyxin B (PxB) induced vesicle-to-vesicle contacts

The unaided rate of fusion of bilayer vesicles is exceeding slow, as also indicated by the fact that the zwitterionic and anionic phospholipid vesicles retain their size and trapped content for several hours and days. However, anionic vesicles tend to aggregate, fuse and become leaky within minutes in the presence of multivalent cations and cationic proteins. Such changes, of course, influence the reaction progress in the scooting mode. Mediators of selective and regulated vesicle-to-vesicle exchange of phospholipids include polymyxins, myelin basic protein, melittin and lung surfactant protein SP-A (Table 4.2).

Table 4.2 Mediators of vesicle-to-vesicle exchange of phospholipids

Peptide/protein	Specificity for the exchanges
Polymyxin B or E	Monoanionic phospholipids
Mastoparan	Anionic phospholipids
Melittin	No specificity
Myelin basic protein	Na-triggered exchange
Lung surfactant protein A	Ca-triggered exchange

4.6.4 Apparent activation due to substrate replenishment

Selective, direct and rapid vesicle-to-vesicle exchange of monoanionic phospho-lipids mediated by polymyxin B is best demonstrated as the rate enhancement due to the substrate replenishment during the reaction progress in the scooting mode (Cajal *et al.*, 1995, 1996a, 1996b). Substrate depletion in the microenvironment of the bound enzyme lowers the observed rate during the reaction progress in the scooting mode, even though substrate is present in the excess vesicles in the reaction mixture. As shown in Figure 4.8, PxB contacts promote exchange of the product

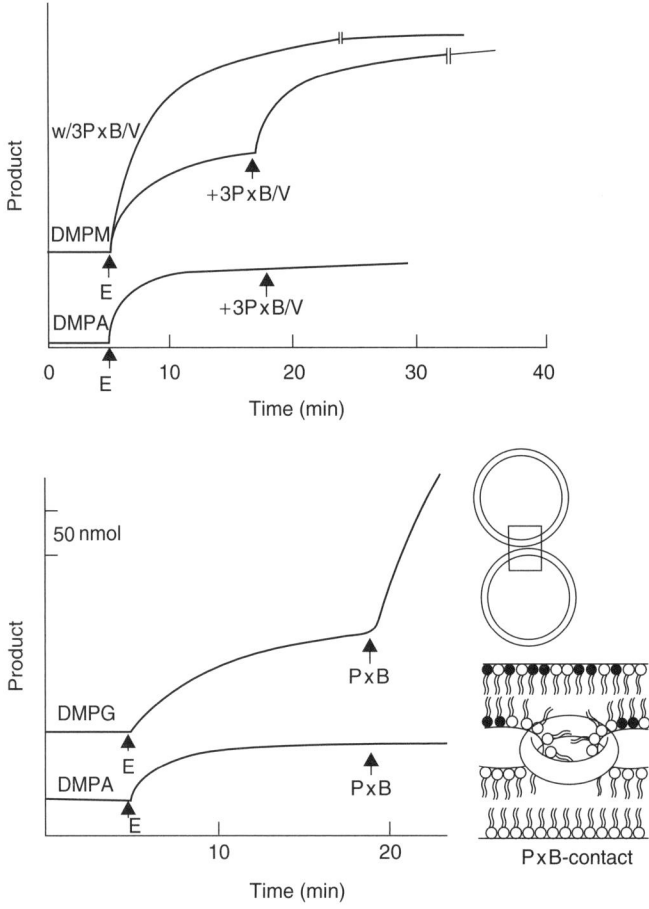

Figure 4.8 (Top) Effect of polymyxin B at 3 polymyxin B molecules per vesicle (3 PxB/V) on the reaction progress for the hydrolysis of DMPM or DMPA vesicles. Note that the order of addition of E or PxB does not make a significant difference in the extent of hydrolysis although the shape of the reaction progress is quite different. Since only a small fraction of vesicles is hydrolyzed, these results show that neither the bound enzyme nor the vesicles in the clusters leave the clusters. (Bottom) Effect of 20 PxB/V on the reaction progress for the hydrolysis of 1,2-dimyristoylphosphatidylglycerol (DMPG) or 1,2-dimyristoylphosphatidic acid (DMPA) by pig pancreatic PLA2. The vesicle-to-vesicle direct exchange of monoanionic phospholipids by PxB is postulated to occur through the PxB contact between the outer layers of vesicles. Reprinted with permission from Cajal *et al.* (1996a, 1996b). Copyright (1996) American Chemical Society

of hydrolysis with DMPM or DMPG. On the other hand, the exchange is not seen with zwitterionic or dianionic phospholipids dispersed alone or codispersed with monoanionic phospholipids. Only the phospholipid molecules present in the outer monolayers of vesicles in contact with each other participate in the exchange. The PxB-mediated selective exchange of phospholipids is clearly distinguishable from the lipid mixing due to fusion or hemifusion.

The kinetic and equilibrium behavior of the PxB-mediated phospholipid exchange suggests that the contacts between the apposed vesicles are stabilized in such a way that neither the peptide nor the vesicles in contact exchange on the time scale of several minutes. This implies that the underlying process is somehow highly cooperative. Analysis of the ensemble behavior of the clusters formed in the presence of PxB shows that their size distribution is bimodal. Thus with an increasing PxB/V ratio the number of clusters increases, but not the number of vesicles in the cluster. It appears that the preferred size is about 40 vesicles per cluster.

4.7 Summary: Ensemble Behavior in a Microscopically Heterogeneous Environment

Reaction progress in the scooting mode provides unequivocal evidence for the interfacial processive catalytic turnover during which the enzyme does not leave the vesicle interface. Thus the kinetic path is unequivocally defined by the species at the interface. Under suitable conditions the ensemble of vesicles and enzyme is amenable to ensemble-averaging, which is critical for the analysis of the observed reaction progress as developed in the following chapters. Such a well-characterized assay system with an unequivocally defined kinetic path permits a full and rigorous analysis of the scooting mode reaction progress, as well as the kinetic characterization of substrate, inhibitors, cofactors and allosteric modulators of an enzyme.

The properties of vesicle dispersions in the scooting mode reaction progress provide information about the ensemble behavior of the distribution of enzyme over vesicles. Such studies take advantage of the tight binding of the enzyme to the vesicle interface. The situation is considerably simplified because the substrate and products also do not exchange between the vesicles unless induced to do so with additives. The ensemble behavior is monitored with the properties of the reaction progress in the scooting mode where the enzyme-to-vesicle ratio is a critical variable related to the observed extent of hydrolysis through the Poissonian distribution. Such results are useful for obtaining the size of the catalytically functional unit of the enzyme at the interface.

5 Analysis of the Processive Reaction Progress

Starting with the assumption that interfacial enzymology is the study of the catalytic behavior at the lipid–water interface, the processive turnover during the reaction progress in the scooting mode permits a detailed kinetic analysis of the catalytic turnover cycle at the interface. Under these conditions the microenvironment of the bound enzyme during the reaction progress is unequivocally defined. Standard steps of a kinetic path for the turnover at the interface are shown in Figure 5.1. In the pre-steady state the enzyme binds irreversibly to the interface, and none of the interacting species leave the interface during the course of the reaction progress.

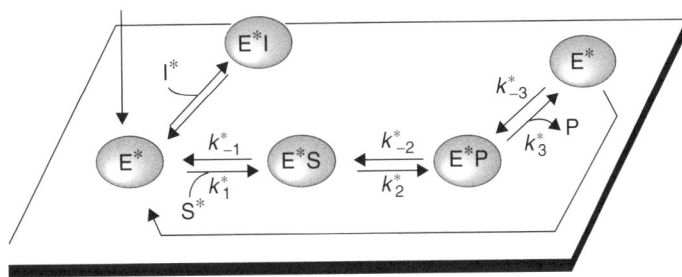

Figure 5.1   The kinetic path for the processive turnover at the interface

5.1 Michaelis–Menten Equation for the Turnover in the Interface

The basic reaction path of the interfacial catalytic turnover cycle is restated in Figure 5.2. Consider the three equilibria for the reversible steps of the turnover cycle as encountered by the enzyme bound to the interface of an aggregate of substrate molecules. The enzyme is assumed to access the substrate only from the interface to which it is bound, and the bound enzyme does not leave the interface. All parameters with an asterisk (*) refer to properties, parameters and reactions taking place in the interface. The rate of substrate binding to the enzyme will be proportional to the rate constant, k_1^*, to the concentration of the substrate accessible in the interface and to the probability that the enzyme is free, P_f. The concentration that the enzyme 'sees' is the surface density in the interface, i.e. the number of substrate molecules per surface area. This is related to the mole fraction of substrate in the interface, X_S^*, which is a more convenient variable. The choice of concentration units was discussed before (Theory Box 3.2 and Section 4.2). At the steady state the flux across each step in the reaction diagram is the same:

$$j = k_1^* X_S^* P_f - k_{-1}^* P_{bS} = k_2^* P_{bS} - k_{-2}^* P_{bP} = k_3^* P_{bP} - k_{-3}^* X_P^* P_f \qquad (5.1)$$

where P_{bS} and P_{bP} refer to the probabilities that the enzyme is in state E*S and E*P, respectively.

The steady state flux, j, can be calculated (Theory Box 5.1) as

$$j = \frac{\dfrac{k_{catS}^*}{K_{MS}^*} X_S^* - \dfrac{k_{catP}^*}{K_{MP}^*} X_P^*}{1 + \dfrac{X_S^*}{K_{MS}^*} + \dfrac{X_P^*}{K_{MP}^*}} \qquad (5.2)$$

The parameters in this equation are the interfacial MM parameters that are determined by the primary rate constants given in the scheme (Figure 5.2 and Equations 3 and 4 in Theory Box 5.1). Here k_{catS}^* corresponds to the apparent first-order rate from state E*S to E* + P, while $k_{catS}^* X_S^*/K_{MS}^*$ is an apparent second-order rate from E* + S* to E* + P*.

The maximum flux over the enzyme occurs when $X_S^* = 1$ and $X_P^* = 0$, so that the *turnover number* is

$$j_{max} = j_0 = \frac{k_{catS}^*}{1 + K_{MS}^*} \qquad (5.3)$$

Thus, in contrast to the solution case, k_{catS}^* is *not* the same as the maximum rate at saturating substrate concentration. Since the concentration that the enzyme 'sees',

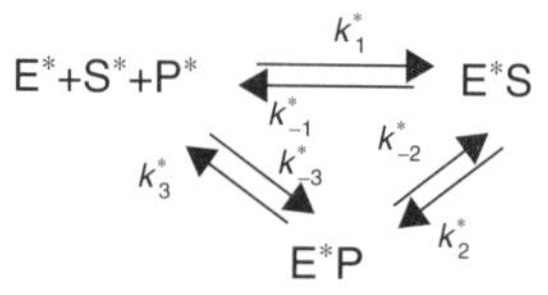

Figure 5.2 Relations of a complete set of equilibria that define the primary parameters of the catalytic turnover cycle at the interface (restated from Figure 5.1)

X_S^*, cannot exceed mole fraction 1, the maximum concentration will be saturating only if $K_{MS}^* \ll 1$. Clearly, if the initial condition is that only substrate is present, $X_S^* = 1$, j_{max} is also equal to the *initial rate*, j_0. Equilibrium is reached when $j = 0$, i.e. when

$$\frac{X_P^{*eq}}{X_S^{*eq}} = \frac{k_{catS}^* K_{MP}^*}{k_{catP}^* K_{MS}^*} = \frac{k_1^* k_2^* k_3^*}{k_{-1}^* k_{-2}^* k_{-3}^*} = K_{eq}^* \tag{5.4}$$

This is the equilibrium constant for the S* to P* transformation in the interface. It can be noted that this is not necessarily the same as K_{eq} for the S to P transformation in solution.

The situation for the maximum substrate concentration at the interface is analogous to a case in solution where the substrate is sparingly soluble and K_M is around the solubility limit (Figure 2.6). In solution kinetics one would manipulate the substrate concentration and measure the initial rate j_0 at different concentrations to deduce k_{catS} and K_{MS} individually. In the interface, however, X_S^* can be manipulated only by adding some other component that reduces the surface density, and hence X_S^*. Of course, the substrate concentration cannot be varied unless a two-dimensional solvent, a *diluent*, is introduced in the interface. In an ideal situation, as developed in Chapter 3, a diluent may not impart new or different properties to the system. Like a buffer in solution enzymology, an ideal diluent changes only the surface density of the components but has little affinity for the active site of the bound enzyme, and it does not change the catalytically relevant properties of the interface. However, as described below, one can also take advantage of the integrated reaction progress where the substrate and product mole fractions change simultaneously, while $X_P^* + X_S^* = 1$ through the course of the reaction.

THEORY BOX 5.1 Steady state flux in the interfacial reaction loop

At the steady state the flux across each step in the reaction diagram is the same:

$$j = k_1^* X_S^* P_f - k_{-1}^* P_{bS} = k_2^* P_{bS} - k_{-2}^* P_{bP} = k_3^* P_{bP} - k_{-3}^* X_P^* P_f \tag{1}$$

where P_f, P_{bS} and P_{bP} refer to the probabilities that the enzyme is in state E*, E*S and E*P, respectively. These probabilities must sum to one: $P_f + P_{bS} + P_{bP} = 1$. These relations give four equations in four unknowns and the flux j can be solved by elimination. One finds

$$j = \frac{\dfrac{k_{catS}^*}{K_{MS}^*} X_S^* - \dfrac{k_{catP}^*}{K_{MP}^*} X_P^*}{1 + \dfrac{X_S^*}{K_{MS}^*} + \dfrac{X_P^*}{K_{MP}^*}} \tag{2}$$

The parameters in this equation are the interfacial MM parameters that are determined by the primary rate constants of the scheme as:

$$k_{catS}^* = \frac{k_2^* k_3^*}{k_2^* + k_{-2}^* + k_3^*} \tag{3a}$$

$$k_{catP}^* = \frac{k_{-1}^* k_{-2}^*}{k_{-1}^* + k_2^* + k_{-2}^*} \tag{3b}$$

$$K_{MS}^* = \frac{k_{-1}^* k_3^* + k_{-1}^* k_{-2}^* + k_2^* k_3^*}{k_1^* \left(k_2^* + k_{-2}^* + k_3^*\right)} \tag{4a}$$

$$K_{MP}^* = \frac{k_{-1}^* k_3^* + k_{-1}^* k_{-2}^* + k_2^* k_3^*}{k_{-3}^* \left(k_{-1}^* + k_2^* + k_{-2}^*\right)} \tag{4b}$$

These results, Equations (2) to (4), are exactly congruent in form to those from the corresponding three-step scheme for solution kinetics (Segel, 1975, Equation II:17). One notable difference is the concentration units. In solution kinetics, concentrations can often be manipulated freely. In the interface, on the other hand, the maximum concentration is mole fraction = 1. Furthermore, if there are no other components in the interface, $X_S^* + X_P^* = 1$ must hold.

It is instructive to dissect the flux equation in its component parts. It consists of two terms, the difference between the total forwards flux and the total backwards flux: $j = j_S - j_P$, where

$$j_S = \frac{k_{catS}^* X_S^* / K_{MS}^*}{1 + X_S^* / K_{MS}^* + X_P^* / K_{MP}^*} = \frac{k_{catS}^*}{K_{MS}^*} X_S^* P_f = k_{catS}^* P_{bS} \tag{5}$$

The total backwards flux, j_P, is analogous with subscripts S and P interchanged everywhere in Equation (5); j_S is the flux from state $E^* + S^*$ to $E^* + P^*$. Due to the equilibrium relation (Equation 5.4), it must hold that

$$\frac{j_P}{j_S} = \frac{X_P^*}{X_S^*} \frac{1}{K_{eq}^*} = \frac{X_P^*}{X_S^*} \frac{X_S^{*eq}}{X_P^{*eq}} \tag{6a}$$

Thus, the net product flux, j, can be written as the forward flux, j_S, times a factor that simply expresses the departure from equilibrium:

$$j = j_S \left(1 - \frac{X_P^*}{X_S^*} \frac{X_S^{*eq}}{X_P^{*eq}}\right) \tag{6b}$$

A departure-from-equilibrium factor of this form, $1 - [P]/([S]K_{eq})$, always appears as the thermodynamic driving force in out-of-equilibrium reactions. The flux from $E^* + S^*$ to $E^* + P^*$, j_S, can also be expressed as the rate of formation, $k_1^* X_S P_f$, of E^*S from $E^* + S^*$ times the probability of going from E^*S to $E^* + P^*$ without substrate dissociation. Thus, this probability can be expressed as

$$P\left(E^*S \longrightarrow E^* + P^*\right) = \frac{k_{catS}^*}{k_1^* K_{MS}^*} = \frac{k_2^* k_3^*}{k_2^* k_3^* + k_{-1}^* k_{-2}^* + k_{-1}^* k_3^*} \tag{7}$$

Equation (5) also contains the probabilities for the three states of the enzyme:

$$P_f = \frac{1}{1 + \dfrac{X_S^*}{K_{MS}^*} + \dfrac{X_P^*}{K_{MP}^*}} \tag{8a}$$

$$P_{bS} = \frac{X_S^*}{K_{MS}^*} P_f \tag{8b}$$

$$P_{bP} = \frac{X_P^*}{K_{MP}^*} P_f \tag{8c}$$

Each term in the denominator of Equation (2) corresponds to each possible state of the enzyme such that $1 = P_f/P_f = [E^*]/[E^*]$, $X_S^*/K_{MS}^* = P_{bS}/P_f = [E^*S]/[E^*]$ and $X_P^*/K_{MP}^* = P_{bP}/P_f = [E^*P]/[E^*]$. Thus, in terms of binding probabilities, K_{MS}^* and K_{MP}^* serve the same role at the steady state as do the dissociation constants at equilibrium.

5.1.1 The initial rate conditions with microscopic steady state

A more pervasive difference from the solution kinetics comes from the nature and properties of the vesicle and the properties of its interface within the constraints outlined above for the interfacial turnover path in Figures 5.1 and 5.2. Equations (5.1) to (5.3) refer to the expected reaction flux for a single enzyme. The total amount of substrate available to each enzyme depends on the size of each aggregate and the number of enzymes bound to it. Assume that an aggregate initially has N_S^0 substrate molecules in the accessible interface and n_E enzymes bound to it ($n_E \ll N_S^0$). Initially, with $X_S^* = 1$ and $X_P^* = 0$, the initial rate per enzyme is given by j_0 from Equation (5.3). If product immediately desorbs from the interface, these initial conditions will last as long as there still remains any substrate in the aggregate. In this case the initial rate will last for a time given by

$$\tau_0 = \frac{N_S^0}{n_E j_0} \tag{5.5}$$

When the product remains in the interface, the initial rate can only last for times much shorter than τ_0.

Thus the duration for the initial rate at $X_S^* = 1$ increases with the vesicle size, but not with the bulk concentration of the vesicles in the reaction mixture (except to the extent that the number of enzymes per vesicle, n_E, decreases). In contrast to the solution case, increasing the bulk substrate concentration does not necessarily prolong the duration of the initial rate. Of course, as shown in the previous chapter, the slow-down due to substrate depletion and product accumulation can be circumvented, for example, by promoting the substrate replenishment through the polymyxin B mediated contacts between vesicles of monoanionic phospholipids. These conditions are well suited with PLA2 for determining the dependence of the initial rates on the enzyme, inhibitor and the cofactor concentrations.

5.2 Catalytic Parameters from the Integrated Processive Reaction Progress

The analysis above offers unequivocal methods for determining the interfacial rate and the substrate affinity parameters from the reaction progress in the

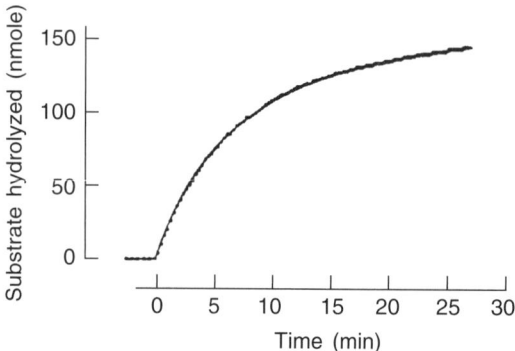

Figure 5.3 The integrated reaction progress for the hydrolysis of dimyristoylphosphatidyl-methanol (DMPM) vesicles by pig pancreatic PLA2. The fit (smooth line) to the data points is based on the integrated Michaelis Equation (5.8). Reprinted with permission from Berg *et al.* (1991). Copyright (1991) American Chemical Society

scooting mode. For the case of at most one enzyme per vesicle, the reaction progress shown in Figure 5.3 provides three pieces of information. As developed in the previous chapter, at least for PLA2, the extent of hydrolysis per enzyme, N_S^0, is a measure of the number of substrate molecules present in the outer layer of the target vesicle. As developed below, the other two parameters, as combinations of K_{MS}^*, K_{MP}^* and k_{catS}^*, are obtained from the analysis of the integrated Michaelis–Menten equations for the processive turnover at the vesicle interface.

5.2.1 Integrated rate equation

If product is retained in the interface, the initial rate j_0 in Equation (5.3) can last only as long as there is no significant change in $X_S^* = 1$ and $X_P^* = 0$ on the enzyme-containing vesicles. Since the initial rate of change of X_S^* is

$$\frac{dX_S^*}{dt} = -\frac{dX_P^*}{dt} = -\frac{n_E j_0}{N_S^0} = -\frac{1}{\tau_0} \tag{5.6}$$

this requires times much shorter than τ_0. For small aggregates, τ_0 may be small and the initial rate not readily observable. In this case, it will be more informative to look at the integrated rate equation for the aggregate. Starting with N_S^0 substrate molecules in an aggregate carrying n_E enzymes, the number of product molecules increases by $n_E j$ per second and the mole fraction X_P^* changes by

$$\frac{dX_P^*}{dt} = \frac{n_E}{N_S^0} j = \frac{n_E}{N_S^0} \frac{k_{catS}^*}{1 + K_{MS}^*} \frac{1 - X_P^*/X_P^{*eq}}{1 + \frac{X_P^*}{K_{MP}^*} \frac{K_{MS}^* - K_{MP}^*}{1 + K_{MS}^*}} \tag{5.7}$$

In the last equality, the flux from Equation (5.2) and the equilibrium relation from Equation (5.4) have been entered. This rate equation can be integrated to give an implicit expression for the reaction progress $X_P^*(t)$ as a function of t. One finds

$$k_i t = -\ln\left(1 - \frac{X_P^*(t)}{X_P^{*eq}}\right) + \left(\frac{N_S^0 X_P^{*eq} k_i}{n_E j_0} - 1\right)\frac{X_P^*(t)}{X_P^{*eq}} \tag{5.8}$$

This result is very similar in form to the integrated MM equation for the two-step solution kinetics (Equation 3.8). The first term on the right-hand side is a first-order contribution which by itself would give a reaction progress with an exponential decline, $X_P^*(t)/X_P^{*eq} = 1 - e^{-k_i t}$, where the relaxation rate is

$$k_i = \frac{n_E}{N_S^0 X_P^{*eq}} \frac{k_{catS}^*}{K_{MS}^*\left(1 + \frac{X_S^{*eq}}{K_{MS}^*} + \frac{X_P^{*eq}}{K_{MP}^*}\right)} \tag{5.9}$$

The second term on the right-hand side of Equation (5.8) is a zero-order contribution. When $X_P^*(t) \ll X_P^{*eq}$, Equation (5.8) gives approximately a linear increase of product in time corresponding to the initial rate: $X_P^*(t) = (n_E/N_S^0)j_0 t$ in accordance with Equation (5.6). Studying the whole reaction progress, fitting $X_P^*(t)$ versus t gives information about the initial rate per enzyme, j_0, and the relaxation rate, k_i, if the other parameters (i.e. n_E, N_S^0 and X_P^{*eq}) in Equation (5.8) are known. The product enters the relaxation rate through X_P^{*eq} and K_{MP}^*; this accounts for product inhibition as well as for the reverse P* to S* reaction. When there is no reverse reaction, $X_P^{*eq} = 1$, K_{MP}^* accounts for product inhibition only.

The reaction progress (Equation 5.8) is calculated as resulting from n_E enzymes on an aggregate of size N_S^0, i.e. with well-defined numbers of both substrates and enzymes in the subsystem. Unless special precautions are taken, these numbers will vary from aggregate to aggregate and from enzyme to enzyme. In this case, all enzymes in the reaction mixture will not 'see' the same substrate concentration, X_S^*, during the reaction progress. Enzymes that happen to be bound to small aggregates where there are many other enzymes bound will see a much faster depletion of the local substrate concentration than those bound at large aggregates with few other enzymes.

The relations above simplify in the limit when the equilibrium is far to the product side, $X_S^{*eq} = 0$ and $X_P^{*eq} = 1$. Furthermore, in what follows we will focus on the limit where there is at most one enzyme per vesicle, $n_E = 1$. Then Equations (5.8) and (5.9) reduce to

$$k_i t = -\ln\left[1 - X_P^*(t)\right] + \left(N_S^0 k_i/j_0 - 1\right)X_P^*(t) \tag{5.10a}$$

$$N_S^0 k_i = \frac{k_{catS}^*}{K_{MS}^*\left(1 + \frac{1}{K_{MP}^*}\right)} \tag{5.10b}$$

Studying the total interfacial reaction progress in this way provides only a limited access to the primary interfacial parameters. The initial rate gives $j_0 = k_{catS}^*/(1 + K_{MS}^*)$

from Equation (5.3) while the ratio of the initial rate and the relaxation rate gives

$$\frac{N_S^0 k_i}{j_0} = \frac{1 + \dfrac{1}{K_{MS}^*}}{1 + \dfrac{1}{K_{MP}^*}} \tag{5.10c}$$

Thus, the integrated rate equation cannot by itself resolve the three interfacial parameters, k_{catS}^*, K_{MS}^* and K_{MP}^*.

5.2.2   Relations in the reaction progress

The integrated rate equation can be plotted in a number of ways. Consider a situation where there is at most one enzyme molecule per vesicle, $n_E = 1$, and where the equilibrium is far to the right so that $X_P^{*eq} = 1$. Free parameters in Equation (5.10a) that can be used are k_i, N_S^0 and j_0. Figure 5.4 shows the number of product molecules, $N_S^0 X_P^*$, produced per enzyme molecule as a function of $j_0 t$ for some values of N_S^0 and $k_i N_S^0/j_0$. As the time axis is scaled by j_0, the initial slope in all curves will be equal to one. After long times, when all substrates have been consumed, the curves level off at N_S^0. The initial rate would be followed for a longer time when N_S^0 is large or when $k_i N_S^0/j_0$ is large (Figure 5.4). At the time when the slope has decreased by the fraction δ from its initial value, one finds

$$X_P^* = \frac{1}{1 + \dfrac{1 - \delta}{\delta} \dfrac{j_0}{N_S^0 k_i}}$$

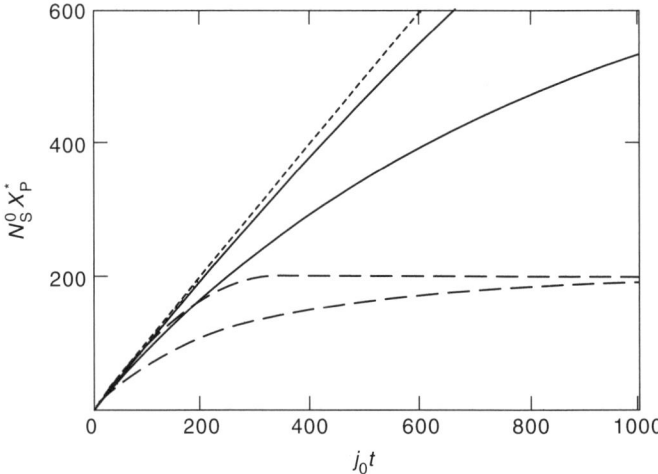

Figure 5.4   Product produced per enzyme as a function of time. The time axis has been scaled by the initial rate, j_0, so that the initial slope for all curves equals one. Solid curves are for $N_S^0 = 1000$ and dashed curves are for $N_S^0 = 200$. The upper curve in each set is for $k_i N_S^0/j_0 = 5$ and the lower one is for $k_i N_S^0/j_0 = 0.5$. The dotted line is the true initial rate (slope 1)

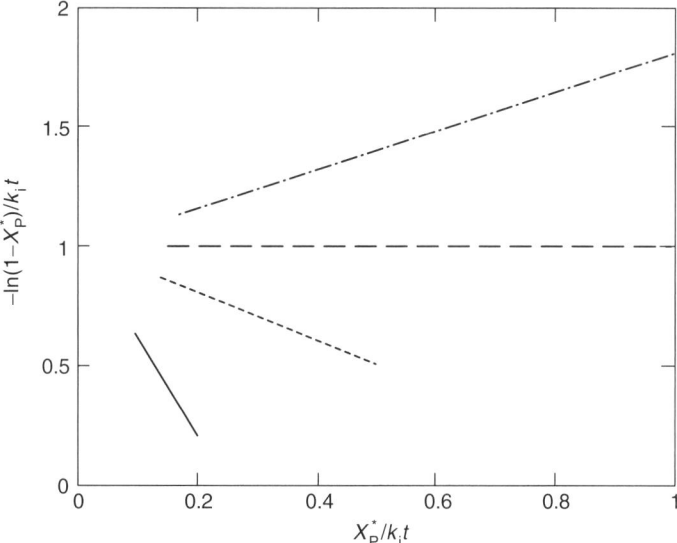

Figure 5.5 Linear plots of the reaction progress. Curves are for $k_i N_S^0/j_0 = 0.2, 1, 2, 5$ (from upper to lower). The time has been scaled by k_i so that the y intercept for all curves equals one and the slope equals $1 - k_i N_S^0/j_0$

If $\delta \ll 1$ and $\delta \ll j_0/N_S^0 k_i$, this will happen at time $\tau_\delta = k_i(N_S^0/j_0)^2\delta$. From Equation (5.10c), it is seen that $k_i N_S^0/j_0$ is large for small K_{MS}^* and large K_{MP}^*, which corresponds to the zero-order limit when the reaction rate per enzyme is constant and largely independent of X_S^* throughout the reaction progress.

Alternatively, to get a linear plot, one can plot $-\ln(1 - X_P^*)/t$ versus X_P^*/t as in Figure 5.5. The extrapolated y intercept in this diagram determines k_i and the extrapolated x intercept is $k_i/[k_i N_S^0/j_0 - 1]$. If $k_i N_S^0/j_0$ is close to 1, the whole reaction progress is nearly exponential and k_i is readily determined (horizontal line in Figure 5.5). From Equation (5.10c), this would be the case if both K_{MS}^* and K_{MP}^* are large or if they are roughly equal. This is the first-order limit where the reaction rate remains proportional to residual X_S^* throughout. The shape of these curves are independent of the size, N_S^0, of the vesicles, except that N_S^0 must be known for the calculation of $X_P^*(t)$ from the observed amount of product produced. N_S^0 in turn is determined as the mean total number of products produced per enzyme after long times.

In the limit described above of at most one enzyme per vesicle, the curves (Figures 5.4 and 5.5) are independent of the enzyme concentration. If the curves are different at different enzyme concentrations, this suggests that the condition is not met.

5.2.3 Extended initial rate by rapid substrate replenishment

As emphasized in Chapter 4, the overall reaction progress must be taken as an appropriate average over each enzyme. In the limit where the number of enzymes is much smaller than the number of aggregates in solution, the probability that there

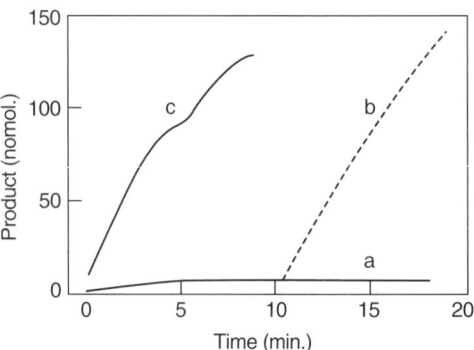

Figure 5.6 The initial rate of the PLA2 catalyzed hydrolysis of clusters of DMPM vesicles with vesicle-to-vesicle contact induced by polymyxin B: curve a, no PxB; curve b, with PxB added at 10 min; curve c, PxB added at $t = 0$. Reprinted with permission from Jain *et al.* (1991c). Copyright (1991) American Chemical Society

will be more than one enzyme bound to an aggregate is small. In this case, $n_E = 1$. Furthermore, if all aggregates are of the same size, the substrate depletion in each affected aggregate will be the same and each enzyme will see the same concentration throughout. When there is no exchange of substrate, product or enzymes between aggregates, this is the limit that is required for the reaction progress to mimic the solution results that are based on perfect mixing of all components at all times between all coexisting interfaces. Another possibility occurs when aggregates are very large and each carries a large number of enzymes. In this case, the relative variation in the number of enzymes per aggregate is expected to be small and all enzymes will again see virtually the same substrate concentration. This situation probably prevails during the PLA2 catalyzed hydrolysis of the PxB-mediated clusters of vesicles (Figures 4.9 and 5.6).

5.3 Additional Constraints for the Analysis of the Interfacial Turnover Events for PLA2

Based on the additional constraints, outlined below and developed later in this and the next chapter, the turnover events for the pig pancreatic PLA2 at the interface are described by three primary events:

$$E^* + S^* \rightleftharpoons E^*S \longrightarrow E^*P \rightleftharpoons E^* + P^*$$

The identification of the individual rate constants is helped by the following observations:

(a)   Analysis of the kinetic isotope effect and the isotope exchange shows that the reverse reaction for the chemical step, k^*_{-2}, is immeasurably slow (Ghomashchi *et al.*, 1991a; Yu *et al.*, 1998)

(b)   The chemical step is the rate-limiting step in k^*_{catS}, i.e. $k^*_2 < k^*_3$, during the turnover cycle (Jain *et al.*, 1992b).

(c)   Calcium is a catalytic cofactor that is obligatorily required for the binding of substrate and mimics to the active site, as well as for the chemical step.

5.3.1   The Michaelis parameters

Observation (a) considerably simplifies the interpretation of the kinetic results. By setting $k^*_{-2} \approx 0$ in Equations (3) and (4) in Theory Box 5.1, we find

$$k^*_{\text{catS}} = \frac{k^*_2 k^*_3}{k^*_2 + k^*_3} \approx k^*_2 \tag{5.11}$$

$$\frac{k^*_{\text{catS}}}{K^*_{\text{MS}}} = \frac{k^*_1 k^*_2}{k^*_{-1} + k^*_2} \approx k^*_1 \tag{5.12}$$

$$K^*_{\text{MP}} = \frac{k^*_3}{k^*_{-3}} = K^*_{\text{P}} \tag{5.13}$$

Observation (b) above suggests that the product dissociation from E*P is rapid, and therefore $k^*_{\text{catS}} \approx k^*_2$ in Equation (5.11). Equation (5.13) shows that when there is no back reaction, K^*_{MP} can be replaced by the equilibrium product dissociation constant, K^*_{P}, which can be measured independently (e.g. see Figure 6.10). Furthermore, from Equation (5.10c), we get

$$\frac{j_0}{N^0_{\text{S}} k_{\text{i}}} = \frac{1 + 1/K^*_{\text{P}}}{1 + 1/K^*_{\text{MS}}} \tag{5.14}$$

Analysis of the interfacial kinetics provides values for j_0, N^0_{S} and k_{i}. Thus K^*_{MS} can be calculated if K^*_{P} is independently obtained (see Chapter 6). When both K^*_{MS} and K^*_{P} can be measured separately, Equation (5.14) can serve as a control. As also developed in Chapter 9, the ratio relates to the difference in the competitive inhibition kinetics under the zero-order versus the first-order reaction progress.

From the steady state initial rate ($X^*_{\text{S}} = 1$, $X^*_{\text{P}} = 0$) and an independently obtained value of K^*_{MS}, one can calculate

$$k^*_{\text{catS}} = j_0 \left(1 + K^*_{\text{MS}}\right) \tag{5.15}$$

K^*_{MS} can be obtained independently (Tables 5.1 and 5.2) by the following protocols described in detail in Chapters 6 and 9:

(a)   From Equation (5.14) if K^*_{P} known.

(b)   From the initial rate when X^*_{S} is varied with a suitable interface neutral diluent that changes only the surface density but not the size and dispersity of vesicles:

$$j_0 = \frac{k^*_{\text{catS}} X^*_{\text{S}}}{X^*_{\text{S}} + K^*_{\text{MS}}} \tag{5.16}$$

(c)   As an obligatory cofactor for the binding of an active-site-directed ligand to the active site, the apparent parameter for the effect of calcium on the initial

Table 5.1 The primary kinetic parameters for the action of pig pancreatic PLA2 on DMPM[a] vesicles. From Berg *et al.* (1991)

$k^*_{catS} = k^*_2 = 400 \text{ s}^{-1}$ (from $j_0 = 300 \text{ s}^{-1}$), the rate constant for the chemical step via E^*S
$k^*_1 = 1350 \text{ s}^{-1}$
$k^*_{-1} = 30 \text{ s}^{-1}$
$K^*_{MS} = k^*_1/(k^*_{-1} + k^*_2) = 0.37$ mole fraction with a neutral diluent (Equation 5.16)
$\phantom{K^*_{MS} = }$ 0.26 mole fraction with $K^*_I = 0.0008$ mole fraction for MJ33
$\phantom{K^*_{MS} = }$ (Equation 5.18)
$\phantom{K^*_{MS} = }$ 0.32 mole fraction with $K^*_P = 0.025$ mole fraction (Equation 5.14)
$\phantom{K^*_{MS} = }$ 0.45 mole fraction with $K^*_{Ca} = 0.28$ mM and $K^*_{Ca}(S) = 0.09$ mM at
$\phantom{K^*_{MS} = }$ $X^*_S = 1$ (Equation 5.17)
O/S ratio $= 10$, the ratio of the k^*_{catS} values for the *sn*-2-oxy ester versus the *sn*-2-thioester of
 DMPM (Jain *et al.*, 1992b)

[a]DMPM, 1,2-dimyristorylphosphatidylmethanol; MJ33, 1-hexadecyl-3-trifluoroethyl-glycero-*sn*-2-phospho-methanol. The K^*_L values are in mole fraction.

Table 5.2 Kinetic parameters for the hydrolysis of the thio-analogs by porcine pancreatic PLA2. From Jain *et al.* (1992a)

Substrate	j_0 s^{-1}	$N^0_S k_i$ s^{-1}	K^*_{MS} (mole fraction)	K^*_P (mole fraction)	k^*_{catS} s^{-1}	k^*_{catS}/K^*_{MS}	$N^0_S k_i$ (calculated)
Dioxy-DMPM	300	34	0.35	0.025	400	1100	28
Monothio-DMPM	280	31	0.15	0.025	300	2000	52
Dithio-DMPM	38	25	0.03	0.044	40	1300	53

rate is given by Equation (9.15) in Chapter 9:

$$K^*_{Ca}(S) = \frac{K^*_{Ca}}{1 + \dfrac{X^*_S}{K^*_{MS}}} \tag{5.17}$$

This allows an independent determination of K^*_{MS} from the Ca dependence of the initial rate.

(d) The effect of a competitive inhibitor (Figure 5.7) is described by Equation (9.6) (with $K^{app}_M(0) = 0$ if all enzymes are in the interface and $[I_T]/[S_T] = X^*_I/(1 - X^*_I)$ when there is no diluent present) as the ratio of the observed initial rates without and with inhibitor present:

$$\frac{j_0}{j_I} = 1 + \left(\frac{1 + \dfrac{1}{K^*_I}}{1 + \dfrac{1}{K^*_{MS}}}\right)\left(\frac{X^*_I}{1 - X^*_I}\right) \tag{5.18}$$

The primary kinetic parameters for pig pancreatic PLA2 are summarized in Table 5.1 along with the values of K^*_{Ca} and K^*_I. For the hydrolysis of DMPM vesicles by pig pancreatic PLA2 under a variety of conditions, the values of K^*_{MS} are in the range of 0.28 to 0.45 mole fraction. The maximum possible X^*_S at the interface is 1, and it is likely to be around 0.5 in the bile-salt dispersions. This is in agreement

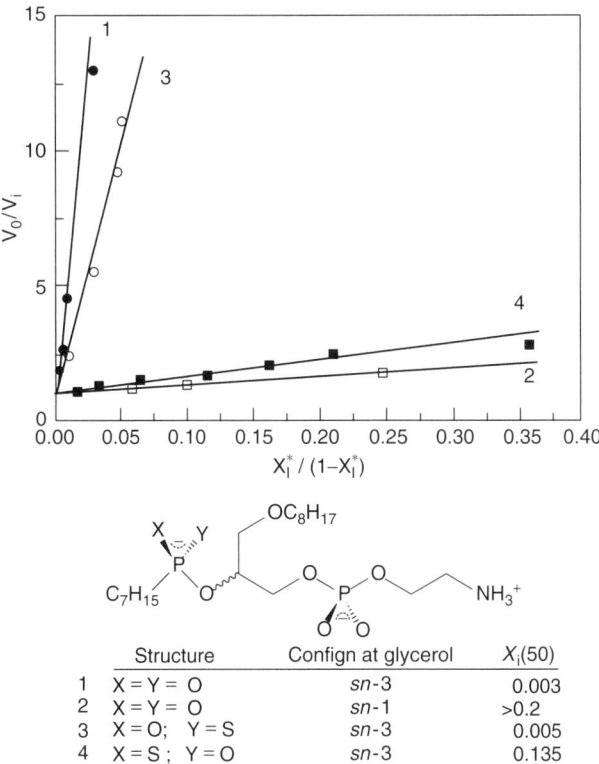

	Structure	Confign at glycerol	$X_i(50)$
1	X = Y = O	sn-3	0.003
2	X = Y = O	sn-1	>0.2
3	X = O;　Y = S	sn-3	0.005
4	X = S ;　Y = O	sn-3	0.135

Figure 5.7 The mole fraction dependence of the inhibitory effect of the four isomeric substrate mimics of 1-octyl-2-phosphoheptyl-*sn*-glycerophosphoethanolamine with the stereochemcially different *sn*-2-phosphonate substituents, see also Figure 9.1 for other examples. Reprinted with permission from Jain *et al*. (1989). Copyright (1989) American Chemical Society

with the general observation that the physiological substrate concentration of most soluble enzymes is usually in the range of K_{MS}.

5.3.2　Product inhibition

The equations developed above are valid for a reaction that creates a single product molecule for each substrate. This is evident in the back reaction where products bind the catalytic site, causing product inhibition. When there are two products, as for the PLA2-catalyzed hydrolysis, this reaction could be proportional to $(X_P^*)^2$, rather than to X_P^*, as assumed here. However, if the product binding is dominated by one of the product molecules, the probability that the catalytic site is occupied by a product is proportional to X_P^*. This seems to be the case for PLA2, where protection experiments with products show proportionality with X_P^* rather than $(X_P^*)^2$, providing an estimate for a monomolecular K_P^* value (see Figure 6.10).

In solution kinetics, product inhibition does not contribute to the initial rate when no products are present. In a heterogeneous system, on the other hand, it is not always possible to observe the true initial rate. For instance, with small vesicles as described above, product can accumulate very quickly in the local environment of the enzyme so that the effects of product inhibition on the observed rate cannot be avoided simply by choosing appropriate initial conditions. The effect of product inhibition on the apparent initial rate will show up in a number of examples, described in the following chapters.

5.4 Uses of the Primary Catalytic Parameters

The primary kinetic parameters for the catalytic turnover of an enzyme with the various reactants provide information about virtually all aspects of the behavior of an enzyme, built into the kinetic path, ranging from the assay to the catalytic mechanism. For example, information about the substrate and cofactor specificity, or the inhibitor selectivity, expressed as characteristic primary parameters gets around the problem of the variables, the local concentration effect and the conditions for the apparent parameters. Certainly, the kinetic and thermodynamic evaluation of a primary event by site-directed mutagenesis is meaningful only through the corresponding primary constant (Yu *et al.*, 1999b; Berg *et al.*, 2001). Additional considerations are developed below.

5.4.1 The chemical step is rate-limiting

Results in Table 5.2 for the oxy/thio effect on the parameters for the hydrolysis of the *sn*-2-oxy versus the -thio substrate by pig pancreatic PLA2 provide significant insight. The reduction in rate between dioxy- and dithio-DMPM is expected primarily in the chemical step k_2^*; hence the listed values of k_{catS}^* are estimates of k_2^* (cf. Equation (5.11). Furthermore, since $k_{-1}^* < k_2^*$ (from the carbonyl isotope effect, see below), k_{catS}^*/K_{MS}^* is an estimate of k_1^* (cf. Equation 5.12). This is largely independent of the oxy/thio ratio, as could be expected.

5.4.2 The carbonyl isotope effect

The ^{14}C-*sn*-carbonyl isotope effect for the hydrolysis of DMPM vesicles by pancreatic PLA2 in the scooting mode is 1.01 ± 0.01 (Ghomashchi *et al.*, 1991a), which shows that the value of k_{catS}^*/K_{MS}^* is unchanged with this isotope replacement. The effect of this carbonyl isotope on the chemical step, k_2^*, is known to be ca 1.14. Thus, from Equation (5.12), one expects that $k_2^* \gg k_{-1}^*$ so that $k_{catS}^*/K_{MS}^* = k_1^*$, and therefore it is possible to obtain the values of k_1^* and k_{-1}^* in Table 5.1.

5.4.3 The solvent isotope effect

The initial rate of hydrolysis of DMPM vesicles in the presence of deuterated water is about 10 % of the rate seen with water (Figure 5.8). This relatively large solvent isotope effect and the bowl-shaped dependence on the mole fraction of D_2O is consistent with the participation of two or more proton transfer steps in the rate-limiting step.

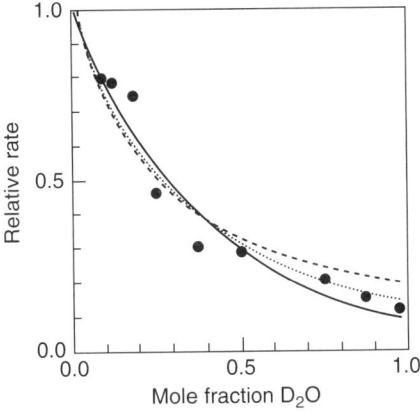

Figure 5.8 Relative rate of hydrolysis of DMPM vesicles by pig PLA2 in the aqueous phase containing varying mole fractions of D_2O in H_2O. The $X_I^*(50)$ values did not change with the solvent, which suggests that the effect is primarily on k_{catS}^*. The solvent isotope effect is large; however, the data points could not be fitted to simple models (Quinn and Sutton 1991). Based on results from Jain *et al.* (1986a), unpublished in this form. A large solvent isotope effect and the bowl-shaped proton inventory curve suggest multiple fractionating transition states; i.e. proton transfer plays a role in two or more substeps of the rate-limiting step

5.4.4 Substrate specificity and substrate conformation

The preference of the various secreted PLA2s for interfaces of the zwitterionic versus anionic phospholipids can be different by several orders of magnitudes (Jain *et al.*, 1982) without a significant difference in the k_{catS}^*/K_{MS}^* substrate specificity in the turnover cycle for the head groups of the naturally occurring phospholipids (Ghomashchi *et al.*, 1991b). The glycerol backbone conformations relative to the chain orientation of different phospholipids are quite different (Figure 4.5), which implies that the substrate specificity does not depend on the crystallographic or ground state conformations. The structure–activity correlation suggests that changes in the *sn*-3 substituent and the chain length also have a significant effect on k_{catS}^* (Jain and Rogers, 1989; Rogers *et al.*, 1996). Since the chemical step remains rate-limiting, in such cases the *sn*-3 substituent apparently influences the position of the substrate binding equilibrium as well as the substrate conformation in the Michaelis complex. This is consistent with the conformation of the substrate mimics in the cocrystals of PLA2, where the gross conformation of the glycerol backbone are moderately different and the *sn*-2-acyl chain can be oriented quite differently in relation to the contacts with residues 2, 3 and 31. At the very least, these results suggest plasticity and flexibility in the substrate binding environment (Yu *et al.*, 1998). Also, the interaction of PLA2 changes the solvation characteristics of the interface that is in contact with the i-face (Chapter 6), and to an extent such changes could depend on the head group structure in the substrate dispersions. Such effects could influence the processivity as well as the distribution of the substrate and mimic in the contact region to possibly influence effective affinities, i.e. the K_{MS}^* or X_I^* (50) values.

5.5 Catalytic Mechanism of PLA2

The His–Asp–calcium ion architecture in the active site provides the structural basis for the chemical step of sPLA2s. Typically, in the cocrystals with the active-site directed mimics, a C=O or P=O oxygen, corresponding to the *sn*-2-ester carbonyl of the substrate, is coordinated to the bound calcium ion of PLA2 (Figure 5.9). A significant part of the stability of the interaction of an *sn*-2-amide or -tetrahedral substituent also comes from the hydrogen bonding with the catalytic residue His-48.

5.5.1 Catalytic triad versus the calcium-coordinated oxyanion mechanism for the chemical step

In analogy with the mechanism for serine proteases, Verheij *et al.* (1981) proposed that Asp-99, His-48 and a water molecule form a 'triad' in which water is 'activated' by His-48 as the general base (Figure 5.10). The *sn*-2-ester carbonyl coordinates with calcium and polarizes it for the nucleophilic attack. This suggestion is consistent with the fact that the chemical step is rate-limiting. However, the rate-limiting transition state presumably occurs during the decomposition of the tetrahedral intermediate in which δN of His-48 is protonated. The decomposition of the tetrahedral intermediate gives an alcoholate which is protonated by δNH^+.

The Ca^{2+}-coordinated oxyanion mechanism (Figure 5.10) has been proposed on the basis of the observations about the flexibility of the calcium coordination environment, and other observations (Yu *et al.*, 1998; Berg *et al.*, 2001). In this mechanism a Ca^{2+}-coordinated oxyanion, the equatorial water (w5) is connected to δN of His-48 through another water molecule (w6). His-48 is protonated during the formation of the tetrahedral intermediate; however, due to the activation of the w5 oxyanion, the activation energy in this step is significantly lower such that the rate-limiting transition state occurs during the decomposition of the tetrahedral intermediate.

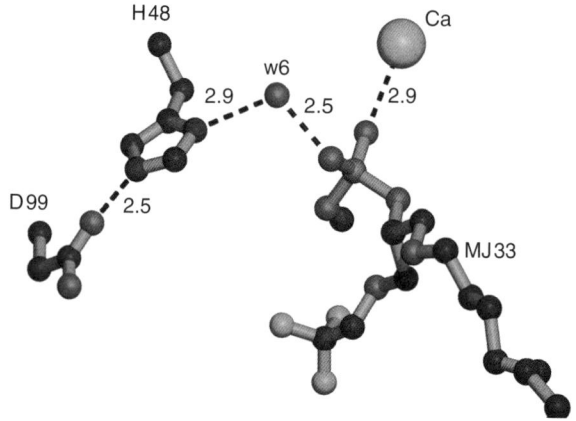

Figure 5.9 The active site interactions of pig pancreatic PLA2 cocrystallized with MJ33 and anions. Reprinted with permission from Pan *et al.* (2001). Copyright (2001) American Chemical Society

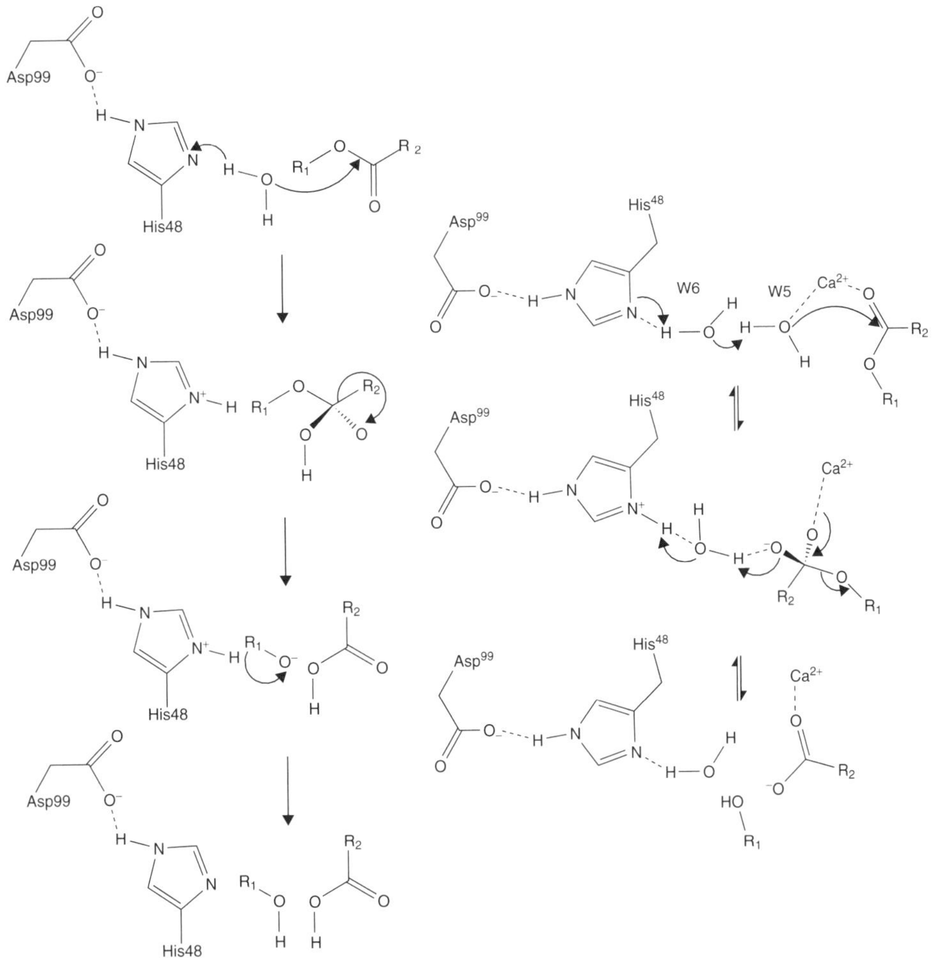

Figure 5.10 (Left) The catalytic triad mechanism (Verheij *et al.*, 1980) and (right) the calcium-coordinated oxyanion mechanism (Yu *et al.*, 1998) for the chemical step of pancreatic PLA2. Reprinted with permission from Berg *et al.* (2001). Copyright (2001) American Chemical Society

There is considerable similarity between the two mechanisms, with the major difference being in the direct role for the calcium cofactor in the coordination of the oxyanion. Such an activation could lower the activation energy for the formation of the tetrahedral intermediate, so that the rate-limiting transition state would lie during the decomposition of the intermediate. As an added feature, in the oxyanion mechanism the Ca-coordination environment changes from seven in the ground state of E*S to eight during the transition to the tetrahedral intermediate. The expanded coordination brings the nucleophile to a better angle of attack to the carbonyl carbon of the *sn*-2-ester. Thus divalent ions that resist such an expansion of the coordination environment are likely to be catalytically inert. This expectation is consistent with the observation that divalent Cd, Zn and Cu cations can only bind the substrate but not catalyze the reaction (Yu *et al.*, 1998). In contrast, the competitive cofactor divalent cations like Co and Ni support the substrate (mimic) binding and also catalyze the reaction as well as calcium does. These roles are consistent with the fact that the smaller cations are likely to have a more flexible or plastic coordination environment.

In the oxyanion mechanism, during the rate-limiting transition state the *sn*-3-phosphate of the substrate replaces the calcium-bound axial water molecule (w12). This is consistent with the observation that the substrate specificity for a variety of *sn*-3-phoshates, phosphonates and arsonates lies in the k^*_{catS} step (Rogers *et al.*, 1996). Since OH of Tyr-69 is also bonded to *sn*-3-phosphate, the Tyr-69–Phe mutant would be k^*_{catS}-impaired, as is found to be case (Yu *et al.*, 1998).

Activation of the oxyanion from calcium is also consistent with the observation that the k^*_{catS} for the Asp-99–Asn mutant is about 5 % of the value for WT PLA2 (Sekar *et al.*, 1997b). In contrast, the Asp-102–Asn mutant of trypsin has only 0.01 % activity, as expected if the attack of the nucleophile is rate-limiting. Thus the 'activating' effect of Asp-99 is lower because it is not as critical for the transfer of proton from δNH^+ to alcoholate.

As discussed above, the oxy/thio effect of 10 in k^*_{catS} suggests that the rate-limiting transition state is unlikely to be during the formation of the tetrahedral intermediate. Also a nucleophilic attack on a thioester should be more favored than on an oxyester. It may be argued that release of a thiolate ion in solution is favored more than that of an alcoholate, which only implies that even with that contribution in the active site the activation energy is still in the proton transfer step from the protonated His–Asp pair. The quantum chemical calculations for PLA2 also support the oxyanion mechanism, albeit without the participation of w6 (Schurer *et al.*, 2000). On the other hand, the modeling and crystallographic evidence (Figure 5.9) shows that the assisting water w6 can be present between a tetrahedral phosphate and His-48 (Berg *et al.*, 2001; Pan *et al.*, 2001).

5.6 The Quality-of-Interface Effects

So far we have focused on the assumption that the interface acts only as a matrix for the substrate accessibility and replenishment. Another issue of general interest is whether the interface also influences the intrinsic kinetic parameters for the bound enzyme. For example, the results in Figure 5.11 clearly show that there is no anomalous effect of the gel-to-fluid phase transition in the DMPM bilayer on the catalytic turnover rate of PLA2. In other words, the sum total of the events of the processive turnover cycle (Equation 5.9) are not influenced by a change in the phase state of the bilayer. This result is also consistent with the observation that the chemical step, and not a physical process, is rate-limiting in the reaction path. As we will see in Chapter 10, this is an important result: anomalous reaction progress at the phosphatidylcholine bilayers is observed in the gel–fluid coexistence region, and such effects are attributed to changes in the E to E* equilibrium.

5.6.1 Lateral phase separation in the bilayer

Anomalous diffusion is expected through the phase boundaries of laterally phase-separated bilayers. The key kinetic issue is whether such restrictions to ideal

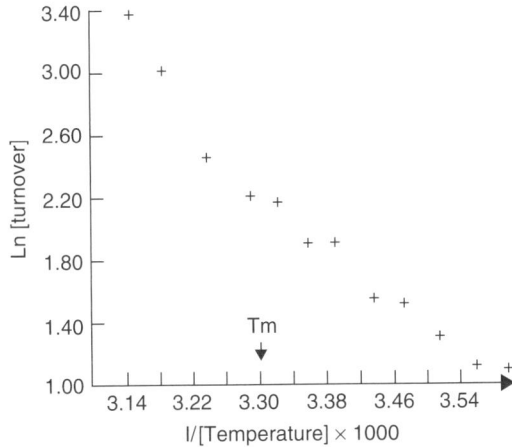

Figure 5.11 The temperature dependence of the first-order processive turnover rate, $N_S^0 k_i$ (on the natural log scale), for the hydrolysis of sonicated vesicles of 1,2-dimyristoyl-3-phosphomethanol (DMPM) by pig PLA2 under the first-order reaction progress (Equation 5.10b) Results not shown here also demonstrate that the bilayer phase properties have no effect on k_2^*. Reprinted from Jain and Rogers (1989). Copyright (1989) with permission from Elsevier Science

mixing of the components in the plane of a bilayer become rate-limiting for the substrate replenishment in the processive catalytic turnover cycles. Evidence based on a significant difference between the rate of hydrolysis of *sn*-2-oxy- versus -thio substrate shows that all such physical processes are not rate-limiting. Since the turnover limited by the chemical step for PLA2 on DMPM vesicles occurs on the time scale of milliseconds, the replenishment rate through the lateral phase boundaries must be much more rapid. This is consistent with the observed values of the lateral diffusion coefficient (Theory Box 5.2). Also, based on the effect of PLA2 on the gel–fluid transition of the bilayer (see Chapter 6), it is clear that the close contact of the i-face with the bilayer creates its own microenvironment in the contact region. Thus, it is quite likely that the interactions that promote lateral segregation in the bilayer are not critical for determining what happens in the bilayer whose interface is in close contact with the bound enzyme.

THEORY BOX 5.2 Time scales for target search by lateral diffusion

The time scale for lateral motions is determined by the lateral diffusion rate constant, D_2. For phospholipids in bilayer, $D_2 = 10^{-7}$–10^{-9} cm^2/s. Consider a spherical vesicle of radius R. The relaxation time for lateral diffusion over the whole surface is

$$\tau_R = R^2/D_2 \tag{1}$$

This is the time it takes to equilibrate an initially nonuniform molecule distribution. The mean time for a single molecule to move one molecular diameter, $2b$, is approximately

$$\tau_1 = 2b^2/D_2 \qquad (2)$$

Phospholipids occupy a surface area of ca. 7×10^{-15} cm^2 ($= \pi b^2$). Thus, $\tau_1 = 4 \times 10^{-6}$–$4 \times 10^{-8}$ s. For a vesicle with N_T phospholipids in the outer surface, $4\pi R^2 = N_T \pi b^2$, so that $\tau_R = N_T \tau_1/8$. Thus, for a vesicle with $N_T = 40\,000$, the diffusional relaxation time is $\tau_R = 0.02$–0.0002 s. Even for a fairly large vesicle, the relevant diffusion times may be short.

Target search by diffusion on a (bilayer) matrix

If there is a single substrate molecule in the surface, the mean time required to reach a particular position is (see Theory Box 12.2 and; Berg, 1985)

$$\tau_a = \frac{R^2}{D_2}\left[2\ln\left(\frac{2R}{b}\right) - 1\right] = \frac{R^2}{D_2}\left[\ln\left(\frac{1}{X_S^*}\right) - 1\right] \qquad (3)$$

This can be interpreted as the diffusion-limited time for a single substrate to reach the position right under the catalytic site of an interfacial enzyme. The last equality in Equation (3) follows from the fact that the mole fraction of substrates in this case is $X_S^* = b^2/(4R^2)$. Equation (3) was calculated as the mean time for a single-substrate single-enzyme encounter on the matrix of a diluent. It also holds very well as the mean time of association of any substrate molecule for the case when there are many substrates uniformly distributed on the surface (except in the limit $X_S^* \to 1$). As the rate of association is $1/\tau_a = k_1^* X_S^*$, the bimolecular association rate constant on the surface can be identified as

$$k_1^* = \frac{1}{\tau_a X_S^*} = \frac{4D_2}{b^2\left[\ln\left(1/X_S^*\right) - 1\right]} \qquad (4)$$

It can be seen that k_1^* is not a true constant as it depends, albeit only weakly, on the surface concentration X_S^*. This is related to the fact that the two-dimensional diffusion-reaction equation does not have a stationary solution, in contrast to the three-dimensional case (see Theory Box 12.2). Nevertheless, Equation (4) provides a good estimate for the apparent rate constant at different surface densities. With the numbers for D_2 and b used above (Equation 2), $k_1^* = 5 \times 10^5$–5×10^7 s^{-1} for $X_S^* = 0.01$ and $k_1^* = 2 \times 10^5$–2×10^7 s^{-1} for $X_S^* = 0.0001$.

The maximum diffusion-limited rate depends strongly on the mobility of the substrates in the interface, but even with the lower estimates the rate is fast. Actually, the enzyme could also have a lateral diffusion constant of the same order of magnitude, and as D_2 should be the sum of the diffusion constants for the substrates and the enzyme, the rate could be even faster.

5.6.2 Lateral diffusion and the target search

Non-idealities in the lateral distribution of the components in a bilayer are likely to emerge from the steric factors or specific interactions for the lateral packing of the amphiphiles in the interface. As such, partitioning driven by the hydrophobic effect would not be significantly influenced by such nonidealities. A target search for a component at low surface density in a multicomponent bilayer could take a significantly longer time if the exchange rate between the coexisting phases is very slow. A delay in the target search by an interfacial enzyme has to be carefully distinguished from the second-order substrate-binding step for the formation of the E*S Michaelis complex. At least in the case of PLA2, the measured rate constant, k_1^*, is much smaller than the estimated diffusion limit (Theory Box 5.2). Also a delay in the target search by an interfacial enzyme on a large surface, such as a monolayer trough (see Chapter 10), is different from the second-order substrate-binding step for the formation of the Michaelis complex or the search through the aqueous phase that involves k_{on} and k_{off} steps for the enzyme binding to the interface.

5.6.3 Changes in the rate-limiting step

Within the constraints of the general interfacial kinetic paradigm several kinetic paths and rate-limiting steps are possible. The magnitude of the oxy/thio ratio is useful for identifying a change in the rate-limiting step. This is supported by certain substitutions in the active site, e.g. Asp-99 and Tyr-69, that change the oxy/thio ratio, suggesting a change in the catalytic mechanism for the chemical step. Also a change in the rate-limiting step for $j_0 = k_{catS}^*/(1 + K_{MS}^*)$ can come about from an increase in K_{MS}^* from <1 to >1; or if K_{MS}^* remains <1, it could come from an increase in the ratio of k_2^*/k_3^* from <1 to >1. While a large (or for that matter a very small) oxy/thio ratio shows that the chemical step is rate-limiting, the element effect close to one is expected for physical processes such as the lateral diffusion for the target search or the substrate binding to the active site of the enzyme. A change in the oxy/thio ratio towards one with a change in the interface would show that the chemical step is no longer rate-limiting, as is the case for the mixed micelles (Jain *et al.*, 1993a).

5.7 Apparent Interfacial Activation

An understanding of the kinetic and structural basis for the coupling between the binding to the interface and the catalytic turnover events constitutes the basis for the allosteric effect of the interface (see Chapter 8) on the catalytic parameters K_{MS}^* and k_{catS}^*. However, to dissect such effects it is necessary to understand the bases for an apparent change in observed rate due to a change in the substrate accessibility and replenishment rate. Some of the interesting results shown in Figures 5.12 and 5.13 are attributed to a decrease in the interfacial processivity.

5.7.1 Reaction progress with slow enzyme exchange

Consider a reaction mixture where the number of enzymes is much smaller than the number of vesicles so that there is no more than one enzyme per vesicle. An enzyme that dissociates from a vesicle will with high probability rebind to a fresh vesicle that has not previously had an enzyme bound. Thus, with some slow rate,

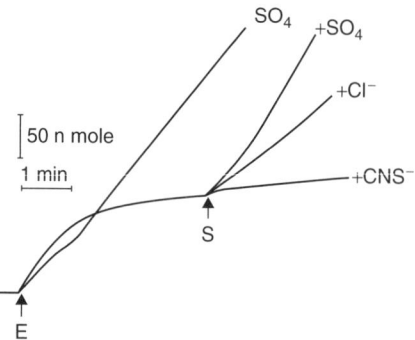

Figure 5.12 Effect of added sodium salts (at S, or added initially) of the indicated anions on the reaction progress of the hydrolysis of DMPM vesicles by pig pancreatic PLA2. Reprinted from Jain *et al.* (1986b). Copyright (1986) with permission from Elsevier Science

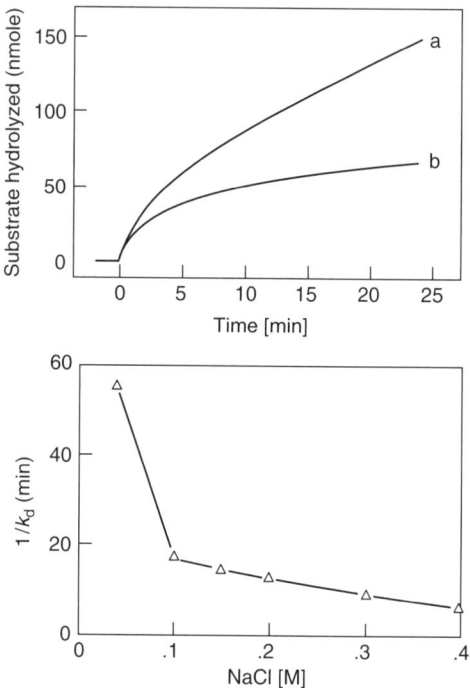

Figure 5.13 (Top) The reaction progress for the hydrolysis of DMPM vesicles with 0.3 M NaCl (curve a) or without NaCl (curve b). Such curves were fitted to Equation (5.19) to obtain $k_{exch} = k_d$ values plotted in the figure (bottom) as a function of [NaCl]. Reprinted with permission from Jain *et al.* (1991c). Copyright (1991) American Chemical Society

enzymes will leave partially or fully depleted vesicles and start anew. The enzyme exchange will extend the scooting mode reaction progress on different vesicles. The exchange rate, k_{exch}, for an enzyme between vesicles equals the dissociation rate constant (see Theory Box 2.1). In the limit where the reaction progress on a single vesicle is exponential, i.e. determined by Equation (5.8) with $j_0 = k_i N_S^0$, one finds that the overall reaction progress is (Jain and Berg, 1989 and Theory Box 5.3)

$$[P^*(t)] = N_S^0 [E_0] \frac{k_i}{k_i + k_{\text{exch}}} \left[k_{\text{exch}} t + \frac{k_i}{k_i + k_{\text{exch}}} \left(1 - e^{-(k_i + k_{\text{exch}})t} \right) \right] \tag{5.19}$$

This has been plotted in Figure 5.14A for some values of k_{exch}/k_i. The result is valid only for as long as a dissociated enzyme is unlikely to rebind a partially depleted vesicle, i.e. for as long as each vesicle has been visited at most once. Furthermore, it requires that the dissociation rate from a fully or partially depleted vesicle is the same as from a fresh substrate vesicle.

The result is much more complicated (Theory Box 5.3) when the reaction progress in each vesicle is not properly exponential, i.e. if there is also a zero-order steady state contribution (Equation 5.9). The overall reaction progress in this case has

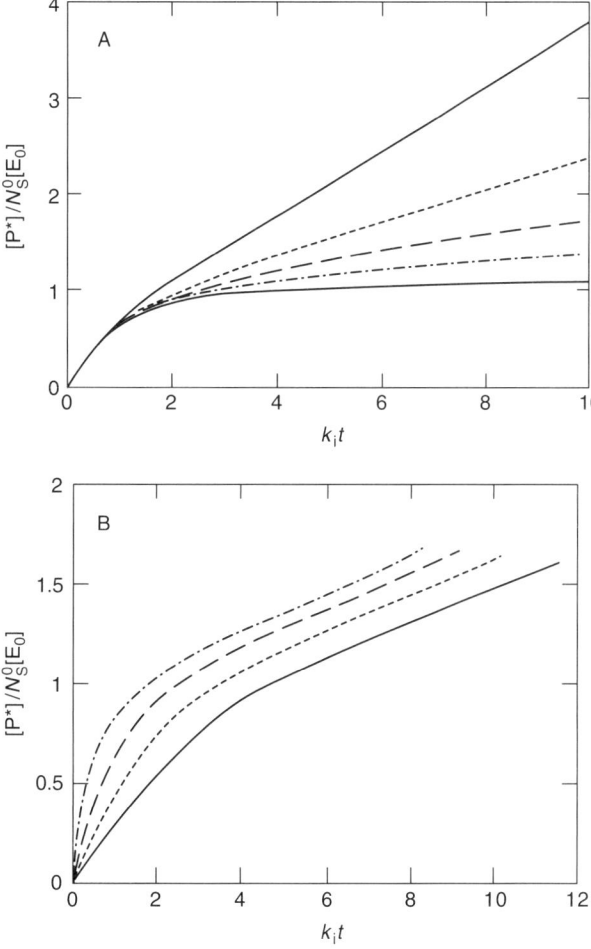

Figure 5.14 Panel A: reaction progress with slow enzyme exchange in the limit $j_0 = k_i N_S^0$ (Equation 5.19), for the first-order reaction progress. From upper to lower curves, $k_{\text{exch}}/k_i = 0.5$, 0.2, 0.1, 0.05, 0.01. Panel B: reaction progress with slow enzyme exchange (Equation 3 in Theory Box 5.3) for $k_{\text{exch}}/k_i = 0.1$ with a zero-order steady state contribution to the reaction progress. From upper to lower curves, $j_0/(k_i N_S^0) = 5$, 1, 0.5, 0.3

been plotted in Figure 5.14B. The shape of the curves is influenced by the value of $j_0/(k_i N_S^0)$, but the slope in the quasi-stationary state remains approximately the same as for $j_0 = k_i N_S^0$. The value of the exchange rate in terms of K_d and aggregate size for enzyme dissociation as a diffusion-limited process is discussed in Theory Box 2.1. In the calculations of Equation (5.19) and Theory Box 5.3, k_{exch} is the effective dissociation rate for all enzyme species in the interface (E^*, E^*S and E^*P) to E, while K_d is defined (see Figure 5.15 later) as the dissociation constant for E^* to E. This fact could make k_{exch} dependent on composition (X_S^*, X_P^*), a complication that has not been accounted for here.

THEORY BOX 5.3 Slow enzyme exchange

An enzyme that has bound a vesicle at time t' will produce product molecules from this vesicle at a rate

$$J_P(t - t') = j_0 \frac{1 - X_P^*}{1 + X_P^*(A - 1)} e^{-k_{exch}(t - t')} \tag{1}$$

at time t, Equation (5.7). Here, the exponential function is the probability that the enzyme remains bound at time t, and $X_P^*(t - t')$ is the expected reaction progress at time t, as given by Equation (5.8) with t replaced by $t - t'$. A is the ratio $j_0/(N_S^0 k_i)$. The fraction of enzymes that exchange to a fresh vesicle during the period t' to $t' + dt'$ equals $k_{exch} dt'$. Thus we can calculate the total product flux as a sum over all enzymes as

$$\frac{1}{[E_0]} \frac{d[P^*]}{dt} = J_P(t) + k_{exch} \int_0^t J_P(t - t') \, dt' = J_P(t) + k_{exch} \int_0^t J_P(t') \, dt' \tag{2}$$

As t is known as a function of X_P^*, (Equation 5.10a), the flux J_P in Equation (1) can also be expressed as a function of X_P^* alone. From Equation (5.7) we also know dt/dX_P^* as a function of X_P^*. Then all integrals over t can be transformed to integrals over X_P^*, and we find from Equation (2) that

$$\frac{[P^*(t)]}{N_S^0 [E_0]} = \int_0^{X_P^*(t)} \left[(1 - z)^{k_{exch}/k_i} e^{-k_{exch}(1 - A)z/(k_i A)} \right.$$

$$\left. + \frac{k_{exch}}{A k_i} \frac{1 + z(A - 1)}{1 - z} \int_0^z (1 - y)^{k_{exch}/k_i} e^{-k_{exch}(1 - A)y/(k_i A)} dy \right] dz \tag{3}$$

This gives the overall reaction progress as a function of the expected mole fraction of product at time t in a vesicle where the enzyme remains bound. As this is known implicitly from Equation (5.10a), the reaction progress can be calculated numerically as a function of t, (Figure 5.14B). In the limit when $A = 1$, i.e. $j_0 = N_S^0 k_i$, the integrals in Equation (3) are elementary and Equation (5.19) results.

5.8 Hopping and the Fast Enzyme Exchange Limit

If the interfacial enzymes 'hop' or the substrates and product exchange very rapidly between aggregates, the substrate depletion in all aggregates will take place at the same rate and all enzymes will see the same substrate and product concentrations, X_S^* and X_P^*. In this limit, the rate of exchange must be much faster than $1/\tau_0$, with $n_E = 1$ from Equation (5.5). Otherwise, each enzyme–aggregate binding event will lead to a substantial substrate depletion.

5.8.1 Reaction progress with fast enzyme exchange

When enzyme exchange between vesicles is faster than the time required to decrease X_S^* significantly, the integrated MM equation is essentially the same as before, (Equation 5.8) if $X_P^*(t)$ is interpreted as the overall average product mole fraction, $X_P^*(t) = [P^*(t)]/[S_0^*]$. Also, the factor $n_E/N_S^0 = [E_0]/[S_0^*]$ must be interpreted as the overall initial enzyme-to-substrate ratio. Thus,

$$k_i t = -\ln\left(1 - \frac{[P^*(t)]}{[S_0^*]}\right) + \left(\frac{k_i\,[S_0^*]}{j_0\,[E_0]} - 1\right)\frac{[P^*(t)]}{[S_0^*]} \tag{5.20}$$

where j_0 is given by Equation (5.3) and k_i from Equation (5.9) as

$$k_i = \frac{[E_0]}{[S_0^*]}\frac{k_{\text{catS}}^*}{K_{\text{MS}}^*\left(1 + 1/K_{\text{MP}}^*\right)} \tag{5.21}$$

giving the ratio

$$\frac{k_i\,[S_0^*]}{j_0\,[E_0]} = \frac{1 + 1/K_{\text{MS}}^*}{1 + 1/K_{\text{MP}}^*} \tag{5.22}$$

Thus, the reaction progress with fast enzyme exchange is exactly the same as for the scooting kinetics with at most one enzyme per vesicle. The difference lies only in the maximum amount of product produced per enzyme: with the scooting kinetics it corresponds to the number of substrates available on the outer surface of a vesicle, N_S^0, while with fast enzyme exchange it is the total accessible substrate-to-enzyme ratio, $[S_0^*]/[E_0]$.

The steady state reaction progress in the quasi-scooting mode (see Chapter 8) satisfies the requirements for rapid exchange of E or S and P.

5.8.2 Not all enzymes are in the interface

If enzymes can exchange between aggregates (Figure 5.15), it is reasonable to consider also the case when they are not bound at all times. In this case the scheme

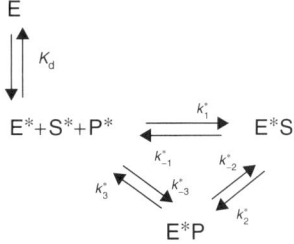

Figure 5.15 Reaction loop for the turnover cycle with exchangeable enzyme

should be modified with an enzyme–aggregate binding step. At the steady state, the flux through the reaction loop (turnover path) for each interface-bound enzyme is determined by the same relations as before, (Equation 5.1). The E to E* binding and dissociation step, on the other hand, is a dead-end that can have no flow through and therefore must be equilibrated at the steady state. As a consequence, the flux per enzyme is found to be

$$j = \frac{\dfrac{k^*_{catS}}{K^*_{MS}} X^*_S - \dfrac{k^*_{catP}}{K^*_{MP}} X^*_P}{1 + \dfrac{X^*_S}{K^*_{MS}} + \dfrac{X^*_P}{K^*_{MP}} + \dfrac{K_d}{[M^*]}} \tag{5.23}$$

[M*] denotes the bulk concentration of accessible molecules in the interface to which the enzyme can bind. The extra term in the denominator, $K_d/[M^*]$, compared to Equation (5.2) is the ratio of [E]/[E*]; in fact, each term in the denominator corresponds to the ratio of each enzyme state to the E* state. The influence on j from any new enzyme states could be introduced in this way simply by adding the corresponding equilibrium ratio as a new term in the denominator, as long as the new states do not interfere with the basic catalytic loop by introducing alternative pathways for the flux (cf. Theory Boxes 5.1 and 6.1).

To calculate the initial rate, we can set $X^*_S = 1$, $X^*_P = 0$ and assume that the concentration of accessible aggregate molecules can be written as $[M^*] = [S^*_0]$, where $[S^*_0]$ is the initial bulk concentration of accessible substrate in the aggregates. For example, for vesicles, the accessible amount is that which is available in the outer layer. Thus the initial rate can be written as

$$j_0 = \frac{k^*_{catS}}{1 + K^*_{MS}(1 + K_d/[M^*])} = \frac{k^*_{catS}[S^*_0]}{[S^*_0](1 + K^*_{MS}) + K_d K^*_{MS}} = \frac{k^{app}_{cat}[S^*_0]}{[S^*_0] + K^{app}_M} \tag{5.24}$$

If the initial bulk concentration of substrate in the interface is manipulated, the initial rate behaves like the solution kinetics result from Chapter 3 (Equation 3.6) with apparent Michaelis–Menten parameters:

$$k^{app}_{cat} = \frac{k^*_{catS}}{1 + K^*_{MS}} \tag{5.25}$$

$$K^{app}_M = \frac{K_d K^*_{MS}}{1 + K^*_{MS}} \tag{5.26}$$

These effective MM parameters can be determined by fitting a hyperbolic curve to the initial rate as a function of $[S^*_0]$ or by using the linear Lineweaver–Burk plot (Figure 3.3); k^{app}_{cat} corresponds to the initial rate when all enzyme is interface bound, while K^{app}_M is the effective dissociation constant from the interface in the presence of substrate binding and catalysis.

The hyperbolic MM relation appears at two levels. When the enzyme is not all in the interface, increasing the bulk concentration of substrate will activate the enzyme by increasing the fraction that is interface-bound. It leads to a product flux that depends on bulk concentration like the solution MM relation (Equation 3.6); for this relation we use the apparent k_{cat} and K_M symbols without the asterisks (Equation 5.24). When the enzyme is all bound to the interface, we have instead a local MM relation (Equation 5.2), where the flux depends on the surface concentration, X_S^*, that the enzyme sees in the interface; for these relations we always use the kinetic parameters with the asterisks.

5.8.3 Reaction progress with fast exchange and not all enzyme in the interface

The flux per enzyme (Equation 5.23) has exactly the same form as Equation (5.2) if K_{MS}^* and K_{MP}^* in Equation (5.2) are replaced by $K_{MS}^*(1 + K_d/[M^*])$ and $K_{MP}^*(1 + K_d/[M^*])$ to get Equation (5.23). As a consequence, the integrated MM equation when not all enzyme is in the interface and with fast exchange is determined by Equations (5.20) to (5.22), with K_{MS}^* and K_{MP}^* replaced in this way. Thus, Equation (5.20) holds with

$$k_i = \frac{k_{catS}^* [E_0]}{K_{MS}^* [S_0^*]} \frac{1}{1 + K_d/[M^*] + 1/K_{MP}^*} \tag{5.27}$$

and

$$\frac{k_i [S_0^*]}{j_0 [E_0]} = \frac{1 + K_d/[M^*] + 1/K_{MS}^*}{1 + K_d/[M^*] + 1/K_{MP}^*} \tag{5.28}$$

When the enzyme binds very weakly to the interface, $K_d \gg [M^*]$, the ratio in Equation (5.28) approaches one and the reaction progress from Equation (5.20) becomes purely exponential; this is the 'hopping' limit with poor enzyme interface binding and low processivity.

5.9 Summary: Variations in the Processivity

Again with a focus on the pancreatic PLA2, a complete analysis of the observed reaction progress in the interfacial processive scooting mode is possible because the reaction path is unequivocally defined, as shown in Chapter 4, only by the steps of the turnover cycle in the interface. The first-order region of the reaction progress dominates the behavior for the hydrolysis of small vesicles largely controlled by product accumulation and inhibition. The steady state zero-order initial rate at $X_S^* = 1$ is discernible during the hydrolysis of large vesicles or when rapid substrate replenishment by vesicle-to-vesicle exchange is induced through the polymyxin B contacts. The kinetic analysis is simplified because the chemical step is rate-limiting and essentially irreversible. Thus the substrate, inhibitor and cofactor specificity under the scooting mode kinetic conditions is fully analyzed in terms of the independently determined equilibrium constants. These results have provided insights for the catalytic mechanism in which a calcium-coordinated oxyanion attacks the *sn*-2-ester carbonyl carbon and the transition state occurs during the decomposition of the tetrahedral intermediate (Figure 5.10).

Exchange of the bound enzyme between vesicles gives rise to a range of kinetic effects that constitute the basis for apparent inhibition or activation. The resulting changes in the residence time of the enzyme influence the turnover processivity. Under suitable conditions such effects can be fully analyzed in terms of the changes in the equilibrium constant for the binding of the enzyme to the interface. Such cases of intermediate processivity may help to account for wide-ranging interfacial kinetic behavior that cannot be fully analyzed without additional information. However, as developed further in Chapter 8, the rapid substrate replenishment conditions also lead to the reaction progress in the quasi-scooting mode, which can be analyzed to obtain primary kinetic parameters with mechanistic significance.

Further Reading

Segel, I.H. (1975). *Enzyme Kinetics: Behavior and Analysis of Rapid Equilibrium and Steady-State Enzyme Systems*, Wiley, New York, 957 pp.

6 Detailed Balance Conditions for Interfacial Equilibria

Life would be infinitely happier if we could only be born at the age of eighty and gradually approach eighteen.

Mark Twain

The general kinetic paradigm for enzymology (Figure 3.1) has several steps that feed into the formation of the Michaelis complex, ES or E*S. Such equilibria control the variables for the substrate accessibility and exchange, processivity, and the steps of the turnover cycle. Each of these steps has characteristic rate and equilibrium constants. As formulated in Figure 6.1, the standard three-dimensional solution equilibria that control the active site occupancy in E in the aqueous phase are defined as K_L and K_A expressed in units of moles per liter. In addition, for the partitioning and binding of solutes and enzyme, there are four other classes of equilibria (Jain *et al.*, 1991a, 1993b; Yu *et al.*, 1993, 1997a, 1997b) that influence processivity and the events of the turnover cycle:

(a)   As discussed before, the dissociation of an amphiphile monomer from the interface is most conveniently defined as K'_L for the partitioning of a solute in the interface (Section 3.5) or for the monomer–micelle equilibrium as the CMC (Section 2.4).

(b)   The dissociation equilibrium between the enzyme bound to the interface versus the aqueous phase, E* to E, is characterized by K_d. Similarly, dissociation of E*L to EL is described by K_d^L. These steps determine the fraction of the enzyme at the interface. For a one-step process, K_d is the free interface concentration, expressed as the aggregated amphiphile concentration, at which half of the total enzyme is bound as E* and the other half is in the E form in the aqueous phase. For the calculations it is more convenient to consider K_d as the dissociation constant from the accessible amphiphiles only, and then only in the limit of a low enzyme-to-amphiphile ratio, so that there are no excluded-surface effects. The same arguments hold for the dissociation of E*L to EL as described by K_d^L.

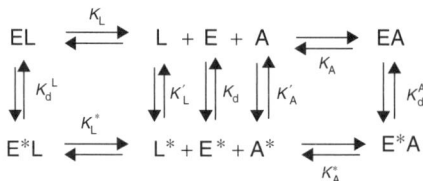

Figure 6.1 Equilibrium relations of an interfacial enzyme (E) with an active-site-directed ligand (L = I, P, S or A as an imperfect diluent) and a neutral diluent amphiphile (A and A*). See Figure 3.1 for the relationship of such equilibria to events of the turnover cycle. Recall that the species marked with an asterisk are in the interface. As shown later in this and the following chapters, the enzyme complexes formed by interaction with cofactors or allosteric effectors can be accommodated within this paradigm

K_d^{eff} describes the effective dissociation constant when 50 % of the total enzyme is in solution and the rest is as the multiple bound species, e.g. E* + E*L.

(c) Dissociation of an active-site-directed ligand from E*L in the interface to L* + E* is characterized as the two-dimensional dissociation constant, K_L^*, expressed in mole fraction units.

(d) The turnover cycle is also influenced by the functionally discrete species of the enzyme with cofactor and allosteric effector. As discussed later, the equilibria to account for their kinetic effects are accommodated with the paradigm of the Scheme in Figure 6.1.

In this chapter we outline properties of the equilibrium relations for the enzyme binding to the interface and the binding of a ligand to the active site. The corresponding equilibrium constants provide insights into the structural features that control the behavior of the enzyme, ligands and the variables related to the interface or the presence of a cofactor. As discussed in the preceding chapter, certain ligand binding constants allow determination of the interfacial catalytic parameters K_{MS}^* and k_{catS}^* in relation to the presence of a cofactor. Operationally, K_d is attributed to the i-face interactions with several amphiphile molecules at the interface. K_L^* is related to the interaction usually of a single ligand molecule with the active site of the enzyme at the interface. Both of these terms relate to the functional difference between the active site versus the i-face, which permits identification of relevant variables. A change in K_d, a property of the i-face, dramatically influences the course of the reaction progress. Thus, not only is the knowledge of the dissociation constant of E* to E important but also the values of K_d^L for the dissociation equilibrium of E*L to EL. Since the equilibria across the interface are explicitly or implicitly coupled, strategies to distinguish the E, E*, EL and E*L species are necessary for the determination of K_d and K_d^L.

6.1 Binding of Ions to the Interface

Interactions of a solute or enzyme with the interface occur within the constraints that are probably never encountered in the classical bulk phases. An appreciation of this environment can be gained from the type of environment encountered by an ionic

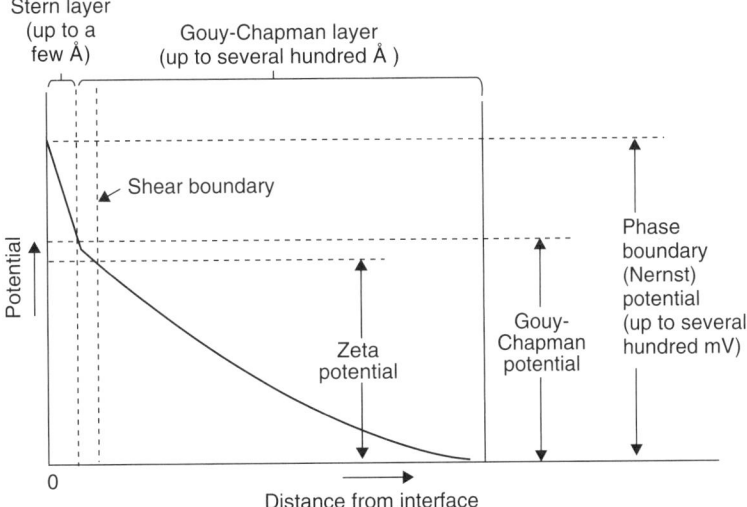

Figure 6.2 A schematic for the surface potential due to the distribution of the counter ion charge density away from the fixed charge at the interface (at distance $= 0$). The diffusing counter ions distributed for up to about 1 nm in the bulk aqueous phase compensate most of the fixed charge

solute in the vicinity of an interface. All interfaces are surrounded by an unstirred aqueous layer whose thickness may extend for several micrometers, depending on the geometry of the interface and the stirring rate (see Theory Box 10.2). As shown in Figure 6.2 for the distribution of counter ions relative to the fixed surface charges, within the unstirred layer near the surface there is a diffuse layer. In this layer, which extends for about 15 Å, the electrostatic pull of the fixed charges on the counter ions is compensated by the thermal motion.

6.1.1 Electrical double layer

Fixed surface charges on an interface give rise to a net electric charge. A net fixed charge on an organized interface is generated due to the dissociation of an ionizable group in the amphiphile. A charged interface is surrounded in the aqueous solution by a diffused layer of counter ions. Gross features of the resulting charge profile are shown in Figure 6.2. Only a part of the fixed charges is balanced by counter ions within the shear plane of the particle in Brownian motion, and the rest is effectively neutralized by more mobile counter ions distributed over a distance away from the interface. Thus the electrostatic potential profile, due to the charge separation at the electrically charged interface, has a characteristic decay that depends on the nature of the interface as well as that of the bulk aqueous phase. Such a surface potential profile exists even at electrically neutral interfaces because the partitioning ability of anions versus cations is different.

The presence of an electrical double layer is best shown by the electrophoretic mobility of the aggregate particles. The smeared-charge profile at the interface is often described by the Gouy–Chapman approximation to the solution of the Poisson–Boltzmann equation for the distribution of the counter ions as point charges (McLaughlin, 1989; Cevc, 1990). The distribution of the counter ions in the double layer is determined by their equilibrium interactions with the interface, the mobility of the particles and the partitioning of the ionic solutes into the bilayer. In the standard model for the electrical double layer, the region of decreasing charge imbalance extends to several (typically less than 10) nanometers away from the plane of the fixed charges. The shape of the charge distribution profile away from the interface depends on the geometry of the interface as well as the hydrodynamic shear boundary of the particle surface.

Binding of cations to an anionic phosphatidylserine (Figure 1.5) vesicle interface has been analyzed from the zeta potential values obtained from the electrophoretic mobility of the vesicles in the presence of the cations (McLaughlin, 1989). The extent of binding was monitored as a change in the electrophoretic mobility as a function of the anion concentration. The surface charge density, σ_S, at the interface increases, with the adsorption of a cation at concentration $[C_+]$ moles per liter, to

$$\sigma_S = \sigma_0(1 + K_+[C_+]) \qquad (6.1)$$

where K_+ is the intrinsic association constant. The model used for the interpretation assumes a Poisson–Boltzmann profile for the distribution of the cations with a plane of shear. Results suggest that the Stern layer plane of the fixed charge with the bound counter ion is of 2 Å thickness. The specific ion association constant for phosphatidylserine was found to be in the range of 0.6 M^{-1} for sodium, 8 M^{-1} for calcium and magnesium to 40 M^{-1} for nickel, corresponding to a specific adsorption potential of −6 to 8 kT for the divalent ions. Electrostatic (Coulombic) interactions of simple ions with the interface occur without desolvation. The contribution per charge to the stability is typically 1 kcal/mole, even when several charged groups are clustered on the same counter ion (Buser *et al.*, 1994).

6.1.2 Electrostatic binding of cationic peptides to anionic interfaces

Long-range nonspecific electrostatic interactions dominate the partitioning or association of cationic peptides to interfaces of anionic phospholipids. The incremental free-energy contribution is typically less than 1 kcal/mole per cationic residue for the binding of pentalysine to 2:1 PC/PS vesicles. The distance dependence of the electrostatic interactions is such that no significant energy gain occurs unless the separation is <4 Å. Only in conjunction with other interactions, which stabilize close contacts through hydrogen-bonding and hydrophobic effects, can such charged motifs contribute significantly to the binding of a protein. Anchoring of an acylated protein by neighboring clusters of cationic residues can make larger contributions (Low, 1989; Murray *et al.*, 2001).

6.1.3 Selective partitioning of anions into zwitterionic interfaces

Although not intuitively obvious, common anions like chloride partition into zwitterionic interfaces more readily than cations like sodium ions (Collins and Washabaugh, 1985). The partitioning ability of the anions generally follow the

Hoffmeister series (SO_4 > F > Cl > ClO_4 > Br > NO_3 > I > SCN), as is also the case for the salting-out effect that lowers the CMC (Figure 2.5). Development of a net negative charge on the interface of phosphatidylcholine vesicles, as a consequence of selective partitioning of anions, is clearly demonstrated, for example, by a change in the electrophoretic mobility of multilamellar liposomes of zwitterionic phosphatidylcholines (Tatulian, 1983). A chemical-trapping method has been developed to estimate interfacial chloride concentration in phosphatidylcholine micelles and vesicles (Berg *et al.*, 1997).

6.1.4 Specific interactions at close contacts

Specific ion binding also influences the charge interactions at close contacts (Chakrabarti, 1993). At the bilayer interface such interactions are mediated by hydrogen bonding (2.8 to 3.5 Å) and ligand coordination (2.3 to 3.0 Å) enhanced by the hydrophobic effect. A remarkable property of the organized interfaces of amphiphiles is the steep gradient of polarity: within a span of about 7 Å, the effective dielectric changes from 80 in the aqueous phase to about 2 in the region where the alkyl chains are segregated. This region has a diffuse-smeared layer of counter ions just outside the hydration layer, possibly containing up to 20 water molecules per phospholipid molecule. For close contacts within 5 Å, the hydration layer must be disrupted to accommodate solutes of wide-ranging polarity, including the residues on the i-face, and participate in specific hydrogen-bonding interactions. Such specific ion-solute binding is possibly operationally best treated as the substitution of water-interface interactions with the solute-interface interactions. For example, specific interactions of ions would involve ligand substitution for the solvated water molecules. An energetically favorable swap would involve the oxygen- and nitrogen-containing functional groups on the interface and the protein brought together within 3 to 4 Å. The local constraints for such short-range contact would certainly involve the local packing constraints of the side chains of the protein as well as the phospholipid head groups at interface. Such specific interactions appear to be at the heart of the higher affinity of the i-face for the anionic interface (Chapters 7 and 8).

6.2 Equilibria for the Binding of the Enzyme to the Interface

The E to E* binding along the i-face is distinct from the steps in which E or E* binds an individual ligand molecule, L or L*, to the active site (Figure 6.1). The presteady state rate (Jain *et al.*, 1988) and the fraction of the bound enzyme is proportional to $[M^*] = [A^*] + [L^*]$, the concentration of interface molecules available for the binding equilibrium (Dam-Mieras *et al.*, 1975; Araujo *et al.*, 1979; Hille *et al.*, 1981; Jain *et al.*, 1982; Jain and Vaz, 1987). For example, the interface of a neutral diluent binds the enzyme in the E* form, because the affinity of A* for the active site of E* is weak or nonexistent (Figure 3.11). As developed below, as a two-dimensional solvent, a neutral diluent is useful for studying the properties of the enzyme at the interface and also for changing the surface concentration of L* for measuring the two-dimensional interfacial equilibrium constants for the occupancy of the active site by a ligand.

6.2.1 The detailed-balance conditions

The thermodynamic information about the molecular interactions between the components of the system lies in the equilibrium constants. Such constants are obtained from the study of the concentration dependence of the various species in the reaction. The equilibria in the scheme in Figure 6.1 can be described by considering the equilibrium between the different states of the enzyme in each branch and then summing over all states to get the known total concentration of enzyme. We will use the E^* state as reference and calculate the concentration of all enzyme states ($E_{state} = E, EL, EA, E^*, E^*L, E^*A$) relative to this. One finds

$$[E] = [E^*] \frac{K_d}{[M^*]} \tag{6.2}$$

$$[E^*L] = [E^*] \frac{X_L^*}{K_L^*} \tag{6.3}$$

$$[E^*A] = [E^*] \frac{X_A^*}{K_A^*} \tag{6.4}$$

$$[EL] = [E^*L] \frac{K_d^L}{[M^*]} = [E^*] \frac{X_L^*}{K_L^*} \frac{K_d^L}{[M^*]} \tag{6.5a}$$

$$[EA] = [E^*A] \frac{K_d^A}{[M^*]} = [E^*] \frac{X_A^*}{K_A^*} \frac{K_d^A}{[M^*]} \tag{6.6a}$$

In the last two equations, [EL] and [EA] were calculated from E^* via E^*L and E^*A, respectively. These concentrations could equally well have been calculated via the state E, giving

$$[EL] = [E] \frac{[L]}{K_L} = [E^*] \frac{K_d}{[M^*]} \frac{[L]}{K_L} \tag{6.5b}$$

$$[EA] = [E] \frac{[A]}{K_A} = [E^*] \frac{K_d}{[M^*]} \frac{[A]}{K_A} \tag{6.6b}$$

Setting each pair of alternative expressions equal and inserting the partitioning relations, Equations (3.11) and (3.12), one finds

$$\frac{K_d^L}{K_d} = \frac{K_L^*}{K_L} K_L' \tag{6.7}$$

$$\frac{K_d^A}{K_d} = \frac{K_A^*}{K_A} K_A' \tag{6.8}$$

These *detailed-balance conditions,* for the equilibrium relations in the so-called thermodynamic box, appear between the equilibrium constants whenever states in a reaction diagram can be reached through different paths.

The choice of units for the constants in Equation (6.7) or (6.8) provides insights into the underlying relations. Both K_d and K_d^L on the left side have the units of moles of interface per liter of aqueous phase. On the other hand, K_L^* is expressed as the mole fraction (unit-less). Thus the units of K_L' (moles of L per liter of aqueous phase) and of K_L (moles of L per liter of aqueous phase) cancel out. Also note that in Equation (6.2), $[M^*]$ is the concentration of amphiphiles in the interface that are available for binding the enzyme. In most cases, we will consider a large excess of amphiphiles over enzyme so that $[M^*]$ is not significantly decreased by enzyme binding.

The equilibrium distribution for the reaction scheme in Figure 6.1 is determined by Equations (3.11) to (3.16) and (6.2) to (6.8). To this we should add the mass-balance requirement that the concentrations of all enzyme states must sum to the total concentration, $[E_0]$, in the system (cf. Theory Box 6.1). The external variables under the experimenter's control are primarily the total concentrations of A and L in the system, $[A_T]$ and $[L_T]$. The relations assume that $[A_T]$ and $[L_T]$ correspond to the total concentrations of 'free' A and 'free' L that are *not* enzyme-bound. This will correspond to the total concentrations in the reaction mixture if A and L are in sufficient excess over E. Furthermore, for the binding of the enzyme to the interface, it is assumed that interface is in large excess so that excluded surface effects are negligible (cf. Theory Box 6.2).

The relations above hold also if the equilibrium constants are not true constants but depend on the composition of the system. It was argued in Chapter 3 that the apparent dissociation constants describing the partitioning of A and L into the aggregates were likely to depend on the composition of the aggregates. Similarly, the enzyme-interface binding may depend on the composition of the interface, e.g. if the enzyme i-face has more (or less) favorable interactions with L than with A. In this case, enzyme binding to the interface will also influence the partitioning equilibria of A and L. Some of these complications are described further in Chapters 10 and 11. If the equilibrium constants do vary significantly in this way, the analysis of the scheme easily becomes intractable. In the following examples it will be assumed that the equilibrium constants are true constants.

THEORY BOX 6.1 Statistical weights of enzyme states

When analyzing extended reaction schemes like the one in Figure 6.1, it is always helpful to use some systematic approach. One possibility is the following. The concentrations of all states of the enzyme are given above (Equations 6.2 to 6.6) relative to the concentration of E^*. We can call the concentration ratio, $[E_{state}]/[E^*] = W(E_{state})$, the statistical weight for each enzyme state, where

$$W(E^*) = 1 \tag{1}$$

$$W(E) = \frac{K_d}{[M^*]} \tag{2}$$

$$W(E^*L) = \frac{X_L^*}{K_L^*} \tag{3}$$

$$W(E^*A) = \frac{X_A^*}{K_A^*} \tag{4}$$

$$W(EL) = \frac{X_L^*}{K_L^*}\frac{K_d^L}{[M^*]} = \frac{[L]}{K_L}\frac{K_d}{[M^*]} \tag{5}$$

$$W(EA) = \frac{X_A^*}{K_A^*}\frac{K_d^A}{[M^*]} = \frac{[A]}{K_A}\frac{K_d}{[M^*]} \tag{6}$$

These weights can be calculated directly by considering the equilibrium relation for each state relative to the reference state, simply 'walking' from E^* along the relevant arrows in the scheme. Each arrow passed contributes a factor to the weight. When doing so, it is always helpful to have the equilibrium constant associated with the correct arrow, e.g. a dissociation constant associated with the dissociation arrow. Then the statistical weight becomes the equilibrium constant in the numerator if the 'walk' is in the same direction as its arrow, and in the denominator otherwise (cf. Equations 2 and 3 where the 'walks' are from E^* to E and E^* to E^*L, respectively). The two different (but equal) weights given in Equations (5) and (6) come from the two possible paths from E^* to EL and EA, respectively. The probability for any state is then determined from

$$P(E_{\text{state}}) = \frac{[E_{\text{state}}]}{[E_0]} = \frac{W(E_{\text{state}})}{\sum\limits_{E_{\text{states}}} W(E_{\text{state}})} \tag{7}$$

where $[E_0]$ is the total enzyme concentration in the system. Thus, the statistical weights for each state appears in the denominator of the probability expression, while the weight for the particular state(s) under consideration appears in the numerator. This is the form that we have been using throughout this text. By using the same reference state, E^*, in all calculations, the form of the probability relations remains the same; all that happens when new enzyme states need to be considered is that new terms describing the weights of these states appear in the denominator. The steady state kinetic results can usually be written on the same form, but the dissociation constants must be replaced by the relevant Michaelis constants (cf. Equations 5.2, 5.23 and 8.1).

6.3   Effective Equilibrium Constants

All the equilibrium constants can be deduced if the concentration of all species in the reaction Scheme of Figure 6.1 can be individually measured. However, in practice only some of the states of the enzymes can usually be determined. Depending on the kind of signals that are available for the various enzyme states, one can study the relations between enzyme in solution and in the interface, $[E + EL + EA]/[E^* + E^*L + E^*A]$, the relation between enzymes with and without ligand in the catalytic site, $[EL + EA + E^*L + E^*A]/[E + E^*]$, or some other combination. These compound equilibria can be described by effective constants that are determined by those of the individual steps.

6.3.1 Effective interface binding of the enzyme

An effective dissociation constant for the interface-bound enzyme to the solution enzyme can be defined from the equilibrium $K_d^{\text{eff}}[E_{\text{all}}^*] = [E_{\text{all}}][M^*]$, involving all species of enzyme in solution ($[E_{\text{all}}] = [E + EL + EA]$) and in the interface ($[E_{\text{all}}^*] = [E^* + E^*L + E^*A]$). In many cases, the E_{all}^* and the E_{all} states can be distinguished by spectroscopic signals. Then the effective dissociation constant can be found by measuring the ratio of unbound to bound enzyme species:

$$K_d^{\text{eff}}(A_T, L_T) = \frac{[E_{\text{all}}]}{[E_{\text{all}}^*]}[M^*] = K_d \frac{1 + \dfrac{(1 - X_L^*)\, K_A'}{K_A} + \dfrac{X_L^* K_L'}{K_L}}{1 + \dfrac{1 - X_L^*}{K_A^*} + \dfrac{X_L^*}{K_L^*}} \qquad (6.9)$$

In the last expression, the concentrations of A and L in solution have been replaced by the total concentrations and X_L^* through the use of Equations (3.11) and (3.12). Finally, replacing X_L^* in Equation (6.9) with the expression from Equation (3.15) gives the effective enzyme interface binding as a function of the total concentrations of A and L. If there is some signal that distinguishes the interface-bound enzyme states from the ones in solution, the effective dissociation constant as a function of X_L^* can readily be measured by applying Equation (6.9) and using Equation (3.16) for $[M^*]$. However, this expression involves far too many equilibrium constants to be useful, except in certain limits.

When there is no A in the system, Equation (6.9) simplifies to

$$K_d^{\text{eff}} = K_d \frac{1 + \dfrac{K_L'}{K_L}}{1 + \dfrac{1}{K_L^*}} = K_d^L \frac{1 + \dfrac{K_L}{K_L'}}{1 + K_L^*} \qquad (6.10)$$

and the effective dissociation constant is a true constant. In the last equality, the detailed balance condition (Equation 6.7) has been entered. If L binds very strongly to the catalytic site, $K_d^{\text{eff}} \approx K_d^L$, while if it binds very weakly, $K_d^{\text{eff}} \approx K_d$, as could be expected.

In the limits when A is a true *neutral diluent*, such that $K_A/K_A' \gg 1$ and $K_A^* \gg 1$, Equation (6.9) simplifies to

$$K_d^{\text{eff}} = K_d \frac{1 + \dfrac{X_L^* K_L'}{K_L}}{1 + \dfrac{X_L^*}{K_L^*}} \qquad (6.11)$$

In this limit, the binding isotherms determined at several X_L^* can give the values of K_L^* and K_L/K_L', if they are sufficiently different and not too large or too small. This relationship shows how the active-site-directed ligand either helps to anchor the enzyme to the interface, if $K_L^* < K_L/K_L'$ (cf. the two lower curves in Figure 6.3), or destabilizes it, if $K_L^* > K_L/K_L'$ (upper curves in Figure 6.3).

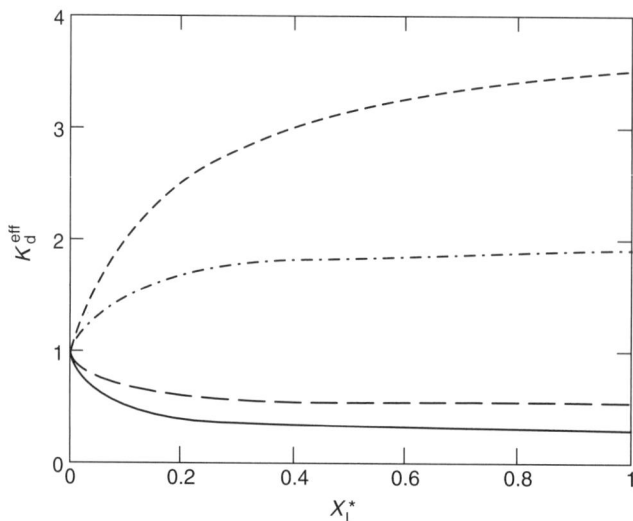

Figure 6.3　Effective dissociation constant with increasing mole fraction of L in the interface, relative to the value at $X_L^* = 0$, from Equation (6.11). (——) $K_L^* = 0.05$, $K_L/K_L' = 0.2$; (– –) $K_L^* = 0.05$, $K_L/K_L' = 0.1$; (–·– ·) $K_L^* = 0.1$, $K_L/K_L' = 0.05$; (- - - -) $K_L^* = 0.2$, $K_L/K_L' = 0.05$

6.3.2　Effective catalytic site binding of the ligand(s)

If there exists a signal (e.g. spectroscopic or, as discussed below, using protection experiments) that can differentiate whether the enzyme has a ligand in the catalytic site or not, one can measure the ratio

$$R^c(A_T, L_T) = \frac{[E^*L + EL + E^*A + EA]}{[E^* + E]} = \frac{\dfrac{X_L^*}{K_L^*}\left(1 + \dfrac{K_d^L}{[M^*]}\right) + \dfrac{X_A^*}{K_A^*}\left(1 + \dfrac{K_d^A}{[M^*]}\right)}{1 + \dfrac{K_d}{[M^*]}}$$

(6.12)

This expression contains too many parameters to be very useful except in certain limits. In the absence of L, and inserting $[M^*] = [A^*] = [A_T] - K_A'$, Equation (6.12) reduces to

$$R^c(A_T, 0) = \frac{1}{K_A^*}\frac{[A_T] - K_A' + K_d^A}{[A_T] - K_A' + K_d}$$

(6.13)

If $R^c(A_T, 0)$ is independent of $[A_T]$, then either $K_d^A = K_d$ or K_d^A and K_d are both $\gg$ or $\ll [A_T] - K_A'$, the latter condition implying that the enzyme is all in the interface. In this limit of large $[A_T]$, the ratio $R^c(A_T, 0)$ gives directly the interfacial catalytic site affinity for A, $1/K_A^*$. If K_d^A and K_d are not equal or are not too large or too small, further analysis of the behavior of $R^c(A_T, 0)$ as a function of $[A_T]$ can give their values as well.

When A is a true neutral diluent, $K_A^* \gg 1$, Equation (6.12) simplifies to

$$R^c(A_T, L_T) = \frac{X_L^*}{K_L^*}\frac{1 + K_d^L/[M^*]}{1 + K_d/[M^*]}$$

(6.14)

The ratio of enzymes with and without an occupied active site can be measured, for instance, by the protection experiments (Equation 6.26), discussed below. In these experiments one monitors the fraction of enzymes whose active site is not protected by ligand binding. In terms of the relations above, this fraction is given by

$$F^c_{unbound}(A_T, L_T) = \frac{1}{1 + R^c(A_T, L_T)} \tag{6.15}$$

6.4 The Cofactor Binding Obligatory for the Substrate Binding

Some enzymes require a cofactor for the binding of a specific ligand to the catalytic site. Consider the case where the cofactor C is in solution and binds the enzyme. The interface is built up from A and L as before, but, for simplicity, assume that A is a true neutral diluent while L cannot bind the catalytic site unless C is already bound. This gives the binding scheme in Figure 6.4, from which we can calculate the effective dissociation constant for C from the enzyme:

$$K^{eff}_C = \frac{[C]([E] + [E^*])}{[E^*C] + [E^*CL] + [EC] + [ECL]} = \frac{K^*_C \left(1 + \dfrac{K_d}{[M^*]}\right)}{1 + \dfrac{X^*_L}{K^*_L} + \dfrac{K^C_d}{[M^*]}\left(1 + \dfrac{X^*_L K'_L}{K_L}\right)}$$

$$\xrightarrow[{[M^*] \gg K_d, K^C_d}]{} \frac{K^*_C}{1 + \dfrac{X^*_L}{K^*_L}} \tag{6.16a}$$

The last step in this equation is the limit when all enzyme is in the interface and expresses how the presence of L will help anchor C to the catalytic site. The fraction of enzymes that have neither C nor L bound at the catalytic site is given by the effective dissociation constant as

$$F^c_{unbound} = \frac{K^{eff}_C}{[C] + K^{eff}_C} \xrightarrow[{[M^*] \gg K_d, K^C_d}]{} \frac{1}{1 + \dfrac{[C]}{K^*_C}\left(1 + \dfrac{X^*_L}{K^*_L}\right)} \tag{6.16b}$$

If binding of C or L to the catalytic site protects it against chemical inactivation, this relationship can be used to find the dissociation constants from the half-times

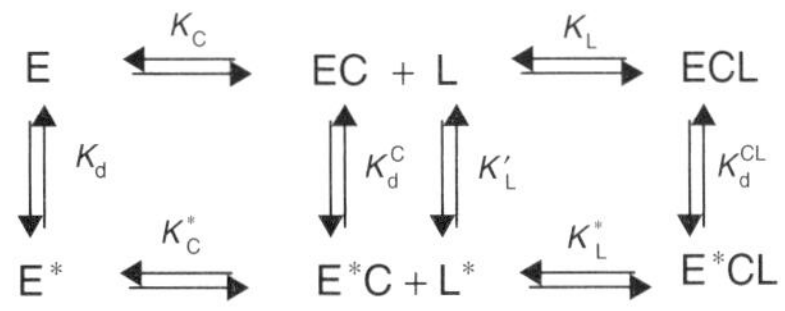

Figure 6.4 The equilibria for the obligatory cofactor (C) requirement for the binding of an active-site-directed ligand (L) in the active site

for inactivation at different concentrations of C and L (Jain *et al.*, 1991a; Yu *et al.*, 1993, 1997b).

Similarly, we can calculate the ratio of ligand-bound enzyme to nonliganded as

$$R^c\left(C, X_L^*\right) = \frac{[E^*CL + ECL]}{[E^* + E^*C + E + EC]} = \frac{\dfrac{[C]X_L^*}{K_C^* K_L^*}\left(1 + \dfrac{K_d^{CL}}{[M^*]}\right)}{1 + \dfrac{[C]}{K_C^*} + \dfrac{K_d}{[M^*]}\left(1 + \dfrac{[C]}{K_C}\right)}$$

$$\xrightarrow[{[C] \gg K_c, K_c^*}]{} \frac{X_L^*[M^*] + K_d^{CL}}{K_L^*[M^*] + K_d^C} \tag{6.17}$$

The last step is the limit when the cofactor binding is saturated. Looking at the ratio of R^c at saturating and nonsaturating C, one finds that

$$\frac{R^c\left(C_{sat}, X_L^*\right)}{R^c\left(C, X_L^*\right)} = 1 + \frac{K_C^*[M^*] + K_d}{[C][M^*] + K_d^C} \tag{6.18}$$

Thus, this ratio is independent of X_L^* and the slope in a plot of this ratio versus the inverse of [C] gives the effective dissociation constant for C in the absence of L (cf. Equation 6.16a in the limit $X_L^* \to 0$).

6.4.1   Detailed balance condition for the cofactor binding

In the Scheme in Figure 6.4, the cofactor could drive either the binding of L to the catalytic site or the binding of enzyme to the interface, or both, depending on the relationships between the equilibrium constants. In the case where cofactor does not influence interface binding, $K_d^C = K_d$ and $K_C^* = K_C$, so that the ratio in Equation (6.18) is independent of [M*]. In the case when cofactor strengthens the interface binding, $K_d^C < K_d$, interface binding will also strengthen the cofactor binding; from the detailed balance condition, $K_d^C/K_d = K_C^*/K_C$. If K_d^C and K_d are sufficiently different and not too small, their values could be deduced by applying Equation (6.18) at different concentrations [M*].

6.4.2   Effect of cofactor partitioning

It was assumed above that the interface-bound enzyme picks up cofactor from solution. In some cases, cofactor will accumulate in the interface and the enzyme could pick up cofactor from there. This does not change any of the results above. Consider the situation where C accumulates in the interface corresponding to a dissociation constant K_C', giving the equilibrium

$$[C] = K_C' \frac{[C^*]}{[M^*]} \tag{6.19}$$

Since $[C^*]/[M^*]$ is the concentration that the enzyme sees in the interface, the E* to E*C equilibrium will be

$$\frac{[E^*C]}{[E^*]} = \frac{1}{K}\frac{[C^*]}{[M^*]} = \frac{1}{KK'_C}[C] \tag{6.20}$$

where K is the appropriate local dissociation constant. Thus, interpreted with the Scheme in Figure 6.4, all that happens is that the dissociation constant can be interpreted as $K^*_C = KK'_C$. For the enzyme equilibria in terms of the concentration of free cofactor in solution, [C], it does not matter whether cofactor accumulates in the interface or not. However, in terms of the total concentration of C added to the system, $[C_T] = [C] + [C^*]$, there will be a difference. From Equation (6.19) we find that

$$[C] = \frac{[C_T]}{1 + \dfrac{[M^*]}{K'_C}} \tag{6.21}$$

so that an increase of interface, $[M^*]$, at constant $[C_T]$ will reduce the concentration of cofactor available for enzyme binding. (Note that C_T, as also with A_T and L_T previously, denotes the total amount except that which is enzyme-bound; only when C is in large excess over enzyme is $[C_T]$ a true total concentration.)

6.5 Detailed Balance Conditions and Local Concentrations for Effective Ligand Binding

It is instructive to consider the probability that the enzyme has bound ligand L as a function of the interface concentration. For simplicity assume that A is a neutral diluent so that the states EA and E*A can be neglected. Then the effective dissociation constant for L can be defined from the thermodynamic box consisting of the five equilibria on the left in Figure 6.1, as

$$K_L^{\text{eff}} = \frac{[L_T]\,([E^*] + [E])}{[E^*L] + [EL]} \tag{6.22}$$

Using the equilibrium relations (Equations 3.14 and 6.2 to 6.7) gives

$$K_L^{\text{eff}} = K_L^* \left([M^*] + K'_L\right)\frac{[M^*] + K_d}{[M^*] + K_d^L} = K_L \left(1 + \frac{[M^*]}{K'_L}\right)\frac{1 + \dfrac{[M^*]}{K_d}}{1 + \dfrac{[M^*]}{K_d^L}} \tag{6.23}$$

The two forms of this expression are equivalent due to the detailed balance condition (Equation 6.7). The first is most useful when enzymes are mostly in the interface and the second one when enzymes are mostly in solution. It can be seen that the effective dissociation constant always increases with $[M^*]$ at sufficiently large $[M^*]$. In fact, $K_L^{\text{eff}} > K_L$ for all values of $[M^*]$ unless

$$\frac{1}{K_d^L} > \frac{1}{K_d} + \frac{1}{K'_L} \tag{6.24}$$

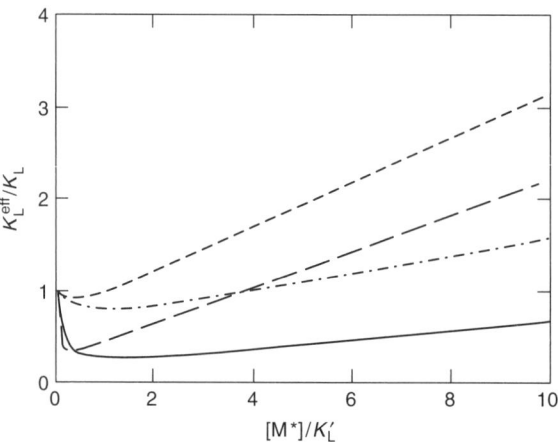

Figure 6.5 Effective ligand binding to the catalytic active site, K_L^{eff}/K_L, as a function of interface concentration (scaled as $[M^*]/K_L'$), from Equation (6.23). (----) $K_d/K_L' = 2$ and $K_d^L/K_L' = 0.5$; (−·−·) $K_d/K_L' = 5$ and $K_d^L/K_L' = 0.5$; (−−) $K_d/K_L' = 1$ and $K_d^L/K_L' = 0.02$; (——) $K_d/K_L' = 2$ and $K_d^L/K_L' = 0.1$

in which case there will be a region of intermediate $[M^*]$ values where $K_L^{\text{eff}} < K_L$ holds, as shown in Figure 6.5. Thus, only if the condition (6.24) holds will the accumulation of the reactants, E and L, in the interface actually increase the binding probability above that in solution; this is in spite of the fact that the local concentration of L in the interface can be very high. The condition implies that the enzyme–ligand complex, E*L, is bound more strongly to the interface than the sum of the binding constants for E* and L* separately. The ratio K_d/K_d^L shows how much stronger the E*L complex is bound to the interface than is E*. Thus it shows how much L contributes to the interface binding of E*L. This could come from an anchoring effect where L retains some interactions with the interface in addition to the binding interactions of E*. It could also come from an allosteric effect where L binding to E* leads to a conformational change in the enzyme that promotes interface binding. If this is the case, then interface binding of E will also promote the conformational change that will increase the affinity for L. This is what the detailed balance condition (Equation 6.7) expresses:

$$\frac{K_d}{K_d^L} = \frac{K_L}{K_L^* K_L'} \tag{6.25}$$

The left-hand side gives the ligand-promoted increase (if >1) in the enzyme–interface affinity, while the right-hand side gives the interface-promoted increase in the enzyme–ligand affinity. The same anchoring effect of a ligand (if $K_L^* < K_L/K_L'$) on the enzyme–interface binding is also evident in Equation (6.11).

6.6   Resolution of the Interfacial Constants for PLA2

The interfacial catalytic turnover parameters in Table 5.1, for the scooting mode reaction progress by pancreatic PLA2 were resolved with the help of the independently determined K_A^*, K_I^*, K_P^* and K_{Ca}^* coupled with the evidence that the enzyme binds to the anionic interface with high affinity. Key results are outlined below.

6.6.1 The high-affinity binding of PLA2 to the anionic interface

Although evident from the scooting mode kinetics (Chapters 4 and 5), the tight binding of pancreatic PLA2 to anionic vesicles and micelles is also shown by direct measurements. Results in Figure 6.6 clearly show the high-affinity binding of PLA2 to DTPM vesicles. The linearity of the signal in both of these plots show that a stoichiometric relationship exists between the bound enzyme and the available

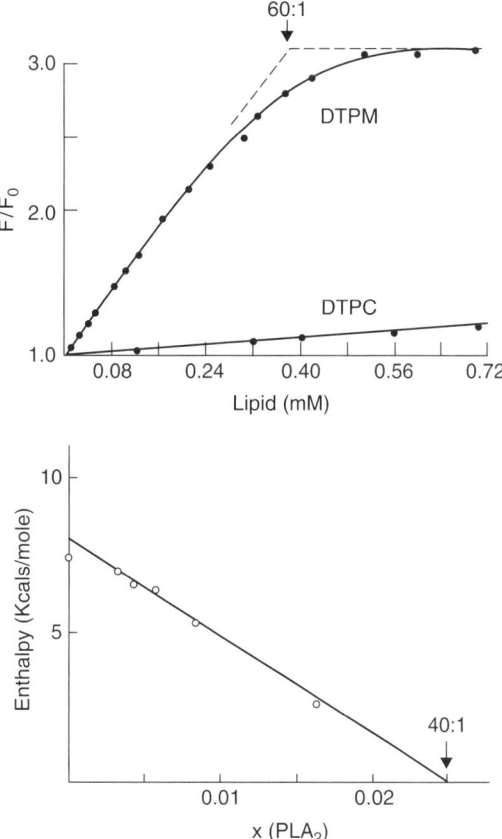

Figure 6.6 (Top) Relative Trp-3 fluorescence of pig pancreatic PLA2 as a function of the added concentration of nonhydrolyzable 1, 2-ditetradecyl-*sn*-glycero-3-phosphomethanol (DTPM) soni-cated vesicles. (Bottom) The change in the enthalpy of the gel–fluid transition of DTPM dispersed in the presence of the PLA2. Both measurements were carried out in the presence of $CaCl_2$ at pH 8.0. The extrapolated 1:60 stoichiometry for the first case is apparently larger because in this case PLA2 binds only to the outer layer of the added vesicles. The samples for the enthalpy measurements were prepared with premixed lipid so that both sides of the bilayer are exposed to the PLA2, and the stoichiometry of 1:40 is consistent with the 2:3 ratio for the number of phospholipid molecules present in the inner to outer layers of sonicated vesicles. Reprinted from Jain *et al.* (1986d). Copyright (1986) with permission from Elsevier Science

interface. Such results show that, depending on the structure of the amphiphile, irrespective of the micellar and vesicle organization, about 25 to 40 amphiphiles are required for the binding of each enzyme molecule. Interpreted in terms of the excluded area effect, possibly each bound enzyme 'covers' the interface occupied by these amphiphile molecules.

Analysis of the titration curve or binding isotherm of the type shown in Figure 6.6 is often problematical for the determination of K_d. Typically, K_d values below 20 μM cannot be accurately measured with about 2 μM protein in the reaction mixture. If $n = 20$–50 amphiphile monomers per enzyme, about 50 μM lipid would be required for the tight binding. This would completely mask the hyperbolic dependence of the titration curve if the dissociation constant is significantly below 50 μM, as is indeed the case for the binding of PLA2 to the anionic interface. Unfortunately no other general methods are available for the determination of low K_d values with vesicle or micellar interfaces. On the other hand, apparent K_d estimated from the vesicle-to-vesicle exchange kinetics of the bound enzyme is in the picomolar range (see Theory Box 2.1). Some possible consequences of the excluded area effects during the high-affinity binding are developed in Theory Box 6.2.

THEORY BOX 6.2 Excluded surface effects in the equilibrium binding of enzyme to large vesicles

Assume that each bound enzyme excludes an area corresponding to n lipids for binding by other enzymes. Then, in the simplest picture, the bound enzymes reduce the accessible surface concentration to $[M^*] - n[E^*]$. The fraction of occupied surface is

$$\Phi = \frac{n[E^*]}{[M^*]} \tag{1}$$

The mass balance at equilibrium could be written as

$$[E]([M^*] - n[E^*]) = K_d[E^*] \tag{2}$$

Replacing the free-enzyme concentration with the total, $[E] = [E_0] - [E^*]$, Equation (2) can be solved for $[E^*]$ as a function of $[E_0]$ and $[M^*]$:

$$\frac{[E^*]}{[E_0]} = \frac{1}{2}\left(1 + \frac{[M^*] + K_d}{n[E_0]}\right)\left(1 - \sqrt{1 - \frac{4[M^*]n[E_0]}{(n[E_0] + [M^*] + K_d)^2}}\right) \tag{3a}$$

Alternatively, we can rewrite Equation (2) as

$$\frac{[E^*]}{[E_0]} = \frac{[M^*]}{[M^*] + K_d/(1 - \Phi)} \tag{3b}$$

Thus, half of the enzyme will be bound at a concentration $[M^*] = K_d/(1 - \Phi)$. This will give a reasonable estimate of K_d only if $\Phi \ll 1$, or when $n[E_0] \ll 2[M^*]$. With a larger surface coverage, it becomes more and more difficult to bind new enzymes, as could be expected when fewer unoccupied sites are available. However, this treatment is a gross oversimplification and would be true only if the surface could be divided into discrete nonoverlapping sites for the independent binding of single enzymes. It may hold for small aggregates that can bind only one or two enzymes.

A more realistic picture for large aggregates would allow bound enzymes to occupy any position on the surface that is not already excluded, thereby allowing partial coverage of a number of 'sites' at the same time. In effect, the bound enzymes would form a two-dimensional gas with hard-sphere exclusions. When one more enzyme binds, it will not just occupy one more 'site', it will effectively reduce the freedom, and therefore the entropy, of all previously bound enzymes. This effect can be accounted for in the two-dimensional scaled-particle treatment of a hard-sphere gas (Lebowitz *et al.*, 1965). As a consequence, the binding relation (Equation 3) will be

$$\frac{[E^*]}{[E_0]} = \frac{[M^*]}{[M^*] + K_d e^{W(\Phi)}/(1 - \Phi)} \tag{4}$$

where $W(\Phi)$ is related to the entropy reduction and is determined

$$W(\Phi) = \frac{3\Phi}{1 - \Phi} + \left(\frac{\Phi}{1 - \Phi}\right)^2 \tag{5}$$

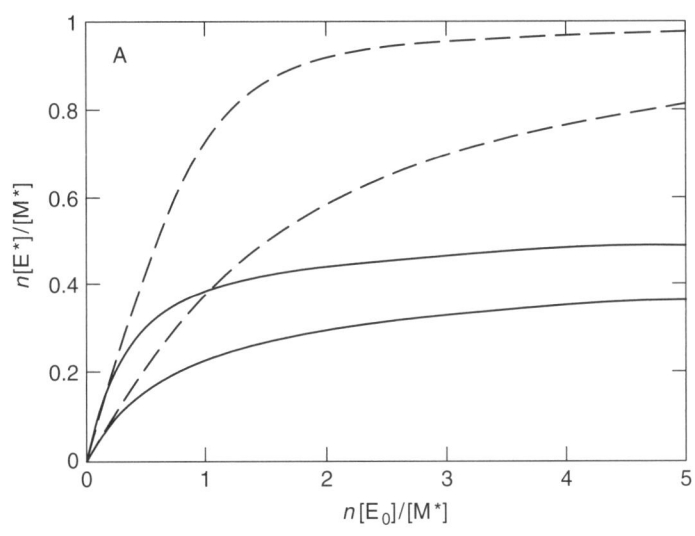

Figure 6.7 Panel A: concentration of bound enzymes as a function of total enzyme concentration, both multiplied by the 'site size' n and divided by $[M^*]$. Thus the y axis shows the surface coverage Φ. $[M^*] = 100$ in all curves. Solid curves are from Equation 4 (with hard-sphere exclusion) using $K_d = 10$ (upper) and $K_d = 100$ (lower); these curves show apparent saturation at ca. 40 to 60 % surface coverage. Dashed curves are from Equation 3 using the same parameters. The initial slopes of these curves are independent of n and equal to $1 + K_d/[M^*]$.

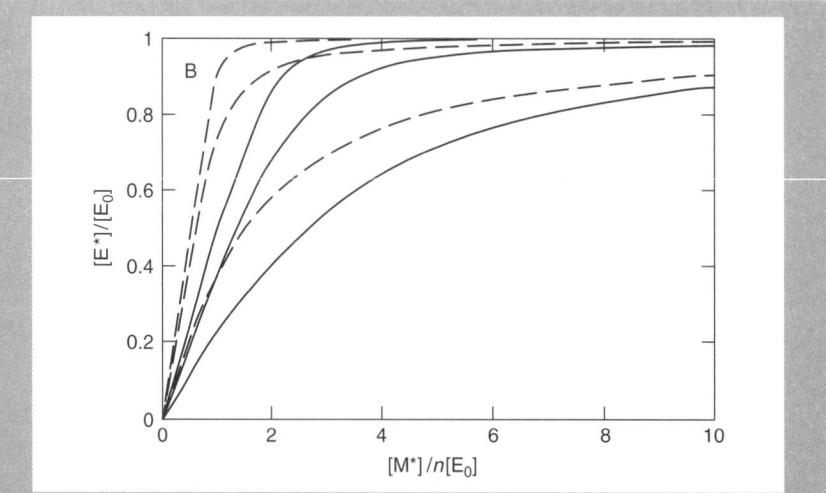

Figure 6.7 (*continued*)   Panel B: fraction of enzymes bound at the interface as a function of interface concentration at constant total enzyme concentration $[E_0]$. Solid curves, using Equation 4 with $K_d/n[E_0] = 0.01, 0.1, 1.0$ (from upper to lower). Dashed curves, using Equation 3 with the same parameters

Since Φ is determined by $[E^*]$ through Equation (1), Equations (4) and (5) give only an implicit relation for the concentration of bound enzymes as a function of $[M^*]$ and $[E_0]$. The results are plotted in Figure 6.7. Figure 6.7A shows that it is extremely difficult with the hard-sphere exclusion (solid curves) to get full enzyme coverage of the surface even if the binding is strong and enzyme is in large excess. As a consequence, the initial slope of $[E^*]/[E_0]$ versus $[M^*]/[E_0]$ (cf. Figure 6.7B) is ca. $(0.4-0.6)/n$, rather than $1/n$ as expected from Equation (3) at strong binding; thus using the initial slope as a measure for $1/n$ can lead to an overestimate of the site size, n, by a factor of 2.

   The calculations above require that each aggregate can bind a large number of enzymes; otherwise one must also account for the finite surface area. Equation (5) does not allow for any attractive enzyme–enzyme interactions that will lead to clustering on the surface. For such cases, the approximation in Equation (3) may give a better description. The behavior of the titration curve (Figure 6.7B) can provide information on the type of binding that is involved; an estimated site size that is larger than the physical size of the enzyme suggests hard-sphere exclusion, while a smaller value suggests enzyme aggregation on the surface.

6.6.2   Dissecting the active site events with the protection method

For the measurements in Figure 6.6, as well under the scooting mode kinetic conditions (Chapters 4 and 5), the active site and the i-face events are not distinguishable, and thus only effective dissociation constants are obtained as in Equation (6.9). If the active site and i-face are structurally distinct, it should be possible to distinguish them functionally in terms of the underlying equilibria in Figure 6.1 with a suitably chosen neutral diluent without affinity for the active site. The experimental criteria

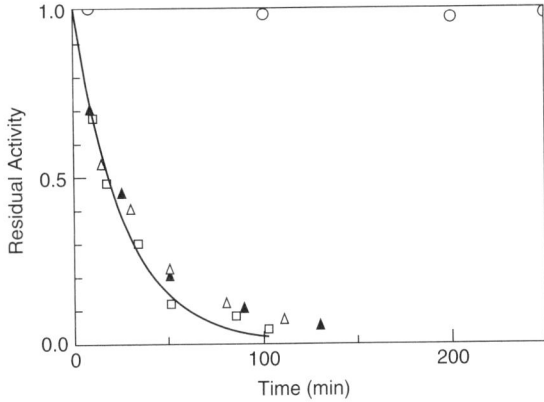

Figure 6.8 Time course of inactivation of pig pancreatic PLA2 (0.03 mM) by *p*-bromophencyl bromide (1 mM) in the aqueous phase (open squares), in 3.3 mM deoxy-LPC(open triangles), in 3.3 mM 2-hexadecyl-glycerophosphocholine (filled triangles), and in 0.5 mM products of hydrolysis of 1,2-dimyristoylphosphatidylmethanol (open circles). The reaction mixture also contained 0.5 mM $CaCl_2$ and 50 mM NaCl in 50 mM cacodylate buffer at pH 7.3 and 23 °C. Reprinted with permission from Jain and Maliwal (1993). Copyright (1993) American Chemical Society

to distinguish an active-site-directed ligand from a neutral diluent for pancreatic PLA2 is shown in Figure 6.8. Here one takes advantage of the fact that the alkylation of the active site residue, His-48 in PLA2, does not occur if the active site is occupied with calcium alone, or with an active-site-directed ligand (Figure 3.11). On the other hand, binding of the enzyme to the interface of a neutral diluent does not protect the enzyme from alkylation.

The role of the micellar surface provided by a neutral diluent is as a two-dimensional interface for the E to E* equilibrium, and it is also a two-dimensional solvent for 'dissolving' or partitioning the ligand for the E* + L to E* L equilibrium (Jain *et al.*, 1991a). In addition, for changing X_S^* during the scooting mode reaction progress the diluent should also mix ideally to retain the bilayer organization. The structural requirements for a useful neutral diluent are rather strict and the same neutral diluent does not necessarily serve well for other PLA2 isoforms. The diluent amphiphile should not have any affinity for the active site, but it should provide the interface for the interaction with the i-face and the interface should also ideally accommodate (mix) the ligands. A list of neutral diluents for certain inter-facial enzymes is given in Table 6.1. Amphiphiles like 1-hexadecyl-propanediol-3-phosphocholine or 2-hexadecyl-glycero-3-phosphocholine in aqueous dispersions

Table 6.1 Examples of neutral diluents

Enzyme	Diluent
PLA2, pancreatic	1-Hexadecyl-propanediol-3-phosphocholine (2-deoxy-LPC)
	2-Hexadecyl-glycero-*rac*-3-phosphocholine
PLA2, synovial	Hexadecylphosphocholine
PLA2, bee venom	3-Hexadecyl-glycero-*sn*-1-phosphocholine (poor diluent)
Lipase, *Humicola*	1-Palmitoyl-2-oleoyl-glycero-*sn*-3-phosphoglycerol

serve as useful neutral diluents for pancreatic PLA2. Based in the criteria outlined above, the qualifier *neutral* is meant to emphasize the *diluent* properties in relation to the equilibria and catalytic turnover.

6.6.3 The dissociation constant for E* or E*L from the interface

The binding of pancreatic PLA2 to an interface is best monitored as a change in the intrinsic fluorescence emission from the only tryptophan present in the molecule, Trp-3 (Figure 6.9). Since several equilibria may be coupled, one often takes advantage of independently obtained information about the individual steps. The apparent dissociation of E* or E*L from the interface becomes possible with the use of a suitable neutral diluent provided that the signal from the appropriate species in equilibrium can be distinguished (Jain *et al.*, 1993b). The isotherm is obtained by changing the bulk concentration of the dispersion of the neutral diluent amphiphile (A*):

$$E + A^* \rightleftharpoons E^*$$

In principle, fitting the concentration of A*, expressed as [M*], dependence of the signal to a hyperbola can give both the apparent dissociation constant, K_d, and the site size n (Theory Box 6.2 and Figure 6.7).

In the example shown in Figure 6.9, the K_d^{eff} for the E* form of pancreatic PLA2 to micelles of dioctylphosphatidylcholine is about 10 mM, and it decreases by at least a factor of 10 in the presence of 4 M NaCl. These measurements are carried out in the presence of EGTA. As discussed later, calcium is an obligatory cofactor for the formation of E*L from E*. Thus in the presence of $CaCl_2$, the K_d^{eff} decreases (if $K_d^L < K_d$) as expressed by Equation (6.11). The hyperbolas drawn in Figure 6.9 were generated without accounting for a possible surface exclusion. As discussed

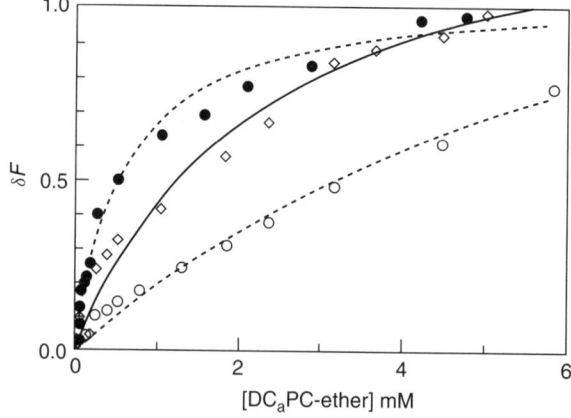

Figure 6.9 Increase in the fluorescence emission intensity, δF, at 333 nm (excitation at 280 nm) from Trp-3 of pig PLA2 as a function of DC_8PC ether concentration (CMC = 0.03 mM): with 1 mM NaCl and 0.2 mM EGTA (open circles), 4 M NaCl and 0.2 mM EGTA (solid circles) or 1 mM NaCl and 10 mM $CaCl_2$ in 10 mM Tris at pH 8.0 (open diamonds). Reprinted with permission from Berg *et al.* (1997). Copyright (1997) American Chemical Society

in Theory Box 6.2, this could lead to an overestimate of K_d unless $K_d \gg n[E_0]$, which could possibly affect the fit to the solid circles (high salt data) in the figure.

6.6.4   The dissociation constant for L* from E*L

The two-dimensional equilibrium dissociation constant, K_L^*, is a measure of the affinity of L* for E*. Based on Equation (6.14), the protection method is useful not only for the identification of a neutral diluent but also for the determination of K_L^* (Jain *et al.*, 1991a; Yu *et al.*, 1997a). As shown in Figure 6.8, the inactivation time for PLA2 does not change significantly in the presence of a neutral diluent; however, in the presence of the products of hydrolysis of DMPM, the inactivation time increases dramatically. Similar increases in the inactivation time are seen if specific inhibitors are added to the neutral diluent micelle to which PLA2 is bound. As shown in Figure 6.10, the increase in the inactivation half-time is proportional to the mole fraction of the ligand in the diluent interface. The inactivation half-times, t_0 and t_L, in the absence and in the presence of L, are inversely proportional to the fraction of enzymes that do not have a ligand in the catalytic site. From Equation (6.15) this gives t_0 proportional to $1 + R^c(A_T, 0)$ and t_L proportional to $1 + R^c(A_T, L_T)$ so that $t_0/t_L = [1 + R^c(A_T, 0)]/[1 + R^c(A_T, L_T)]$. Using Equation (6.12) in the limit where all enzyme is in the interface and replacing A by ND gives the linear relation:

$$\frac{1}{1 - t_0/t_L} = 1 + \frac{K_L^*}{X_L^*} \frac{1 + 1/K_{ND}^*}{1 - K_L^*/K_{ND}^*} = 1 + \frac{K_L^*}{X_L^*} \tag{6.26}$$

If ND is a true neutral diluent, $K_{ND}^* \gg 1$ and can be neglected, giving the last equality. The slope of the line gives K_L^* if K_{ND}^* is independently known. In this case, according to the results in Figure 6.8, $K_A^* (= K_{ND}^*) > 1.8$ mole fraction. If the y intercept is different from one, this suggests that the occupied catalytic site is not totally protected by ligand binding. The y intercept in Figure 6.10 shows that the binding of the ligand in E*L offers more than 97% protection to His-48 from the reaction with the alkylating agent.

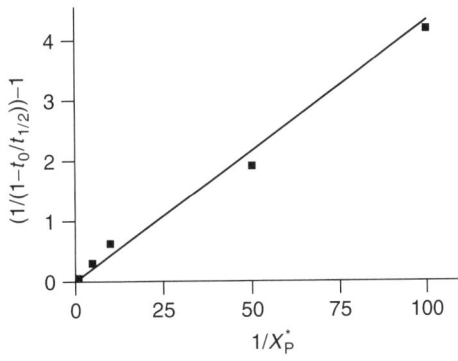

Figure 6.10 The relationship between the inactivation time for the E* form of PLA2 by *p*-bromophenacyl bromide without (t_0) or with an active-site-directed ligand ($t_{1/2} = t_L$) as a function of the mole fraction of the products of hydrolysis of DMPM in micelles of a neutral diluent. Based on the results in Jain *et al.* (1991a)

6.7 Detailed Balance Conditions for PLA2

The interfacial binding parameters for pancreatic PLA2 are summarized in Table 6.2. As related by Equation (6.25), the values of these parameters provide useful information about the nature of the underlying interactions. For example, it is possible to compare the intrinsic ligand binding affinities of EL and E*L, even though values of K_L and K_L^* cannot be directly compared because their units are different. As developed in the next section, the detailed balance condition relates K_L', K_L, K_L^*, K_d and K_d^L as in Equation (6.25), which provides a rational basis for understanding the effect of the local substrate concentration on the observed rate. Intuitively, the concentration of monomeric phospholipid in the aqueous phase is exceedingly low and the local substrate concentration (number density) at the organized interface is high. This apparent advantage will tend to cancel out because the hydrophobic effect organizes the substrate in the interface and increases their number density. On the other hand, for the formation of E*S, the substrate moves from the hydrophobic environment of the interface to that of the active site. Thus the net hydrophobic driving force for transferring L* to the active site of E* will be insignificant. In general, the same thermodynamic force that is used to increase the local concentration in the aggregate must be counteracted again when the ligand is pulled out of the aggregate to bind at the catalytic site. Thus the high local concentration by itself does not have any effect on the intrinsic binding of ligand to E* relative to that of E.

6.7.1 K_L^* activation Factor

K_L is the dissociation constant for $EL \rightleftarrows E + L$ and $K_L' K_L^*$ is for $E^*L \rightleftarrows E^* + L$ (Figure 6.1). The ratio $K_L/(K_L' K_L^*)$ is therefore a direct measure of how much better (or worse if <1) the interface-bound enzyme binds the ligand. If the ratio is equal to one, there is no difference in the interactions between enzyme and ligand

Table 6.2 The activation factor (Equation 6.25) for the binding of a mimic to the active site of pig pancreatic PLA2 at pH 8, 25 °C[a]

Mimic	K_d	K_d^I	K_I	K'	K_I^*
MJ33 (*rac-*)	0.9	0.028	0.008	0.01	0.0026
MJ72 (*rac-*)	0.9	0.028	0.086	0.3	0.002
MG14	0.9	0.028	0.04	1	0.0009
RM3	0.9	0.15	0.07	4	0.0033
RM2	0.9	0.065	0.05	0.25	0.003
DC$_8$PC-ether	10.0	2.5		0.24	0.15
DC$_7$PC-ether	9.7	1.5	0.08	0.3	0.2
DC$_6$PC-ether			1	16	0.4

[a]The interfacial constants with deoxy-LPC ($K_d = 3.7$ mM). All parameters are in units of mM, except that K_I^* is in mole fraction. The constant for PCn-ether are in its own micelles. MJ72 is 1-octyl-3-trifluoroethyl-*rac*-glycero-2-phosphomethanol; MG14, 1-octyl-2-phosphonoheptyl-*sn*-glycero-3-phosphoethanolamine; PCn-ether, 1,2-dialkyl-glycero-*sn*-3-phosphocholine. Results are from Berg *et al.* (1997). RM2 is 1-thio-octadecyl-2-acetyl-2-aminoglycero-*sn*-3-phosphocholine; RM3 is the 1-octyl analog.

at the interface and in the aqueous phase. From the detailed balance condition (Equation 6.25), $K_L/(K'_L K^*_L) = K_d/K^L_d$. This equality simply expresses the reciprocity: if the interface increases the ligand binding by a certain factor, the ligand will increase the enzyme–interface binding by the same factor. Thus, the K^*_L activation factor can be determined equally well by studying the ratio K_d/K^L_d. In Table 6.2 the value of K_I has the largest experimental uncertainty, and for the discussion of the extent of K^*_L activation we have relied more on the K_d/K^L_d ratio.

When the ratio is larger than one, the interactions within E*L are more favorable than those in EL. For example, this could come about at least in two different ways: *allosteric activation* where the interface promotes an enzyme conformation that binds the ligand better or a *scaffolding effect* where the ligand remains partially bound and stabilized by the interface even when it is in the active site (see also Chapter 12, Section 12.2). When the ligand is the substrate, the ratio $K_L/(K'_L K^*_L)$ is a measure of the K^*_S type of interfacial activation factor; i.e. the intrinsic affinity of the bound enzyme for the ligand increases by this factor (Jain *et al.*, 1993b). Values of the activation factor for pig pancreatic PLA2 are in the range of 20 to 50 seen for the inhibitors, and less than 10 for the ether analogs of the short chain phosphatidylcholines (Table 6.2). This effect is apparently not dependent on the length of the hydrophobic tails of the ligands, which suggests that the activation is not due to the scaffolding effect where part of the ligand remains anchored in the interface. The activation factor is about 50 when the active-site-directed inhibitors contain an *sn*-2 substituent that can provide an oxygen ligand for coordination with calcium (Table 6.2). On the other hand, the activation factor is more than tenfold lower for *sn*-2-ether analogs of the substrate. This is consistent with the idea that K^*_S activation may involve the ternary complex and the calcium loop.

At present there is no clear evidence that the magnitude of the K^*_S activation depends on the nature of the interface. Possible contributions of systematic or method-based errors in these constants can be minimized by using kinetically measured values of K^*_L and K'_L/K_L. The rationale is that if K'_L values are different in the different interfaces, we would also expect K^*_L values to be different, but in such a way that the product $K'_L K^*_L$ is invariant. Therefore, if a change in K'_L is due to a stronger interaction of L with the interface, then we expect K'_L to decrease and K^*_L to increase in roughly the same way. Because of the detailed balance condition and because K_L by definition is independent of the interface, K_d and K^L_d would also change at the different interfaces in roughly the same way so that K_d/K^L_d remains invariant. A similar argument also holds for the effect of added NaCl. The magnitude of the salt effect on the CMC ($= K'_L$) is of the same magnitude as the effect on K_L for the monodisperse EL of PLA2, while K^*_L is unchanged, which suggests that the K^*_S activation factor does not change with added NaCl.

6.8 Summary: Primary Equilibrium Parameters for the Kinetic Path

The resolution of the equilibria in the interface and across the interface (Figure 6.1) has provided insights into the nature of the primary events, as well as about the role of the interface in controlling the active site events. These relations permit resolution of virtually all the kinetic constants for the interfacial turnover cycle (Chapter 5). The equilibrium partitioning of a solute (K'_L) or the binding of the enzyme to the interface (K_d and K_d^L) is described by the interface concentration variable related to the amphiphile concentration in the interface, [M*]. On the other hand, the mole fraction of the ligand in the interface is the variable for the two-dimensional equilibrium dissociation of a ligand (substrate, inhibitor, product or imperfect diluent) from the active site of the enzyme at the interface. The K_L^* values are measured at the interface of a neutral diluent doped with the ligand. The apparent K_d values are related to the binding of the enzyme to the interface of the diluent and K_L^* for the sequential step at the interface. It is also clear that further progress in interfacial enzymology would require development of protocols to distinguish the bound forms of the enzyme to facilitate quantitation of the underlying equilibria.

The detailed balance condition relates all the equilibrium constants for the processes in a closed loop or the thermodynamic box. It permits a comparison of the two- and three-dimensional equilibrium constants, such as K_L^* versus K_L. As expected, the effect of the interface on this reaction is compensated by comparable changes in the affinities of E versus EL for the interface. Together, these equilibrium parameters provide clear evidence for the K_L^* allostery, i.e. the enhanced binding of the active-site-directed ligand to the active site of the enzyme at the interface.

7 Rapid Substrate Replenishment in the Quasi-Scooting Mode

...Then Jose Arcadio Buendia threw three doubloons into a pan and fused them with copper filings, orpiment, brimstone, and lead. He put it all to a boil in a pot of castor oil until he got a thick and pestilential syrup which was more like common caramel than valuable gold. In risky and desperate process of distillation, melted with the seven planetary metals, mixed with hermetic mercury and vitriol of Cyrus, and put back to cook in hot fat for lack of any radish oil, Ursula's precious inheritance was reduced to a large piece of burnt hog cracklings that was firmly stuck to the bottom of the pan.

Gabriel Garcia Marquez in *One Hundred Years of Solitude*

The interfacial kinetic paradigm (Figure 3.1) can be constrained to steady state kinetic paths where the substrate in the interface is rapidly replenished. In this limit, the turnover by the bound enzyme occurs at the interface, where the substrate depleted during the turnover is rapidly replenished by the exchange of the product with the excess substrate present in the aqueous phase or other interfaces in the ensemble. For a substrate with K'_S in the micro- to millimolar range, significant levels of X^*_S is achieved (Equation 3.10), while the absorption and desorption rates of the substrate and product are rapid (Theory Box 3.1). Also, k_{on} and k_{off} for the enzyme binding to the interface (Figure 1.8) do not significantly contribute towards the turnover time. Within such constraints, variables for the steady state microenvironment of the bound enzyme are unequivocally obtained from the total concentration values. Three kinetic paths with rapid substrate replenishment are distinguishable:

(a) As shown in Figures 4.8 and 5.6, the peptide-mediated direct vesicle-to-vesicle exchange of the bilayer-forming long chain anionic phospholipids establishes the steady state condition for the scooting mode reaction progress with constant $X^*_S = 1$ for an extended period of time.

(b) In the next chapter we will consider the steady state turnover path where during the turnover the enzyme is bound to the interface of substrate micelles.

During such a quasi-scooting mode reaction progress, the amphiphile provides the interface for the binding of the enzyme, and the same amphiphile is also the fast-exchanging substrate for the interfacial turnover cycle.

(c)   In this chapter we consider the turnover path where the substrate partitions, S to S*, and the enzyme, E to E*, binds to a diluent interface (A*) for the interfacial turnover. A* and S* are different molecular species and the active site of E* accesses only S* (interfacial enzyme).

Note that the successive turnover cycles in path (a) remain demonstrably processive; i.e. the enzyme does not leave the interface (the scooting mode). Ideally in paths (b) and (c) for the reaction progress with rapid substrate replenishment, the microenvironment of the bound enzyme does not change between the successive turnover cycles. Strictly speaking, for the steady state analysis of the quasi-scooting mode reaction progress, it does not matter whether the bound enzyme acts processively or not as long as the initial X_S^* remains constant (Theory Box 7.1).

7.1   Interfacial Catalytic Cycle Turnover in the Quasi-Scooting Mode

Consider the reaction path in Figure 7.1 for the hydrolysis on a diluent interface to which the enzyme is bound and a sparingly soluble substrate is partitioned. When the substrate is replenished and the product is removed very fast, it does not matter if the enzyme exchanges rapidly, slowly or not at all (Theory Box 7.1). The initial steady state in the interface can be upheld as long as there is no significant depletion of bulk substrate. From Equation (5.23) we get the rate per enzyme:

$$j_0 = \frac{k_{catS}^* X_S^*}{X_S^* + K_{MS}^* \left(1 + \dfrac{K_d}{[M^*]}\right)} \tag{7.1}$$

for the case where X_S^* is the mole fraction of substrate in the interface at the steady state and $X_P^* = 0$. The major problem is to identify the concentrations X_S^* and the interface concentration, $[M^*]$, in terms of the variables that are under direct experimental control, i.e. the amount of substrate and neutral diluent added to the system. In the simplest case, the mixture of S and the diluent A in the interface can be considered as an ideal mixture as described in Chapter 3. For simplicity,

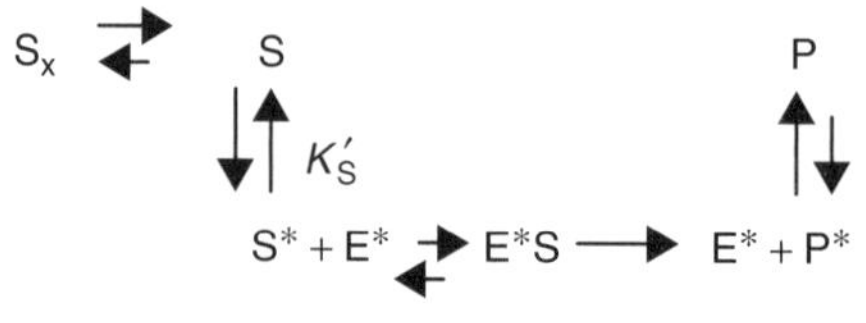

Figure 7.1   Reaction path for reaction progress in the quasi-scooting mode for the hydrolysis of a sparingly soluble substrate partitioned in the interface to which the enzyme is also bound. Note that the substrate exists in three states: S, monodispersed in the aqueous phase; S*, partitioned in the interface; and S_x, precipitated above its solubility limit

let us also assume that A is all in the interface; i.e. $K'_A \ll 1$ can be neglected in Equations (3.16) and (3.17). Then the steady state rate per enzyme will be

$$j_0 = \frac{k^*_{\text{catS}} X^*_S}{X^*_S + K^*_{\text{MS}}\left(1 + \dfrac{K_d\left(1 - X^*_S\right)}{[A_T]}\right)} \tag{7.2a}$$

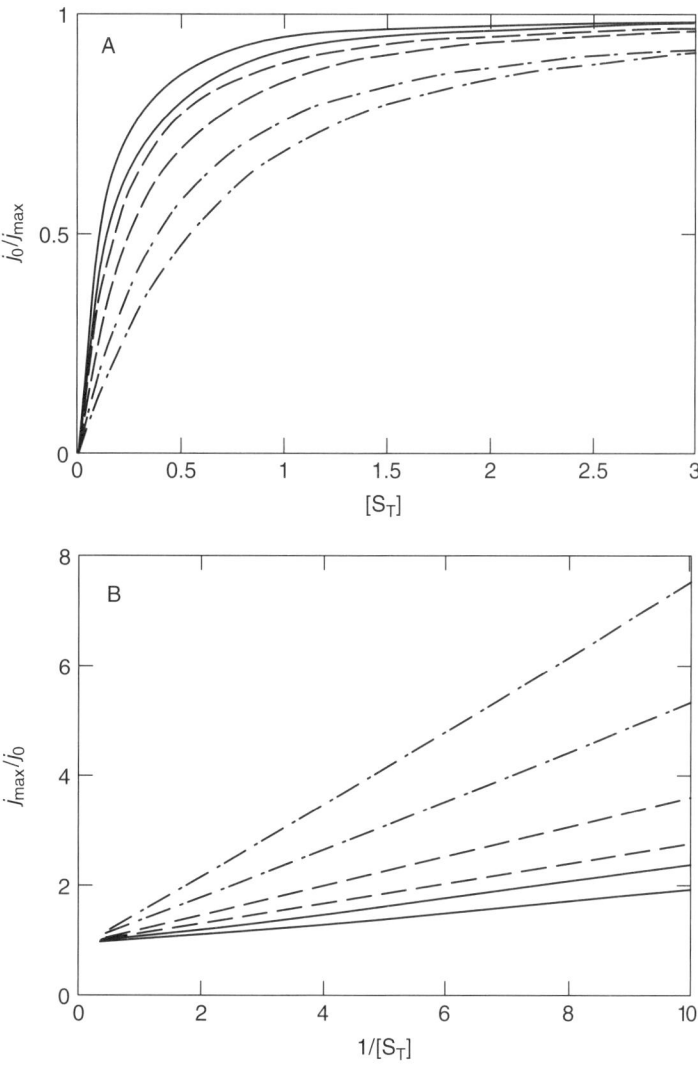

Figure 7.2 Initial rate from Equations (7.2a) and (7.3) relative to the maximal flux (Equation 7.2b), when all enzymes are in the interface, as a function of substrate concentration $[S_T]$ at constant $[A_T] = 0.5$. Panel A is the hyperbolic plot; panel B is the inverse plot (Lineweaver–Burk plot) of the same data as in panel A. Solid curves, $K^*_{\text{MS}} = 0.05$, $K_d = 0.5$; dashed curves, $K^*_{\text{MS}} = 0.1$, $K_d = 0.5$; dash–dot curves, $K^*_{\text{MS}} = 0.1$, $K_d = 2$; for each pair of curves, the upper is for $K'_S = 1$ and the lower for $K'_S = 0.5$ (in panel A, reversed in panel B)

When all enzyme is in the interface and $X_S^* = 1$, the maximum rate is

$$j_{max} = \frac{k_{catS}^*}{1 + K_{MS}^*} \tag{7.2b}$$

The flux j_0 depends on the total amounts of substrate and diluent through the mole fraction as given by Equation (3.15) for $K_A' = 0$:

$$X_S^* = \frac{[A_T] + [S_T] + K_S'}{2K_S'} \left(1 - \sqrt{1 - \frac{4K_S'[S_T]}{([A_T] + [S_T] + K_S')^2}} \right) \tag{7.3}$$

Together, Equations (7.2a) and (7.3) give the steady state flux as a function of the total concentrations of diluent and substrate in the system (Figures 7.2 and 7.3). Despite the square-root relation in Equation (7.3), the initial rate behaves very nearly like the standard hyperbolic MM relation:

$$j_0 = \frac{k_{cat}^{app}[S_T]}{K_M^{app} + [S_T]} \tag{7.4a}$$

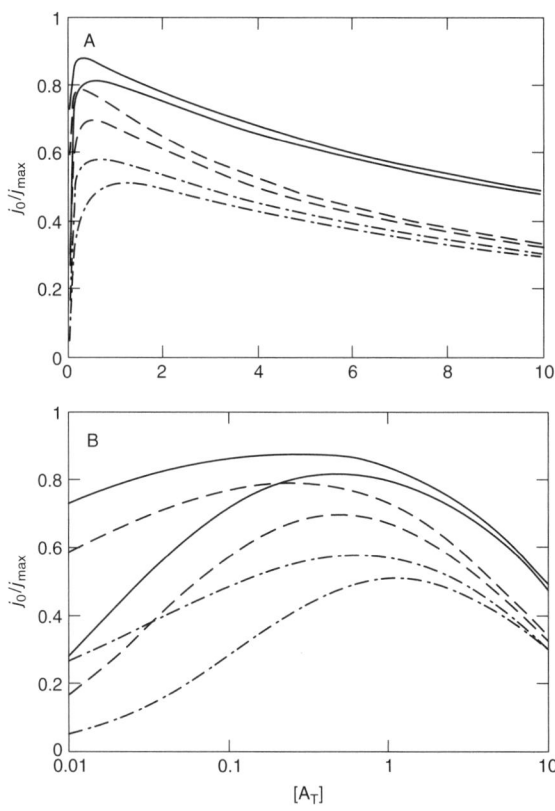

Figure 7.3 Initial rate from Equations (7.2a) and (7.3) relative to the maximal flux (Equation 7.2b), when all enzymes are in the interface, as a function of diluent concentration $[A_T]$ at constant $[S_T] = 0.5$. Panel B is the same as in panel A with the x-axis in logarithmic scale to show the behavior at low concentrations. Solid curves, $K_{MS}^* = 0.05$, $K_d = 0.5$; dashed curves, $K_{MS}^* = 0.1$, $K_d = 0.5$; dash–dot curves, $K_{MS}^* = 0.1$, $K_d = 2$; for each pair of curves, the upper is for $K_S' = 0.5$ and the lower for $K_S' = 1$

as demonstrated by the apparently straight lines in the Lineweaver–Burk plot (Figure 7.2B). In this case, $k_{\mathrm{cat}}^{\mathrm{app}} = j_{\max}$ from Equation (7.2b) and

$$K_{\mathrm{M}}^{\mathrm{app}} = \frac{K_{\mathrm{d}} + [\mathrm{A_T}]}{1 + 1/K_{\mathrm{MS}}^*} \left(1 + \frac{K_{\mathrm{S}}'}{[\mathrm{A_T}]} \frac{1 + 1/K_{\mathrm{MS}}^*}{1 + 1/K_{\mathrm{MS}}^* + 1 + K_{\mathrm{d}}/[\mathrm{A_T}]} \right) \tag{7.4b}$$

corresponds to the substrate concentration $[\mathrm{S_T}]$ for which $j_0 = (1/2)j_{\max}$.

At low concentration of interface, $[\mathrm{A_T}]$, the enzyme activity is low because the enzyme is not bound in sufficient amounts, while at high $[\mathrm{A_T}]$, the rate decreases because the substrate becomes increasingly diluted in the interface. From these kinds of plots, the catalytic parameters k_{catS}^* and K_{MS}^* could be determined if K_{S}' and K_{d} are determined independently.

THEORY BOX 7.1 Requirements for rapid substrate replenishment

The assumption that the partitioning of S is equilibrated at the steady state is equivalent to the requirement that the association–dissociation reactions of substrate to the interface are much faster than the rate of catalytic change. At the steady state, the net association flux of substrate to an enzyme-containing particle must equal the enzyme flux. The association flux of monomers to an aggregate is given by Equation (3) in Theory Box 2.1 as $k_a N_T[\mathrm{S}]$ (molecules per second). As discussed in Theory Box 3.1, the monomer concentration in solution equilibrates rapidly by desorption from aggregates. Thus, $[\mathrm{S}] = K_{\mathrm{S}}' X_{\mathrm{S}}^*$ (Equation 3.11) and the requirement for replenishment to keep up with catalysis is

$$k_a N_T K_{\mathrm{S}}' X_{\mathrm{S}}^* \gg n_E \frac{k_{\mathrm{catS}}^* X_{\mathrm{S}}^*}{K_{\mathrm{MS}}^* + X_{\mathrm{S}}^*} \tag{1}$$

assuming that all enzymes are in the interface and there is no product ($X_{\mathrm{P}}^* = 0$). The value of k_a is given in Theory Box 2.1 for a diffusion-limited association and $k_a N_T = 10^{10} \mathrm{M}^{-1} \mathrm{s}^{-1}$ for an aggregate of size $N_T = 10^3$. Thus

$$K_{\mathrm{S}}' \gg \frac{k_{\mathrm{catS}}^* n_E}{\left(K_{\mathrm{MS}}^* + X_{\mathrm{S}}^*\right) k_a N_T} \tag{2}$$

and K_{S}' in the µM range or above is sufficient if k_{catS}^* is around $10^3 \mathrm{s}^{-1}$ or smaller.

If replenishment is much slower than catalysis, the product flux per enzyme is

$$j = k_a N_T K_{\mathrm{S}}' X_{\mathrm{S}}^* / n_E \tag{3}$$

where n_E is the number of enzymes per enzyme-containing particle. This result requires that the substrate in solution ($[\mathrm{S}] = K_{\mathrm{S}}' X_{\mathrm{S}}^*$) is buffered by desorption from particles not containing enzyme in which X_{S}^* is unchanged while enzyme-containing particles are becoming depleted.

7.2 Sparingly Soluble Substrates

The useful working range for X_S^* attained by the partitioning of a sparingly water-soluble substrate in the interface of a suitable diluent is limited by K_S' and the solubility limits of the substrate and product in the aqueous phase (Figure 7.1). Less soluble substrates are expected to have a higher partition coefficient (or lower K_S'); therefore these two parameters place opposing demands for the working range of the substrate. Since the maximum possible X_S^* for a partitioned substrate would be significantly less than 1, the assay protocol with such limits for the reaction path in Figure 7.1 is useful only for sparingly soluble substrates with low K_{MS}^*.

7.2.1 Effect of solubility limit and other nonidealities on the partitioning

The solubility limit for a sparingly soluble substrate in the aqueous phase establishes not only the maximum aqueous phase concentration, $[S_{max}]$, but also the maximum mole fraction of the substrate in the interface, $X_S^{*\,max}$. If the interface does not undergo a phase change, excess solute beyond the solubility limit would form its own dispersed phase, and thus be excluded from the aqueous phase as well as the diluent interface. In other words, changing concentration of the aggregated or precipitated solute above its solubility limit does not have any influence on the solution or partitioned substrate concentration variables. The solute concentration in the aqueous phase, S, is in equilibrium with S*. S* is the solute partitioned into the interface of a diluent A, and its concentration, X_S^*, is determined by K_S' (Equation 7.3). However, it may also be important to account for a size difference between the solute and the diluent molecules. This can be done by introducing the area correction factor, φ, as defined in Theory Box 3.2 and further elaborated in Theory Box 7.2. In the more general case, φ should be considered as a first-order correction to nonideal mixing properties in the interface (see Theory Box 11.3).

If the solubility limit in the aqueous phase is $[S_{max}]$, $X_S^{*\,max}$ is given by

$$X_S^{*\,max} = \frac{[S_{max}]}{K_S' - \varphi[S_{max}]} \tag{7.5}$$

and the partitioning equation (Equation 2 in Theory Box 7.2) is valid only as long as $X_S < X_S^{*\,max}$ and $[S] < [S_{max}]$. The concentration $[S^*]$ is

$$[S^*] = \frac{[A_T][S]}{K_S' - (1 + \varphi)[S]} \tag{7.6a}$$

which holds as long as $X_S^* < X_S^{*\,max}$. The maximum is

$$[S_{max}^*] = \frac{[S_{max}][A_T]}{K_S' - (1 + \varphi)[S_{max}]} \qquad (7.6b)$$

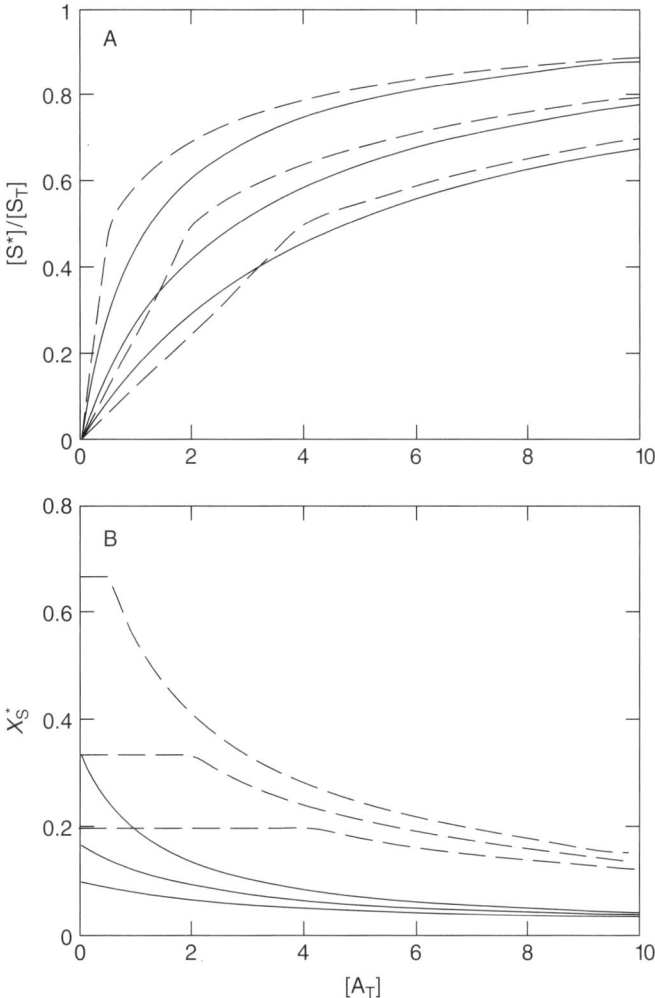

Figure 7.4 Partitioning of substrate as a function of diluent concentration, from Equations (7.3) and (7.6a) with $[S_{max}] = 1$ and $\varphi = 0$. Panel A shows the fraction of substrate, $[S^*]/[S_T]$, that is in the interface. Panel B shows the mole fraction of S in the interface, X_S^*. Solid curves are for $[S_T] = 0.5$ and dashed curves are for $[S_T] = 2$. In each set, $K_S' = 1.5, 3, 5$ from upper to lower curves. The horizontal straight-line behavior at low $[A_T]$ for the dashed curves corresponds to the region where precipitates form; the slope in this region for the curves in panel A equals $1/((K_S'/[S_{max}] - 1 - \varphi)[S_T])$, from Equation (7.6b)

When $K'_S < (1 + \varphi)[S_{max}]$, Equations (7.5) and (7.6) would imply that there is no precipitated S_x, but the substrate prefers to go into the interface regardless of how little A may be present. This is not possible for a substrate that precipitates, and $K'_S > (1 + \varphi)[S_{max}]$ will be assumed in the following. The total concentration of S above which precipitates start to form is

$$[S_T^m] = [S_{max}] + [S_{max}^*] = [S_{max}] \left(1 + \frac{[A_T]}{K'_S - (1 + \varphi)[S_{max}]}\right) \qquad (7.7)$$

The partitioning of substrate in the interface is shown in Figure 7.4. When $[S_{max}]$ and φ are known, the slope of $[S_T^m]$ versus $[A_T]$ can be used to determine K'_S (Figure 7.5). Alternatively, K'_S can be determined from the slopes in Figure 7.4.

Stated simply, as related by K'_S, a sparingly soluble solute partitions between the aqueous phase and the diluent interface. However, excess solute, beyond the maximum solubility, $[S_{max}]$, in the aqueous phase and the interface, is precipitated in the S_x form. The solubility and partitioning parameters put a physical limit on the maximum solute mole fraction that can be incorporated in the diluent interface, which provides significant insights into the nature of the underlying equilibria involved in the steady state catalytic turnover. This analysis is critical for the evaluation of the mole fraction of a sparingly soluble substrate in the interface that the bound enzyme 'sees'.

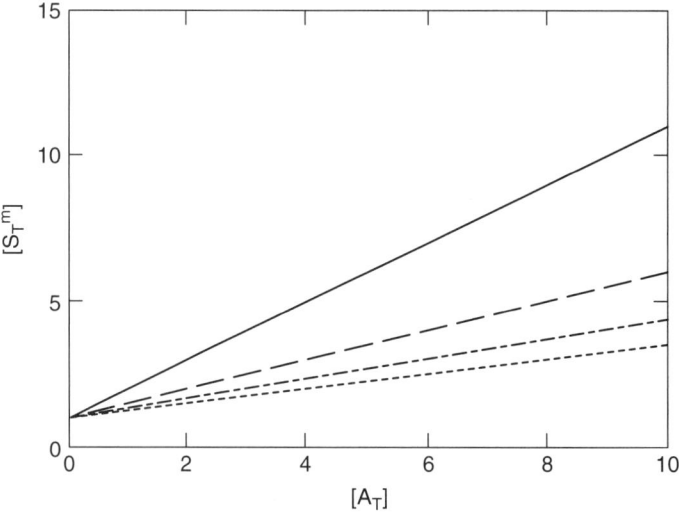

Figure 7.5 The concentration $[S_T^m]$ at which precipitates start to form, from Equation (7.7). The intercept determines $[S_{max}]$ and the slope equals $[S_{max}]/(K'_S - (1 + \varphi)[S_{max}])$. $[S_{max}] = 1$, $\varphi = 0$ and $K'_S = 2, 3, 4, 5$ from upper to lower curves

THEORY BOX 7.2 Molecular area effects on partitioning and binding

In Theory Box 3.2 we introduced the area correction factor for a substrate partitioned into a diluent interface as

$$\varphi = a_S/a_A - 1 \tag{1}$$

As a consequence, the partitioning equilibrium can be written as (see Equation 7 in Theory Box 3.2)

$$[S] = \frac{X_S^* K_S'}{1 + \varphi X_S^*} \tag{2}$$

or, equivalently,

$$X_S^* = \frac{[S]}{K_S' - \varphi[S]} \tag{3}$$

In the general case, the factor $1/(1 + \varphi X_S^*)$ should be considered as an activity coefficient that accounts for all nonidealities in the partitioning to first order in X_S^* (see Theory Box 11.3), in which case φ would be an adjustable parameter. The presence of this correction changes X_S^* from the expression given in Equation (7.3) to that given in Equation (9) in Theory Box 3.2 with L replaced by S. Furthermore, when all A molecules are in the interface, $[S^*] = X_S^*([A_T] + [S^*])$, and one finds, using Equation (3), that

$$\left[S^*\right] = \frac{X_S^*}{1 - X_S^*}\left[A_T\right] = \frac{[A_T][S]}{K_S' - (1 + \varphi)[S]} \tag{4}$$

If [S] or [S*] can be measured, this relation can be used to determine K_S' and φ separately (see Figure 11.6).

If S and A contribute differently to the interface area, the enzyme binding could also be affected. To obtain the enzyme flux expressions (Equations 7.1 and 7.2a), it was assumed that all molecules that are partitioned into the interface contribute equally to the binding of enzyme. In a more general case, all molecules contribute proportionally to their interface area. Then, the concentration of interface area is

$$a_A\left[A^*\right] + a_S\left[S^*\right] = a_A\left[M^*\right]\left(1 - X_S^* + \frac{a_S}{a_A}X_S^*\right) = a_A\left[M^*\right]\left(1 + \varphi_d X_S^*\right) \tag{5}$$

where $\varphi_d = a_S/a_A - 1$ is the same *area correction factor* as for the substrate partitioning (see above and Theory Boxes 3.2 and 11.3). The concentration of interface molecules is $[M^*] = [A_T] + [S^*]$ as before, assuming that all A is in the interface. If the enzyme binding is proportional to the interface area, the equilibrium distribution of free and interface-bound enzyme will be

$$\frac{[E]}{[E^*]} = \frac{K_d}{[M^*]\left(1 + \varphi_d X_S^*\right)} = \frac{K_d}{[A_T]}\frac{1 - X_S^*}{1 + \varphi_d X_S^*} \tag{6}$$

When $X_S^* = 0$, this gives the binding to the pure diluent interface with the same dissociation constant K_d as before. The same limit is reached when $\varphi_d = -1$, i.e. when the substrate does not contribute to the surface area; this would be the case when the substrate partitions into the interior of the bilayer without expanding the surface. Thus the area effect corresponds to the introduction of an X_S^*-dependent dissociation constant $K_d/(1 + \varphi_d X_S^*)$ to replace the K_d used before.

In the case when the substrate in the interface contributes differently to the interactions with enzyme than the diluent molecules, the enzyme binding would have a more complicated dependence on X_S^* than in Equation (3). This is the case for PLA2, for instance, when anionic molecules S are partitioned into a zwitterionic interface (Chapter 10). To account for this more general situation, φ_d should be considered as an adjustable parameter describing these effects to first order in X_S^*. Nonideal effects in the enzyme–interface binding can manifest themselves at higher enzyme concentration through excluded surface effects (Theory Box 6.2); this effect is not taken into account here.

The area effect in Equation (6) caused by shifting the amount of bound enzyme leads to a correction of the flux in Equation (7.2a):

$$j_0 = \frac{k_{catS}^* X_S^*}{X_S^* + K_{MS}^* \left(1 + \dfrac{K_d (1 - X_S^*)}{[A_T](1 + \varphi_d X_S^*)}\right)} \tag{7}$$

where X_S^* is given by Equation (9) in Theory Box 3.2 and Equation (7.5) as before. Except when φ_d is a large number (e.g. see Chapter 10), this effect on the enzyme flux is difficult to substantiate and has not been considered further.

7.3 Reaction Rate with the Partitioned Substrate

The initial enzyme flux, j_0, is determined by Equations (7.2a) with X_S^* from Equation (9) in Theory Box 3.2 for X_S^* below $X_S^{*\,max}$ given by Equation (7.5). When all enzymes are in the interface and $[S_T]$ is so large that $X_S^* = X_S^{*\,max}$, the maximal flux is reached:

$$j_m = \frac{k_{catS}^* X_S^{*\,max}}{X_S^{*\,max} + K_{MS}^*} = \frac{k_{catS}^* [S_{max}]}{[S_{max}](1 - \varphi K_{MS}^*) + K_S' K_{MS}^*} \tag{7.8}$$

As shown in Figure 7.6, the hyperbolic increase in j_0 versus $[S_T]$ (Equation 7.4a) is interrupted when the maximum amount of S has partitioned.

7.4 Interfacial Turnover by Triglyceride Lipase

A large number of serine hydrolases with substrate specificity towards acylglycerols (as in the glycerides of dietary fat) include many industrially and biologically important evolutionary families of lipases. It was noted quite early that lipases

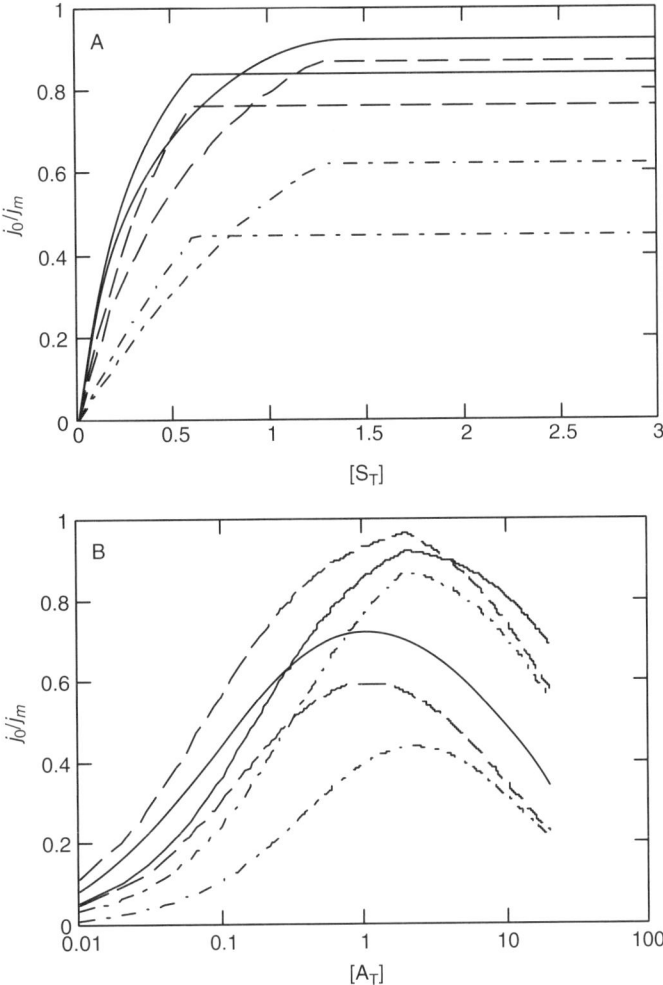

Figure 7.6 The initial enzyme flux from Equation (7.2a) relative to the maximal from Equation (7.8) as a function of $[S_T]$ (panel A) and of $[A_T]$ (panel B); $K'_S = 3$ and $\varphi = 0$ everywhere; solid curves $K^*_{MS} = 0.05$, $K_d = 0.5$; dashed curves $K^*_{MS} = 0.1$, $K_d = 0.5$; dash–dot curves $K^*_{MS} = 0.1$, $K_d = 2$. Panel A: $[A_T] = 0.5$ for all curves; for each pair of curves, $[S_{max}] = 1$ (high plateau) and $[S_{max}] = 0.5$ (low plateau). Panel B: $[S_{max}] = 1$ for all curves; for each pair of curves, the upper is for $[S_T] = 2$ and the lower is for $[S_T] = 0.5$

exhibit a preference for the substrate at the interface (Sarda and Desnuelle, 1958). As shown in Chapter 3, it is difficult to establish whether or not the hydrolysis observed below the solubility limit is at interfaces or through the monomer path. It is also difficult to characterize the interfacial path for lipases because the polymorphism and organizational state of glycerides in the aqueous dispersions is difficult, if not impossible, to control. Nonpolar glycerides are virtually insoluble in water. They have very weak polar groups and therefore glycerides are not amphiphilic enough to form organized dispersions. When suspended in water they adsorb on extraneous surfaces, or phase-separate to form films and droplets of uncontrolled dispersity, as in an oil–vinegar mixture. To overcome such difficulties, dispersed emulsions of glycerides are often used for the assay of lipases. In one such assay tributyrin is used in a froth emulsion stabilized by gum Arabic, analogous to packaged salad dressing. Considering such biophysical realities of emulsions (Chapter 2), it is not surprising that the kinetic path for only one lipase has been characterized so far.

7.4.1 Hydrolysis of PNPB in POPG vesicles by tl-lipase from *Thermomyces lanuginosa*

The kinetic path in Figure 7.1 has been useful for the analysis of tl-lipase catalyzed hydrolysis of a sparingly soluble substrate, such as tributyrin or *p*-nitrophenylbutyrate (PNPB), below its solubility limit (Berg *et al.*, 1998). The initial rate of hydrolysis of PNPB by WT (Figure 3.5) or the W89 mutant (Figure 7.7) of the tl-lipases increases dramatically in the presence of the small unilamellar vesicles (Figure 2.3) of anionic POPG. In contrast, little or no increase in the rate of hydrolysis is seen in the presence of zwitterionic POPC-suv, or with the large unilamellar vesicles (luv) of POPG. The hydrolysis of PNPB occurs at the same site as that of a triglyceride, and tributyrin (tributylglycerol) acts as a competitive substrate with PNPB. The hydrolysis of both of these substrates is eliminated by the substitution of the catalytic Ser-146 by alanine in the S146A mutants.

7.4.2 Assumption of tl-lipase as an interfacial enzyme

The enhanced rate of hydrolysis by tl-lipase in the presence of POPG-suv is attributed to the fact that the enzyme can only pick up the substrate bound to the interface (Figure 7.1). Results described below can also be interpreted as the

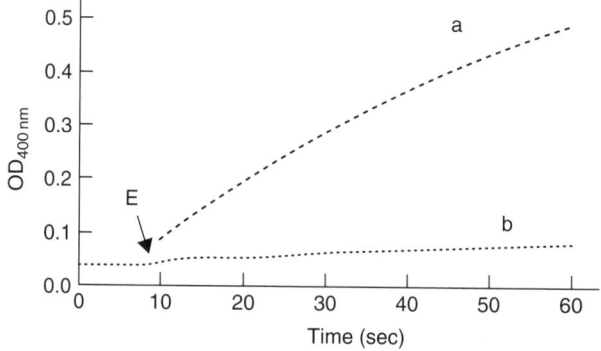

Figure 7.7 Reaction progress for the hydrolysis of unstirred 0.312 mM PNPB solution by the mutant of *Thermomyces* tl-lipase (from which all but Trp-89 are deleted) in the presence of (a) small unilamellar vesicles (30 nm diameter) of 0.054 mM POPG. (b) No significant hydrolysis is seen in the absence of enzyme, or with S146A mutant, or in the presence of POPC-suv, or in the presence of large unilamellar vesicles (80 nm diameter) of POPG. Note that virtually identical effects are seen with WT (Figure 3.5). Adapted with permission from Cajal *et al.* (2000b). Copyright (2000) American Chemical Society

Table 7.1 Interfacial kinetic parameters for the hydrolysis of PNPB or tributyroylglycerol by tl-lipase. From Berg *et al.* (1998) and Cajal *et al.* (2000b)

Parameter (units)	PNPB	Tributyroylglycerol
φ (Theory Box 7.2)	−1	0
K_S' (mM)	2.0	5.3
K_{MS}^* (mole fraction)	0.4	0.047
k_{catS}^* (s^{-1})	2600	3900
$[S_{max}]$ (mM)	1.2	0.75
K_d (POPG-suv) (mM)	0.053	0.05
Spectrosopic K_d (mM)		
POPG-suv	0.065	
POPG-luv	0.367	
POPC-suv	0.308	

activating effect of the interface on the intrinsic catalytic turnover parameters of the lipase as a matrix enzyme that accesses its substrate from the aqueous phase. As discussed in Chapter 3, for the soluble substrates the steady state kinetics cannot distinguish the interfacial turnover path of an interfacial enzyme from the aqueous turnover path for an interfacially activated matrix enzyme. This quandary also applies to the quasi-scooting mode turnover as long as substrate partitioning is equilibrated on the time scale of the observed reaction progress. Based on the observations below, the kinetics of hydrolysis of PNPB by the tl-lipase has been analyzed for the interfacial path to obtain the parameters summarized in Table 7.1. If the lipase functions as a matrix enzyme accessing substrate from aqueous solution, the parameters should be reinterpreted such that its K_M value towards the solution concentration will be $K_{MS} = K_{MS}^* K_S'$; thus the results in Table 7.1 can easily be translated to the case of a matrix enzyme.

7.4.3 Partitioning of PNPB into POPG-suv

Based on the results in Figure 7.8, the K_S' value for PNPB partitioned into POPG vesicles is 2.0 mM. This is consistent with the theoretical expectations in Figure 7.4, as well as with the steady state kinetic results.

7.4.4 Substrate dilution on POPG-suv

The change in the initial rate of hydrolysis with increasing amount of POPG vesicles is biphasic (Figure 7.9). These results are consistent with the substrate accessibility for the bound enzyme directly from the interface (interfacial enzyme) or through the aqueous phase (matrix enzyme). In the minimum model (Figure 7.1) the observed rate increases initially as the fraction of the bound enzyme increases. At the higher POPG concentrations, X_S^* decreases (Figures 7.9) when all enzymes are at the interface and active. For the reasons developed in Chapter 3, the biphasic POPG concentration dependence is consistent with the interfacial as well as the

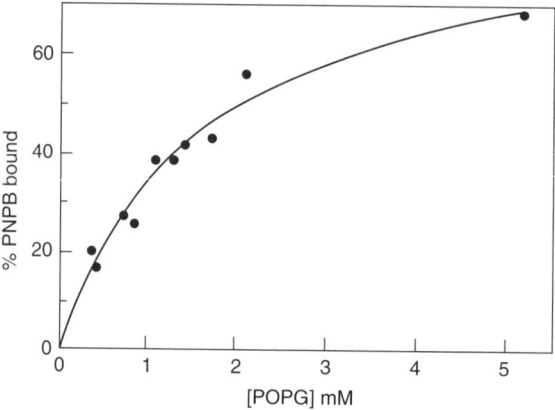

Figure 7.8 The partitioning behavior of 0.28 mM PNPB in POPG vesicles as described by the fitted curve for Equations (7.3) and (7.6a) to obtain K'_S. Reprinted with permission from Berg *et al.* (1998). Copyright (1998) American Chemical Society

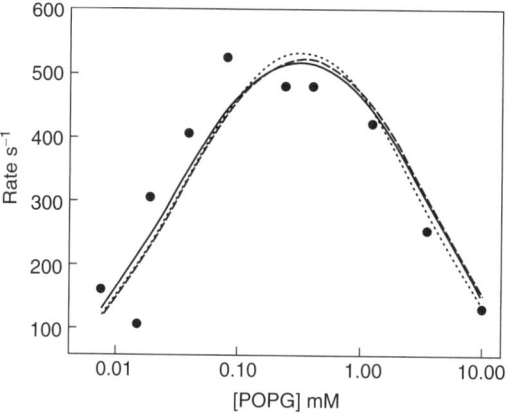

Figure 7.9 Dependence of the initial rate of hydrolysis of 0.353 mM PNPB by *Thermomyces* tl-lipase on POPG concentration as small sonicated vesicles (POPG-suv). These results are consistent with the expectations shown in Figure 7.6. The solid line is the best fit to Equation (7.2a) for j_0 with Equation (9) in Theory Box 3.2 for X^*_S, using fixed $\varphi = -1$, $K_d = 0.05$ mM and $K'_S = 2$ mM, to obtain $K^*_{MS} = 0.4$ mol fraction and $k^*_{catS} = 2000 s^{-1}$. The dashed line is the best fit to a hyperbolic approximation, which gives $K^*_{MS} = 0.5$ mole fraction and $k^*_{catS} = 2900$ s^{-1}. Note that X^*_S remains below 0.15 mole fraction through the whole concentration range; therefore such fits show a strong covariance. Reprinted with permission from Berg *et al.* (1998). Copyright (1998) American Chemical Society

matrix mechanism because at constant S, both [S] in the aqueous phase and X^*_S for PNPB partitioned in the interface would also decrease with an increasing amount of POPG in the reaction mixture.

7.4.5 The substrate concentration dependence

The solubility limit for PNPB in the aqueous phase is 1.2 mM. As shown in Figure 7.10, the observed rate of hydrolysis in the presence of 0.05 mM POPG

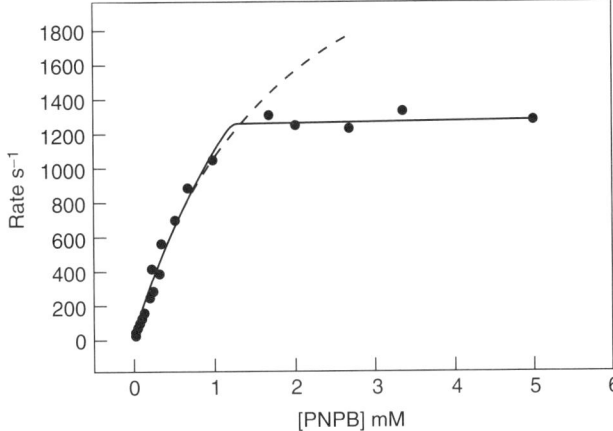

Figure 7.10 The initial rate of hydrolysis of PNPB by *Thermomyces* tl-lipase in the presence of 0.05 mM POPG vesicles. The apparent rate parameters were obtained by curve-fitting to Equations (7.2a), (7.8) and (9) in Theory Box 3.2. with $\varphi = -1$, $K'_S = 2.0$ mM and $K_d = 0.05$ mM. Reprinted with permission from Berg *et al.* (1998). Copyright (1998) American Chemical Society

$(= K_d)$ increases up to the solubility limit. As expected (Equations 7.4 to 7.6), neither X_S^* nor the concentration of the monodisperse substrate will change above the solubility limit. Here again the result cannot resolve the substrate dilemma because the insoluble substrate is not accessible for the binding to either form of the enzyme.

The parameters summarized in Table 7.1 are based on the assumption that tl-lipase accesses its substrate from the interface. While the assay system provides a set of very reproducible conditions for the study of the enzyme, the choice of the interfacial versus matrix, turnover path for the accessibility of the substrate cannot be made from the steady state kinetic measurements. An argument can be made in favor of the interfacial path for most if not all lipases. The maximum observed rate of hydrolysis with PNPB by tl-lipase is in the same range as with suitable emulsions of virtually insoluble triglycerides. The aqueous phase concentration of the long chain triglycerides is likely to be in the subnanomolar range. Thus $[S] < 10^{-9}$ M and with a maximal diffusion-limited association rate of $k_1 = 10^8$ M^{-1}s^{-1}, the maximal enzyme rate for a matrix enzyme would be smaller than 0.1 s^{-1}. The observed turnover rate of several hundred per second with glyceride emulsions could therefore not be supported by a direct access of the substrate in the aqueous phase by a matrix enzyme bound to the interface.

The affinity of POPG diluent for the active site is considerably smaller than it is for PNPB or tributyroylglycerol, i.e. $K_{ND}^* \gg K_{MS}^*$. Also, the same value of K_d is obtained from steady state kinetics and the fluorescence changes associated with the binding to POPG vesicles. These observations together suggest that the occupancy of the active site has little effect on the apparent K_d obtained under the kinetic conditions. The observation that K_d does not change for the active enzyme, or the inactive S146A mutant, suggests that the interface binding for tl-lipase to POPG vesicles is independent of the occupancy of the active site. The basis for the equilibrium underlying the K_d to POPG-suv vesicles is for the activated form of the enzyme at the interface, which may include lid opening and other events at the interface (see below).

7.5   Lid on tl-Lipase Active Site

A diluent is useful not only for the kinetic and equilibrium studies but also for the structural characterization of the enzyme at the interface. The tl-lipase binds with high affinity to POPG vesicles as well as to POPC vesicles; however, the changes in the fluorescence spectra depend on the nature of the interface. The changes in the Trp-89 fluorescence emission seen on the binding of tl-lipase to the interface (Figure 7.11) are consistent with a two-state model where the enzyme is in an active or an inactive form at the interface (Figure 7.12). As discussed below, a comparison with the crystallographic structures (Brzozowski *et al.*, 2000) suggests that the active state of the enzyme with $K_d = 0.053$ mM, has the open-lid conformation, and K_d for

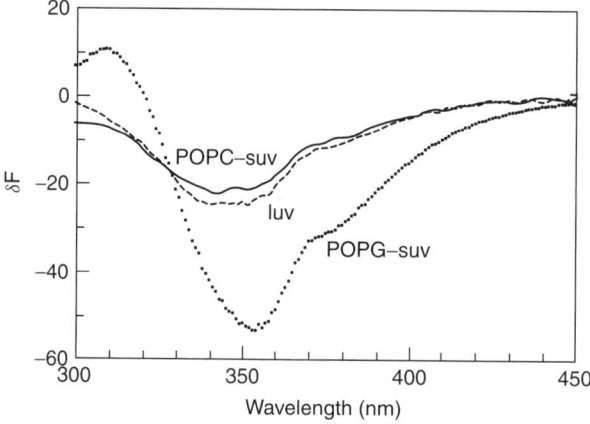

Figure 7.11   The change in the fluorescence emission of W89-lipase mutant on the addition of POPG-suv (dotted line), POPC-suv (solid line), or POPG-luv (dashed line). POPC-suv and POPG-luv induce essentially identical changes. In this S146A tl-lipase mutant only Trp-89 is present, and the three other tryptophan residues are substituted as 117-Phe, 221-His and 260-His. Adapted with permission from Cajal *et al.* (2000b). Copyright (2000) American Chemical Society

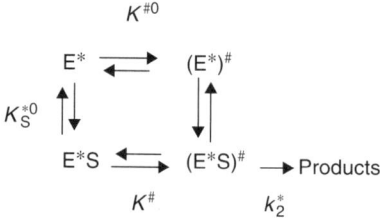

Figure 7.12   A kinetic model for interfacial activation. The interface-bound enzyme exists in two forms, E* and (E*)#. If the E* form binds S very poorly, or not at all, the effective affinity for S can be modified by shifting the equilibrium towards (E*)#. Such a shift can be caused by properties (e.g. charge) of the interface, the 'quality of interface', and corresponds to K_S^* activation. One possible example is the 'lid' opening of lipase discussed in this chapter. Alternatively, both forms bind substrate, but the inactive form E*S is favored. In this case, properties of the interface can shift the equilibrium towards the active form (E*S)#, thereby increasing enzyme activity, which corresponds to the two-state model for k_{catS}^* activation in Chapter 8 (Figure 8.7 and Theory Box 8.2)

the closed-lid inactive state is about 0.35 mM, as seen with POPG-luv or POPC-suv (Table 7.1).

7.5.1 Relationship of the 'lid' to the active site

Lipases are very widely distributed. The tl-lipase belongs to a subfamily of lipases from microorganisms. Lipases have been structurally characterized by biochemical methods, mutagenesis and X-ray crystallography. Crystal structures of several types of lipases (carboxyesterase or triglyceride hydrolases) have been determined. They have the CXSXG motif for the serine hydrolases with a Ser–His–Asp(Glu) triad of the catalytic active site. As a class, lipases belong to the α/β-fold family of proteins with 20 000 to several hundred thousand molecular weight. The catalytic triad of tl-lipase consists of Asp-201, His-258 and serine-146 in the consensus catalytic sequence GHS^{146}LG. The catalytic triad is accessible from the surface of the protein only through a pocket with a lid. Although the lid structure is not found in all lipases, the lid-containing lipases appear to be quite widespread.

The tl-lipase belongs to the subgroup that has a structural feature that putatively acts as a lid to cover the active site. Trp-89 is on the lid helix. Results in Figure 7.11 show that the indole environment is different under the conditions where the lid is expected to be open or closed. This correlation provides a structural interpretation for the activation scheme. The fluorescence measurements are carried out on a diluent where the active site is not occupied, and also in the W89-mutant of S146A mutant. These results show that the structural change on the binding occurs with the E^* form of the enzyme in which the active site is not occupied. In this conceptualization of a diluent for the lipase the functionally active form is present only on POPG-suv, but not on POPC (suv or luv) or even on POPG-luv. Such observations at different interfaces suggest that the 'open-lid' form (Figure 7.13), which accepts the substrate into the active site, is present only at the highly curved anionic interface.

7.5.2 Cationic residues as the 'hinge' for the interfacial activation by the anionic charge

The crystallographic structures with an 'open' and 'closed' lid could be related to the active and inactive $(E^*)\#$ and the E^* forms in Figure 7.12. This requires, as supported by the kinetic and fluorescence results with the W89-lipase, a control of the lid by the properties of the interface. The requirement for the anionic surface charge at the interface for the lid opening implies a role for certain suitably placed cationic residues on the lipase structure. The catalytic site lined with hydrophobic residues is buried under a lid of a two-turn (residues 86 to 93) amphiphilic α-helix containing Trp-89 on the hydrophobic face:

$$F_{80}\text{RGSR}\underline{\text{SIENWI}}_{90}\text{GNLNFDLKEI}_{100}$$

Based on the kinetic and binding properties it is suggested that in this sequence Ser-83 and Asp-96 serve as pivots for the movement of the lid. Interactions of

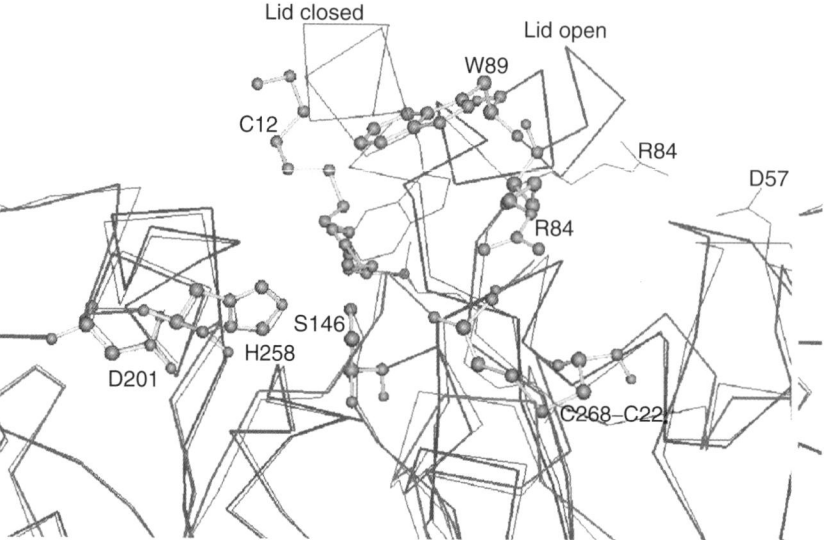

Figure 7.13   The catalytic site, lid and hinge region of tl-lipase. The thinner line represents the backbone of the closed-lid form and the thicker line shows the open-lid form of the dodecyldiethylphosophonate (C12) inhibitor complex. The ball and stick representations for the residues in the open form include Arg-84 (with the side chain for the closed form) and the isomeric forms of the Cys268–Cys22–disulfide bond. This view was kindly provided by Professor Marek Brzozowski (University of York) based on his crystallographic work (Brzozowski *et al.*, 2000)

one or more of the cationic residues with the interface could leave the lid in the open position to connect with the β-turns in the tl-lipase structure. In fact, the cationic residues K73, R80, R84 and K98 on both sides of the lid helix could play a role in controlling the hinges of the lid. The fact that the surface curvature plays a role implies that additional factors are also at work here, possibly related to specific interactions between the groups at the interface with those on the enzyme.

Structures of nine different crystal forms suggest that tl-lipase can assume two or three discrete sets of structures with distinct positions for the lid and that Arg-84 acts as a hinge for the changes in the position of the lid (Brzozowski *et al.*, 2000). Depending on the crystallization conditions the lid can be in the open (active) or closed state, and in some cases an intermediate state is also stabilized. The open conformation is favored in media of low dielectric or low ionic strength, and therefore possibly by the lipid interface. Calculations suggest an activation energy barrier of 10.6 kcal/mole between the open and closed states with an enthalpy difference of about 3 kcal/mol. The opening of the lid is accompanied by a 120° change in the side chain of Arg-84 and the hinge region. At least two other regions undergo a major change: *cis–trans* isomerization of the 22–268 disulfide and the lid helix which swings open to make the substrate cavity accessible. These changes are stabilized by the isomerization. The roll-over of the lid exposes a hydrophobic patch whose function is not related to the binding of the enzyme to the interface.

7.6 Interfacial Allostery for the Quality-of-Interface Effects

Qualitatively all the available results are consistent with a model in which the tl-lipase could exist in an active or an inactive form, which would potentially result in four different forms of the interface-bound enzyme (Figure 7.12). The kinetic, spectroscopic and structural evidence is consistent with a model in which the bound enzyme exists at least in two states that may correspond to the open and closed forms stabilized with the Arg/disulfide switch activated by the anionic interface. In addition to the rate enhancement by the surface charge, the vesicle size also influences the observed rate, which suggests that other factors are also at work. A complete analysis of the activating effect of the interface on the bound enzyme is not available yet. However, considerations outlined below show that the lid on the active site of the bound enzyme assumes a range of states with the changing interface.

7.6.1 Defining the problem of lid opening versus interfacial activation

As a guide to understanding interfacial activation and allostery in terms of the Scheme in Figure 7.12, certain points can be made on the basis of available evidence. The fact that the rate of hydrolysis of monodisperse substrate is virtually zero and that a rate enhancement is seen with POPG-suv suggests that the E form of the enzyme is not functional. This could be because either the lid remains closed or the E form is intrinsically inactive for the catalytic turnover. Lid opening to control the access of the substrate to the bound enzyme is only one of the many possible ways in which the turnover rate at the interface can be influenced. In terms of the interfacial activation scheme in Figure 7.12 a likely effect of the lid opening is that only the fraction of the enzyme in the form capable of binding the substrate has changed. In this case, the scheme reduces to

$$
\text{E} \underset{K_d^0}{\rightleftharpoons} \text{E}^* \overset{K^{\#0}}{\rightleftharpoons} \left(\text{E}^*\right)^{\#} + \text{S}^* \underset{K_S^{*\#}}{\rightleftharpoons} \left(\text{E}^*\text{S}\right)^{\#} \longrightarrow
$$

Thus the effect of the charge on the interface could be on the second step, but would not influence the stability of the Michaelis complex. The effect of the interface on either the first or the second step would mean that only the enzyme binding to the interface is influenced by the properties ('quality') of the interface. If the dissociation constant in the first step is K_d^0 and the equilibrium constant for the second step is $K^{\#0}$ (cf. Figure 7.12), the effective dissociation constant for $\text{E}^* + (\text{E}^*)^{\#}$ to E would be $K_d^0/(1 + K^{\#0})$. Thus the apparent binding of the enzyme to the interface can be influenced by the properties of the interface, as appears to be the case with the tl-lipase (Table 7.1), even if only the second step is affected. This second step could be the opening of the lid of the bound enzyme at the interface. The enzyme flux in this scheme would be

$$
j = \frac{k_2^* X_S^*}{X_S^* + K_S^{*\#}\left[1 + \dfrac{1}{K^{\#0}}\left(1 + \dfrac{K_d^0}{[\text{M}^*]}\right)\right]}
\tag{7.9}
$$

A comparison with Equation (5.24) shows that this corresponds to the effective parameters

$$K_{MS}^* = K_S^{*\#}\left(1 + 1/K^{\#0}\right) \tag{7.10}$$

$$K_d = K_d^0 / \left(1 + K^{\#0}\right) \tag{7.11}$$

Thus, changes in the equilibrium constant $K^{\#0}$ correspond to K_S^* activation.

From the steady state kinetics one could dissect the effect of the interface on the catalytic turnover parameters, K_{MS}^* or k_{catS}^*. This would require the evidence that the bound enzyme is in the catalytically competent form at the interface with its lid open and that none of the first two steps are rate-limiting in the turnover cycle, or that they are part of the pre-steady state.

7.7　Motifs for Close Contact of Proteins with the Interface

Crystallographic structural information has been interpreted to identify putative membrane contact regions of interfacial enzymes (Table 7.2). The kinetic significance of such motifs and their relationship to the catalytic cycle events remain to be evaluated. General features of the membrane contact regions are introduced below.

7.7.1　Hydrophobic residues

Defining the hydrophobic drive of the amino acid residues of a protein is a challenging problem. The balance of the hydrophobic drive comes into play on the basis of the region of a molecule exposed to the aqueous phase. As summarized in Table 7.3, several approximations have been made to take into account the restricted environment at the interface. For example, a nonpolar helix that spans a bilayer is usually identified via the hydropathy plot constructed from the sequence information. The optimized energy for the transfer of a residue in a helix from the aqueous phase to the membrane interior for the hydropathy plot is given in column 2 of Table 7.3. The values in column 3 are for the transfer to bilayer of a residue in a pentapeptide that is most likely to be localized in the glycerol backbone region. The values in column 4 are for the transfer of N-acetyl amino acid amides to octanol and corrected for the difference in the size of the solute and solvent molecules. These

Table 7.2　Interface and membrane binding domains of membrane-associated enzymes

Motif (residues, ion)	Enzyme
i-face	Secreted PLA2
C2 domain	cPLA2 mammalian PI-PLC, pancreatic lipase
C1 domain (150 residues)	Protein kinase (PKCγ, δ and μ) diacylglycerol kinases
PH (120 residues)	Mammalian PI-PLC
FYVE (Zn and 75 residues)	PI-phosphate 3 kinases
Transmembrane helics	Bacteriorhodopsin, photosynthetic reaction center
C terminus (50 residues)	Cycloxygenase

Table 7.3 Free energy (kcal/mole) for the transfer of X residue from water to the indicated environment (pH 8)

X residue	α-Helix[a]	Bilayer[b]	n-Octanol[c]	V[d]	A[e]
Ala	−1.60	0.17	−2.03	231	347
Arg	12.32	0.81	−1.78	318	468
Asn	4.81	0.42	−1.40	264	379
Asp	9.21	1.23	−1.09	261	378
Cys	−2.00	−0.24	−4.19	258	360
Gln	4.11	0.58	−2.48	297	408
Glu	8.21	2.02	−1.78	290	405
Gly	1.00	0.01	−1.01	198	323
His	3.01	0.17	−2.95	297	432
Ile	−3.11	−0.31	−5.87	332	402
Leu	−2.80	−0.56	−5.89	341	399
Lys	8.80	1.0	−2.01	329	445
Met	−3.40	−0.23	−4.80	316	425
Phe	−3.71	−1.13	−6.03	342	429
Pro	0.20	0.45	−3.51	300	364
Ser	−0.60	0.13	−1.59	233	352
Thr	−1.20	0.14	−2.53	263	370
Trp	−1.90	−1.85	−7.14	369	470
Tyr	0.701	−0.94	−4.90	342	447
Val	−2.61	0.07	−4.51	300	375

[a]Estimate for the residue transfer in α-helix (Engelman *et al.*, 1986).
[b]From the partition coefficients in mole fraction units for the residue transfer as Ac–Trp–L–X–LL peptide from water to bilayer (Wimpley and White, 1996).
[c]Values for the molarity partition coefficient for *N*-acetyl amino acid amide between ocatnol and water corrected for the solute–solvent size differences (Sharp *et al.*, 1991). Comparable values for transfer to cyclohexane or the gaseous phase are also given in this paper.
[d]Volume as cm^3/mole.
[e]molecular area as $Å^2$.

estimates suggest that the actual values are likely to be strongly dependent on the microenvironment of the residue. The volume and exposed area for the residues are given in columns 5 and 6. Such information may be useful for the evaluation of the changes in the local environment of the protein induced by site-directed mutagenesis.

7.7.2 Hydrophobic segments

The hydrophobic surfaces of the membrane-incorporated proteins consist of α-helix and a barrel of β-strands in contact with the acyl chains of phospholipids. The procedure for the construction of a hydropathy plot to ascertain the transmembrane propensity of a helix or strand consists of calculating the energy of transfer of a hypothetical helix of 19 to 21 amino acid residues from the aqueous phase. The choice of this number actually depends on the thickness of the hydrophobic region of the membrane. Thus, starting from the 1–20 segment, the procedure is repeated for 2–21, 3–22, 4–23, ... until n–20 to n. This moving average of the energy for transfer of a helix is plotted against the first residue of the helix. Typically, a peak of 20 kcal/mole correlates with a stable insertion of the membrane spanning helix.

7.7.3 Amphipathic helix

Typically, the amphipathic helix motifs have significant differences in the polarity of the opposing faces along the length. Several methods have also been developed to identify an amphipathic helix. For example, in the helical wheel diagram, the side-chain orientation is projected on to a plane perpendicular to the long axis of the helix. Using these methods seven classes of amphipathic helices have been identified in peptides that putatively interact with phospholipid interfaces (Segrest *et al.*, 1990). Each class contains 3.6 residue per turn, but the clustering of hydrophobic and polar residues on the helix may be significantly different. Thus amphipathic helixes found in apolipoproteins, lytic peptides and globular proteins differ significantly. The ratio of the polar and nonpolar regions provide differing constraints on the contact surfaces which allow the helix to lie on the membrane surface, or form a helix bundle with hydrophobic contacts, or form a transmembrane bundle with a polar pore in the middle. One of the significant differences is in the mean angle subtended by the polar face, which may be less than 100° to more than 320°, and thus the ratio of the hydrophobic to polar residues and the hydrophobic moment per residue differ significantly.

7.7.4 Membrane anchors

A number of acylated proteins have been identified with a variety of hydrophobic anchors (Low, 1989). Often such anchors are specific, such as myristoyl, palmitoyl, farnesyl, geranyl–geranyl, or phosphatidylionositol chains attached to specific residues on a protein through a Cys thiol or through the C or the N terminus. Since attachment of anchors appears to have a regulatory role in the critical cellular process, including signal transduction and cell division, the interfacial enzymes that remove or place these anchors must be well regulated. It appears that *N*-myristoyltransferase requires myristoyl-CoA, and the reaction occurs co-translationally.

7.7.5 Putative membrane-binding domains

Protein kinase C (C1 and C2), pleckstrin (PH) and a host of other homology domains have been identified in lipases and other proteins involved in eukaryotic signal transduction and membrane trafficking (Hurley and Misra, 2000). A dissection of the catalytic and membrane-binding domains may have an evolutionary origin. For example, the catalytic domains of PI-specific phospholipase C have a remarkably similar $(\beta\alpha)_8$ – or TIM – barrel fold. However, the PH and C2 membrane-binding domains found in the mammalian enzymes are not present in the Gram-positive bacterial PI-PLC with poor affinity for the interface and a somewhat different substrate specificity (Griffith and Ryan, 1999).

Only in some cases is the catalytic site a part of the membrane-binding domain, and often the catalytic site is accessible from the face that binds to the interface. Some membrane-binding domains show specificity for the phospholipid head group and often bind a divalent ion for the membrane-binding function. For example, the C1 domain of protein kinase C consists of about 50 residues in two small β-sheets and a short C-terminal α-helix around two 3-Cys-1-His zinc binding clusters. C1 may also be a part of the allosteric activation site, and the diacylglycerol and phorbol ester-binding site is formed at one tip of the domain. Type IIβ phosphatidylinositol kinase binds to anionic vesicles, and most of the binding is reversed in the presence of 1 M NaCl or by mutating three lysines (72,76,78) to glutamate. Its crystallographic structure has been interpreted to identify a possible membrane-binding domain.

7.7.6 Colipases

Lipases, bile salts and a 10 kDa colipase protein are co-secreted by pancreas, along with other hydrolytic enzymes. The lipase does not require colipase for the catalytic activity, but their cocomplex may have a regulatory role. In the crystallographic structure, colipase is bound to the C-terminus β-sheet of the lipase (Tilbeurgh *et al.*, 1999). This structural fold, also noted in the colipase structure, has a superficial resemblance to the eukaryotic C2 domain mentioned above. This noncatalytic region of about 100 residues is on one edge of the active site slot and the 'lid' is on the opposite edge. The lid is 'closed' in the complex, as it is for the lipase alone. However, in the cocrystals of colipase with inhibited lipase, with the catalytic serine covalently modified with undecanoylphosphonate, the lid is 'open'. This change brings the lid into contact with the N-terminus part of colipase. The suggestive role of the lipase–colipase interaction in the stabilization of the 'open' conformation is also consistent with the observation that the interaction between lipase and colipase is weak in solution, but increases in the presence of the interface. Also, colipase increases the binding of the lipase to the interface. The functional significance of such interactions in terms of the primary events of the catalytic cycle remains to be established.

7.8 Summary: Multiple States of the Enzyme at the Interface

Steady state catalytic turnover by the bound enzyme on the substrate partitioned in the interface of a diluent can be analyzed to obtain the primary kinetic parameters. Kinetic strategies based on the use of the substrate partitioned in an organized interface not only provide a measure of the interface concentration but also eliminate many of the problems associated with the anomalous effects of the extraneous surfaces. Such conditions eliminate, or at least reduce, the possible number of uncontrolled variables to a point that the steady state kinetics can be characterized in terms of primary parameters with well-established functional significance. Also, many of the nonidealities of the effect of the solute partitioned in the interface can be quantified. For example, the value of the area correction factor, φ, may provide a meaningful measure of the distribution of the solute in the interface. Such kinetic studies do not distinguish between the interfacial versus the interface-activated matrix mechanisms. However, circumstantial evidence for tl-lipase supports the conclusion that it is an interfacial enzyme that directly accesses the substrate from the interface.

The model for interfacial activation is based on multiple states of the enzyme in the interface. The bound tl-lipase can exist in active or inactive forms that have different substrate accessibility to the active site. As suggested by the kinetic and spectroscopic studies, the crystallographic results show that a 'lid' controls the access of the substrate to the active site. In addition, the change from the closed to the open lid form is facilitated by the charge interactions between the anionic interface and the Arg–disulfide switch. It is still intriguing that such interfacial activation is seen only on small anionic vesicles or micelles but not with large anionic vesicles.

Significant progress has been made in the characterization of the crystal structure of lipase as well as many other interfacial enzymes. The active sites in many of these proteins have been characterized with reasonable certainty. On the other hand, our current knowledge of the membrane and interface contact regions of virtually all such proteins is based on circumstantial evidence. The problem is compounded by the fact that the membrane contact region consists of scores of amino acid residues. Therefore, evaluation of the incremental contributions of the structural features by site-directed mutagenesis studies requires methods capable of resolving and quantifying modest changes in specific equilibrium and kinetic parameters for a kinetic path. In the absence of suitable kinetic protocols for a defined kinetic path, such a structure function correlation is unlikely to be meaningful.

8 Interfacial Allostery

> *We are old-fashioned monkeys and futuristic apes. We are sympathetic, canny, crude, and dazzling. We are profoundly aggressive, and we have many loci of control over that aggression. We feel our way to the narthex of love and think our way down its nave. We are like nothing else that has even appeared on this threshing blue planet, and we will become, in the next few centuries, like nothing we can fathom now. And we will do it wearing our same old Stone Age genes.*
>
> **Natalie Angier** in *Woman*

The activating effect of the interface on the observed rate has been the phenomeno-logical grail of interfacial enzymology for over five decades. For example, as shown in Figure 8.1, the observed rate of the pig pancreatic PLA2 catalyzed hydrolysis of diheptanoylphosphatidylcholine (DC_7PC) increases above the critical micelle concentration, and a further rate increase is seen with added NaCl (de Haas *et al.*, 1971; Pieterson *et al.*, 1974). The standard kinetic paradigm for soluble enzymes does not readily account for these results. The rate increase above the CMC is attributed to interfacial activation, with the suggestion that the activity of PLA2 is significantly higher on the micellar substrate than it is on the monomeric substrate. Such a rationalization in terms of Scheme I is the starting point for defining the kinetic path that can be analyzed to obtain insights into the underlying processes.

In this chapter we describe analytical protocols that are useful for identifying the kinetic path in the quasi-scooting mode under rapid substrate replenishment conditions. This permits interpretation of the observed effects of the interface-related variables in terms of the parameters related to the primary events. A complete analysis of the phenomena in Figure 8.1 shows that several effects are at work here (Berg *et al.*, 1997). For example, analysis of the rate increase above the CMC shows that the catalytic turnover occurs at the interface of substrate micelles. The fraction of the total enzyme at the interface increases with the concentration of the micellar interface. The catalytic turnover in the quasi-scooting mode by PLA2 bound to the interface of short chain phosphatidylcholines (DC_nPC) micelles occurs with rapid substrate replenishment. In addition, an allosteric effect of the NaCl-induced anionic charge at the interface on the chemical step of the interfacial turnover cycle is attributed to the charge compensation of Lys-53,56,120.

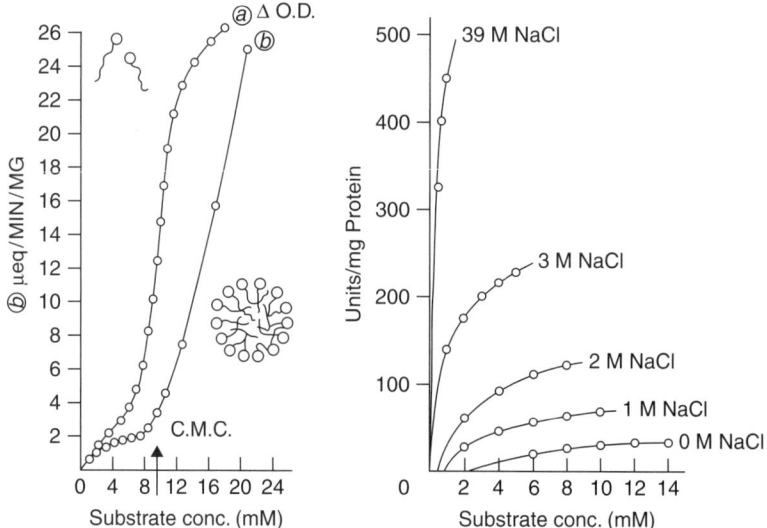

Figure 8.1 (Left) DC$_6$PC concentration dependence for the change in turbidity (curve a) or the rate of hydrolysis by pig pancreatic PLA2 (curve b). (Right) Effect of [NaCl] on [DC$_7$PC] dependence of the hydrolysis by PLA2. The specific activity (ordinate) is 4.2 times the flux s^{-1}. As shown in Figure 2.5, the CMC of DC$_7$PC is 1.5 mM in the absence of NaCl and decreases in the presence of added NaCl. Reprinted with permission from Pieterson *et al.* (1974). Copyright (1974) American Chemical Society

The monomer rate versus the micellar rate

The catalytic and exchange fluxes in the general kinetic paradigm (Figure 3.1) are not always readily resolved. The protocols outlined in Figure 3.6 show that the observed monomer rate below the CMC in the absence of extraneous surfaces, if any, is <0.1 s^{-1} for pig pancreatic PLA2 (Yu *et al.*, 1999a). Significantly larger monomer rates seen below the CMC in Figure 8.1 are attributed to the reaction at the surfaces of air bubbles and vessel walls in the stirred assay mixtures.

The rate above the CMC is attributed to the reaction on the micellar interface of the zwitterionic short chain phosphatidylcholines. For example, under these conditions the reaction begins immediately after the addition of the enzyme and the rate does not change with stirring and shaking. The initial rates are linear for several minutes and decrease gradually until all the micellar substrate is consumed. The enzyme activity remains undiminished because hydrolysis begins again when a fresh aliquot of the micellar or vesicle substrate is added to the reaction mixture. Results show that with the micellar substrate all the substrate is accessible to the enzyme, and the contribution of the reaction on extraneous surfaces is negligible (Yu *et al.*, 1999a).

Defining the kinetic path for the reaction progress in quasi-scooting mode

Micelles of the rapidly exchanging substrate amphiphiles provide the interface for the reaction progress in the quasi-scooting mode. The kinetic path in Figure 8.2 is

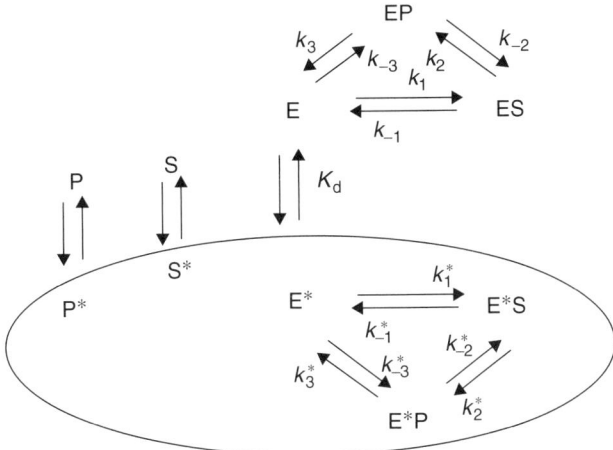

Figure 8.2 The catalytic turnover loops for the monomer and the micellar paths in the quasi-scooting mode. Note that the rapid substrate replenishment condition (Theory Box 8.1) eliminates the kinetic contributions from the product accumulation on the enzyme-containing micelle. According to the constraints of the detailed balance condition, the species in the interface are in rapid equilibrium with the corresponding species in the aqueous phase

useful for the detailed analysis in this limit of rapid substrate replenishment because the microscopic steady state condition for the successive turnover cycles remains invariant during the reaction progress by the bound enzyme. The variables for the steady state turnover are defined as follows:

(a) The bulk concentration of phospholipid present as micelles, $[S^*] = [S_T] -$ CMC, determines the fraction of the enzyme bound to the interface (E^*). The fraction of the total enzyme present as E^* increases with $[S^*]$. Thus the fraction of interface-bound enzyme, as well as the observed rate, shows a hyperbolic dependence on $[S^*]$ (Equation 5.24). This $[S^*]$ dependence of the rate is described by the parameters K_M^{app} and k_{cat}^{app} obtained from the hyperbolic fit. As analyzed below, only under suitable constraints are these parameters related to the primary rate and equilibrium constants for the interfacial turnover path.

(b) The number density (concentration) of the substrate at the micellar interface, approximated as $X_S^* = 1$, determines the substrate concentration for the formation of the E^*S complex. This requires very fast product dissociation from the interface (Theory Box 8.1). The substrate replenishment from excess micelles must be rapid on the time scale of the turnover event. This is a critical condition to assure that the steady state turnover path does not change with the reaction progress. Rapid substrate replenishment ensures that the microscopic steady state condition for the turnover cycle of an enzyme bound to a small micelle remains the same as the average of the total reaction mixture. The same result for the apparent parameters could be achieved if enzyme exchange between micelles is very rapid.

8.1 Interfacial Catalytic Turnover in the Quasi-Scooting Mode

For the general kinetic path in Figure 3.1, there would be net steady state fluxes for the different enzyme states between solution and interface, i.e. across the steps $E \rightleftarrows E^*$, $ES \rightleftarrows E^*S$ and $EP \rightleftarrows E^*P$, requiring a knowledge of the six rate constants for these steps rather than just the equilibrium constants. These rate constants would then enter the effective Michaelis–Menten parameters for the scheme. The description is considerably simplified if the catalytic fluxes are decoupled, e.g. as in Figure 8.2, where the dissociation of an enzyme with a substrate or product in the catalytic site is blocked. This would be the case, for instance, if the ligands help anchor the enzyme to the interface. It should be stressed that the equilibrium constants for the two reactions $ES \rightleftarrows E^*S$ and $EP \rightleftarrows E^*P$ are well defined as determined by the detailed-balance conditions, Equation (6.7) with L replaced by S and P respectively; only the corresponding rates are assumed slow.

The steady state rate for the kinetic path in Figure 8.2 can easily be calculated as outlined below. Since there is just a single connection for enzyme in the interface and solution, there can be no flux across this branch. Each steady state catalytic flux can be calculated independently (Theory Box 5.1) and then normalized to the total enzyme concentration by coupling across the equilibrated $E \rightleftarrows E^*$ step. The result is

$$j = \frac{\frac{k^*_{catS}}{K^*_{MS}} X^*_S - \frac{k^*_{catP}}{K^*_{MP}} X^*_P + \frac{K_d}{[M^*]} \left(\frac{k_{catS}}{K_{MS}}[S] - \frac{k_{catP}}{K_{MP}}[P] \right)}{1 + \frac{X^*_S}{K^*_{MS}} + \frac{X^*_P}{K^*_{MP}} + \frac{K_d}{[M^*]} \left(1 + \frac{[S]}{K_{MS}} + \frac{[P]}{K_{MP}} \right)} \tag{8.1}$$

As in the previous expressions (see Equation 5.23), each term in the denominator corresponds to each possible enzyme state relative to E^*, e.g. $(K_d/[M^*])([S]/K_{MS}) = [ES]/[E^*]$. When all enzymes are in the interface at the saturating interface concentration, $K_d \ll [M^*]$, Equation (5.2) is recovered. When all enzymes are in solution, $K_d \gg [M^*]$, one gets the Michaelis–Menten equation for the soluble enzyme reaction path. The interfacial turnover parameters are given in terms of the rate constants in Theory Box 5.1, and the corresponding solution parameters (k_{catS}, K_{MS}, k_{catP} and K_{MP}) look exactly equivalent after removal of the * superscripts.

Consider the case when (a) there are no back reactions in the catalytic steps, $k_{-2} = k^*_{-2} = 0$, which gives $k_{catP} = k^*_{catP} = 0$; (b) no product is present initially, $[P] = 0$; (c) the solution concentration of S equals the CMC, $[S] = CMC = K'_S$; and (d) the interface concentration is given primarily by the substrate, $[M^*] = [S^*] = [S_T] - CMC$. With these assumptions, Equation (8.1) simplifies to

$$j = \frac{\frac{k^*_{catS}}{K^*_{MS}} X^*_S[S^*] + \frac{k_{catS}}{K_{MS}} K_d K'_S}{[S^*]\left(1 + \frac{X^*_S}{K^*_{MS}} + \frac{X^*_P}{K^*_P}\right) + K_d\left(1 + \frac{K'_S}{K_{MS}}\right)}$$

$$= \frac{k^{app}_{cat}[S^*] + k^{mono}_{cat} K^{app}_M}{[S^*] + K^{app}_M} \tag{8.2}$$

The last equality follows after identifying the apparent constants

$$k_{\text{cat}}^{\text{app}} = \frac{k_{\text{catS}}^* X_{\text{S}}^* / K_{\text{MS}}^*}{1 + \dfrac{X_{\text{S}}^*}{K_{\text{MS}}^*} + \dfrac{X_{\text{P}}^*}{K_{\text{P}}^*}} \xrightarrow{X_{\text{P}}^* = 0} \frac{k_{\text{catS}}^*}{1 + K_{\text{MS}}^*} \tag{8.3}$$

$$K_{\text{M}}^{\text{app}} = \frac{K_{\text{d}}\left(1 + K_{\text{S}}'/K_{\text{MS}}\right)}{1 + \dfrac{X_{\text{S}}^*}{K_{\text{MS}}^*} + \dfrac{X_{\text{P}}^*}{K_{\text{P}}^*}} \xrightarrow{X_{\text{P}}^* = 0} \frac{K_{\text{d}}\left(1 + K_{\text{S}}'/K_{\text{MS}}\right)}{1 + 1/K_{\text{MS}}^*} \tag{8.4}$$

$$k_{\text{cat}}^{\text{mono}} = \frac{k_{\text{catS}} K_{\text{S}}'}{K_{\text{S}}' + K_{\text{MS}}} \tag{8.5}$$

These apparent MM parameters derive from variation of substrate concentration [S*]. Therefore $K_{\text{M}}^{\text{app}}$ corresponds to the enzyme–interface dissociation constant relating all enzyme species in the interface (E*, E*S and E*P) to all in solution (E, ES and EP). These 'constants' still contain the interface concentrations X_{S}^* and X_{P}^*. If product desorbs very fast and is replaced by substrate, the limits of $X_{\text{S}}^* \approx 1$ and $X_{\text{P}}^* \approx 0$ will hold on the enzyme-containing micelles, as indicated by the arrows in Equations (8.3) and (8.4). This requires that product exchange is faster than the effective catalytic rate in the interface, i.e. faster than $k_{\text{catS}}^*/(1 + K_{\text{MS}}^*)$, and a bound enzyme will always see the local concentration $X_{\text{S}}^* \approx 1$ regardless of the rate of enzyme exchange. Alternatively, the rapid substrate replenishment condition would also hold if enzyme exchange is sufficiently fast, so it moves to a fresh micelle before there is any significant depletion. If product exchange is not quite as fast as this, but still fast enough to uphold the integrity of the enzyme-bound micelle, there will be a nonzero steady state presence of product which is determined by the ratio between the exchange rate and the catalytic rate (Theory Box 8.1). Thus, in a pre-steady state process, the enzyme binds a micelle very quickly and establishes a microscopic local depletion of substrate. In this case there could be product inhibition in the apparent initial (steady state) rate, even if product is not present initially. In what follows, we will mostly assume that substrate/product or enzyme exchange is sufficiently fast so that Equations (8.2) to (8.5) with $X_{\text{S}}^* \approx 1$ and $X_{\text{P}}^* \approx 0$ are valid. In this limit, the actual rates of exchange do not enter the results, only their equilibrium constants.

Based on the Michaelis–Menten Equation (8.2), it is also possible to calculate the integrated reaction progress. This is most useful when the apparent initial rate does not last long enough to be accurately measured. In the absence of a monomer rate (or premicellar aggregates, wall reactions, etc.), the reaction would stop when [S$_{\text{T}}$] equals the CMC.

THEORY BOX 8.1 Exchange-limited kinetics: activation by product desorption

Consider the change in the number, N_{P}, of products in a micelle that has one enzyme bound. Assume that a product leaves the micelle with rate k_{d}^{P}

and is replaced by a substrate from solution. Then the rate of change of N_P is

$$\frac{dN_P}{dt} = -k_d^P N_P + \frac{k_{catS}^* X_S^*}{X_S^* + K_{MS}^* \left(1 + X_P^*/K_P^*\right)} = 0 \quad (1)$$

which equals zero at the steady state. Identifying $N_P = N_T X_P^* = N_T(1 - X_S^*)$, the steady state flux per enzyme (apparent initial rate) can be written as

$$j_{ss} = k_d^P N_T X_P^* = \frac{k_{catS}^* \left(1 - X_P^*\right)}{1 + K_{MS}^* - X_P^* \left(1 - K_{MS}^*/K_P^*\right)} \quad (2)$$

Replacing X_P^* everywhere by $j_{ss}/(k_d^P N_T)$ gives a quadratic equation for j_{ss} that can be solved as

$$\frac{j_{ss}}{j_0} = \frac{\beta(1+\alpha)}{2}\left[1 \pm \sqrt{1 - \frac{4\alpha}{\beta(1+\alpha)^2}}\right] \quad (3)$$

where

$$\alpha = \frac{k_d^P N_T \left(1 + K_{MS}^*\right)}{k_{catS}^*} = \frac{k_d^P N_T}{j_0}, \qquad \beta = \frac{1 + K_{MS}^*}{1 - K_{MS}^*/K_P^*}$$

As shown in Figure 8.3, j_{ss} approaches the maximal initial flux (true initial rate), $j_0 = k_{catS}^*/(1 + K_{MS}^*)$, only when $k_d^P N_T \gg j_0$, unless $K_{MS}^* \ll 1$ and $K_P^* \gg K_{MS}^*$ when it is sufficient that $k_d^P N_T > j_0$. With strong product inhibition, $K_P^* < K_{MS}^*$, $\beta < 0$, the more stringent requirement is $k_d^P N_T \gg j_0$ to avoid exchange limitation. Thus when $k_d^P N_T < j_0$, the enzyme flux j_{ss} is determined mostly by the exchange rate and the catalytic properties of

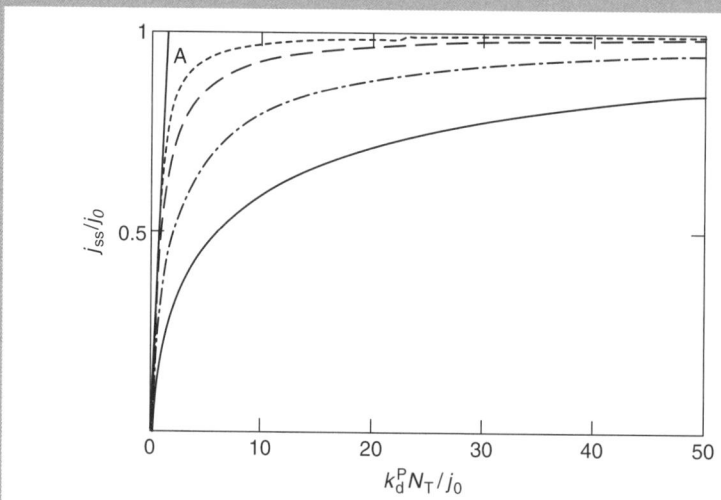

Figure 8.3 Activation by product desorption. The steady state rate, j_{ss}, relative to the true initial rate, j_0, as a function of $k_d^P N_T/j_0$ for some values of K_{MS}^* and K_P^*. Panel A: the straight solid line is for $K_{MS}^* = 0$ where $j_{ss} = k_d^P N_T$ for $k_d^P N_T < j_0$ and $j_{ss} = j_0$ otherwise; all curves from upper to lower, $\beta = 1, 1.5, 5, -0.5, -0.1$.

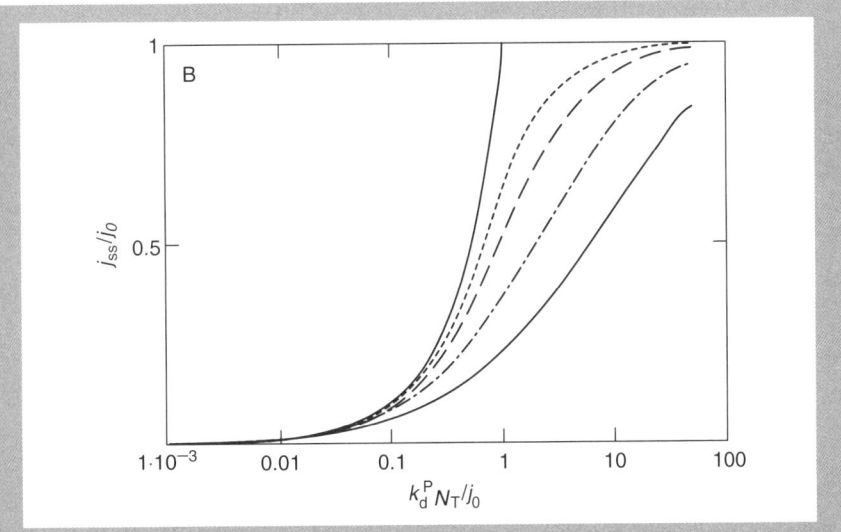

Figure 8.3 (*continued*)   Panel B: same data as in panel A with a logarithmic scale on the *x* axis. All parts of the curves that are lower than 1 show a significant dependence on the product exchange kinetics. The part of the curves that follow the line for $\beta = 1$ are exchange limited with $j_{ss} \approx k_d^P N_T$

the enzyme have little impact. Increasing the exchange rate k_d^P leads to activation by replenishment.

For micelles of short chain phosphatidylcholines the order of magnitude estimates support the assertion that the observed rates are not affected by product accumulation: $K_{MS}^* = 1$, $K_P^* = 0.1$, giving $\beta = -0.2$. Also, the other stringent condition is satisfied because with $K_P' > 1$ mM and the diffusion-limited exchange rate, k_d^P is estimated (Theory Box 2.1) to be greater than 10^4 s^{-1}, compared to $j_0 = 1000$ s^{-1} and $N_T = 100$.

8.2 The Apparent Rate Parameters for the Pig Pancreatic PLA2

The experimental results of the type shown in Figure 8.1, interpreted in terms of the model outlined above in Figure 8.2, allow an analysis of the interfacial turnover events:

(a)   Values of K_M^{app} and k_{cat}^{app}, obtained from the bulk micellar substrate concentration dependence of the interfacial rate (Table 8.1), are related to K_d, K_{MS}^*, K_S', K_{MS} and k_{catS}^* (Equations 8.3 and 8.4). The dissociation constant for E*, K_d, is independently measured using dispersions of a neutral diluent (Chapter 6). Values for K_{MS}^* (Table 8.1) are obtained from the inhibition kinetics (Equation 9.5 in Chapter 9).

(b)   Even though the branch ES $\rightleftarrows$ E*S has been blocked in the catalytic scheme (Figure 8.2), the relationship between the E, ES, E* and E*S species are

Table 8.1 The kinetic parameters for the hydrolysis of micelles of short chain phospholipids by pig pancreatic PLA2

Substrate[a]	in [NaCl]	K_M^{app} (mM)	k_{cat}^{app} (s^{-1})	K_{MS}^* (MF)	k_{catS}^* (s^{-1})
DC$_8$PC	1 mM	1.44	250	0.035	270
	4 M	0.06	2200	0.07	2350
DC$_8$PM	1 mM	0.16	1260	0.015	1450
	4 M	0.18	900		
DC$_7$PC	1 mM	2.3	16	0.07	16
	0.1 M	2.0	28	0.05	28
	4 M	0.25	660	0.13	660
DC$_7$PM	1 mM	0.1	700	0.05	700
DC$_6$PC	1 mM	>10	5.5		
	4 M	0.85	265	0.2	300
DC$_6$PM	1 mM			0.15	250

[a]DC$_n$-, 1,2-diacyl-*sn*-glycero-3-; PC, phosphocholine; PM, phosphomethanol.

determined from the detailed-balance condition (Equation 6.7). This provides a basis for analyzing K_S^* activation as discussed in Chapter 6.

(c) The effect of added NaCl on the apparent parameters shows that the interfacial charge has a direct effect on the chemical step, i.e. the k_{catS}^* activation.

8.2.1 Resolution of the anomalous contribution of the monomer rate

The hyperbolic dependence of the observed rate at the subsaturating concentrations of the micellar substrate is shown in Figure 8.4. These conditions, where the maximum rate is low, are chosen to illustrate the strategy to eliminate the anomalous contributions to the observed rate with the monodisperse substrate. According

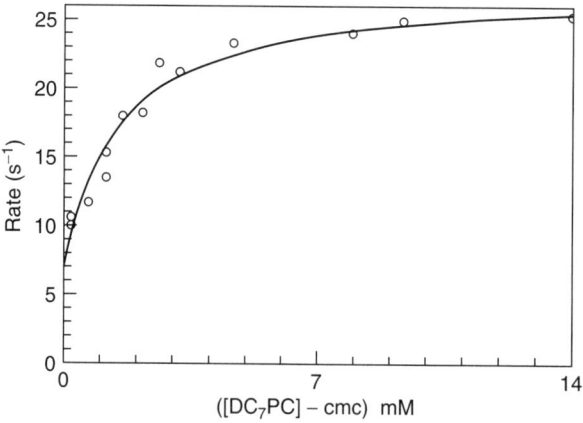

Figure 8.4 The micellar substrate concentration dependence of the rate of hydrolysis of DC$_7$PC by pig pancreatic PLA2 in 0.1 M NaCl. The curve represents the three-parameter fit to Equation (8.2) to obtain k_{cat}^{app}, K_M^{app} and k_{cat}^{mono}. Reprinted with permission from Berg *et al.* (1997). Copyright (1997) American Chemical Society

to the general kinetic scheme the observed rate with micellar dispersions has two components, the interfacial turnover and the 'monomer' turnover. The k_{cat}^{mono} contributions from the rates due to the turnover outside the micellar interface in the aqueous phase and the micellar interface together are related to the apparent turnover parameters at the micellar interface as given by Equation (8.2).

As shown in Figure 8.4, the apparent stirred rate at the CMC of the substrate is significant and increases modestly with the increasing micellar substrate concentration at 0.1 M NaCl. Of course, k_{cat}^{mono} calculated from such fits is the maximum premicellar rate through all possible kinetic paths including the reaction through monodisperse ES, through the premicellar complexes and the reaction at the extraneous interfaces. Obviously, other methods have to be devised to find out the kinetic origins of k_{cat}^{mono}, but the analytical strategy in Equation (8.2) is useful to eliminate all such apparent monomer rate contributions.

The contributions to the activating effect of NaCl on k_{cat}^{app} can be dissected (Figure 8.5). The fact that the unstirred rate with monodisperse substrate (Figure 3.6) is negligible shows that the pancreatic PLA2 functions most efficiently on aggregate substrate. As summarized in Table 8.1 for several phosphatidylcholines, k_{cat}^{app} at the zwitterionic interface increases 30- to 100-fold in the presence of 4 M NaCl. Recall that NaCl lowers the CMC (Figure 2.5), and a similar salting-out effect of added NaCl is seen on the total substrate concentration dependence of the PLA2 catalyzed rate (Figure 8.1). In all such cases, irrespective of the NaCl concentration in the reaction mixture, the rate increase is always seen above the CMC. Since added NaCl does not have a significant effect on the unstirred rate seen with monodisperse substrate, the effect of NaCl on the rate with the zwitterionic micellar interface is due to an effect on the intrinsic parameter for the turnover at the interface. The analysis described in Section 8.4 shows that indeed the interfacial anionic charge has an allosteric effect on the chemical step of the turnover cycle. Results with some interesting limits are outlined below.

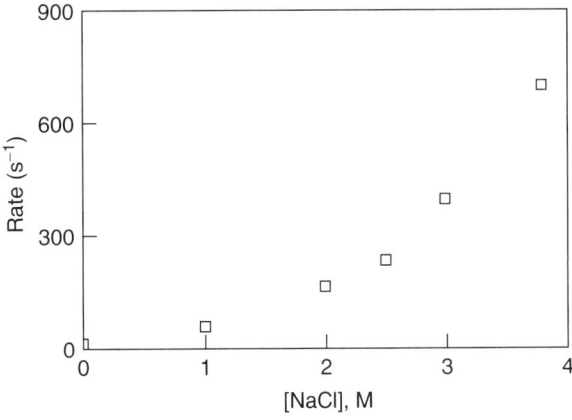

Figure 8.5 Effect of added NaCl on the maximum rate of hydrolysis at the saturating concentration of micellar DC₇PC. Reprinted with permission from Berg *et al.* (1997). Copyright (1997) American Chemical Society

8.2.2 Hydrolysis of monodisperse anionic substrate via premicellar aggregate

Hydrolysis of anionic phospholipids by pancreatic PLA2 does not exhibit many of the anomalous kinetic characteristics observed with the zwitterionic phospholipids (Van Oort *et al.*, 1985a, 1985b; Jain and Rogers, 1989; Rogers *et al.*, 1996). For example, as summarized in Table 8.1, the rate of hydrolysis of anionic short chain phosphatidylmethanol does not change with added NaCl. In fact, as shown in Figure 8.6, the rate of hydrolysis of short chain anionic phospholipids increases monotonically with the substrate concentration. Only a barely detectable increase in the rate is seen at the CMC, and most of the substrate concentration-dependent increase in the observed rate is seen below the CMC. As compared in Table 8.1, the maximum rate with the anionic substrates is virtually the same as with the zwitterionic substrate at 4 M NaCl. The CMC effect is not seen during the hydrolysis of structurally diverse anionic phospholipid substrates by pancreatic PLA2. Also certain snake venom PLA2 show a higher premicellar rate with the phosphatidylcholine (Van Eijk *et al.*, 1983). Whether a monomer kinetic path operates under any set of conditions with any interfacial enzyme remains to be demonstrated.

The higher rates of hydrolysis by PLA2 observed under certain conditions with a monodisperse substrate are not inconsistent with the interfacial mechanism (Figure 3.7). Even when the bulk of the anionic amphiphile in the assay mixture is monodisperse, many, if not all, secreted PLA2 form premicellar aggregates with anionic substrate amphiphiles. Certain PLA2s also form premicellar aggregates with the zwitterionic amphiphiles. Independent evidence shows that a premicellar aggregate of the enzyme with the substrate or mimic amphiphiles is formed in all cases where the monodisperse substrate is hydrolyzed. Although the structural details of such premicellar aggregates are not established, their kinetic behavior is virtually identical with that at the preformed interfaces. A detailed analysis of the interfacial turnover in the premicellar aggregates would require an understanding of the organization, partitioning and exchange behavior of the aggregates.

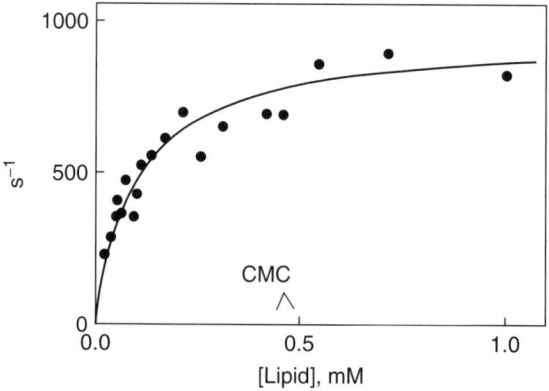

Figure 8.6 The total substrate concentration dependence of the rate of hydrolysis of dioctanoylphosphatidylmethanol by pancreatic PLA2. Reprinted with permission from Rogers *et al.* (1996). Copyright (1996) American Chemical Society

8.2.3 [S*] dependence due to fusion–fission of micelles

Increasing the concentration of [S*] would also increase the rate of fusion–fission of micelles. Consider the possibility that a minimum number of S* bound to E* are not accessible to the active site of E*. In the current model (Equations 8.1 to 8.4), enzyme-containing micelles will be quickly depleted of the accessible substrate before a fusion event takes place. Then at sufficient micelle-to-enzyme excess, most productive fusion–fission events will be when a depleted enzyme-containing micelle meets a fresh one. Assuming that fusion–fission occurs with rate $k_f[S^*]$ and that each such event brings in N_S new substrate molecules, the total enzyme rate would be $j = N_S k_f[S^*]$, if limited by replenishment. Only at sufficiently high substrate concentration will the rate be limited by the catalytic turnover if the replenishment through the monomer path is not sufficient. A more detailed model of replenishment by the fusion–fission process is analyzed in Theory Box 11.2.

8.3 K_S^* Allosteric Effects of the Interface

In standard enzymological terms there are two possible allosteric effects of the interface on the events of the interfacial catalytic turnover cycle. In this section we examine the consequences of the K_S^* allosteric effect of the interface on the substrate binding. As developed in the next section, the anionic charge at the interface also increases k_{catS}^* by transition-state stabilization. Based on the detailed balance condition, for the occupancy of the active site developed in Chapter 6, the interface has a K_L^* allosteric effect on the pancreatic PLA2. K_S^* activation with short chain substrate is unlikely to come from the scaffolding effect because the substrate chains will not extend out of the active site into the interface.

8.3.1 Quasi-equilibrium for the catalytic turnover in micelles

The activation factor in K_L^* (Section 6.7) cannot be applied to K_{MS}^* unless there is a true quasi-equilibrium ($k_{-1}^* \gg k_2^*$ so that $K_{MS}^* = K_S^*$; Equation 8.6 below) for the steps feeding into the rate-limiting chemical step. The compound parameter K_M^{app} (Equation 8.3) is a measure of the effective dissociation constant of the enzyme from the interface. Since it includes all the interface-bound forms of the enzyme, E*, E*S and E*P, it also includes all the steps of the catalytic cycle in the interface:

$$E^* + S^* \underset{k_{-1}^*}{\overset{k_1^*}{\rightleftharpoons}} E^*S \overset{k_2^*}{\longrightarrow} E^*P \underset{k_{-3}^*}{\overset{k_3^*}{\rightleftharpoons}} E^* + P^*$$

with the substrate–enzyme association and dissociation with rate constants k_1^* and k_{-1}^*, the chemical step k_2^* and product–enzyme dissociation and association rate constants k_3^* and k_{-3}^*. The contribution of k_{-3}^* is negligible when the product rapidly leaves the interface. Based on the magnitude of the oxy/thio element effect, k_2^* is rate-limiting in k_{catS}^*, which requires that $k_3^* \gg k_2^*$, as discussed in Chapter 5. In this limit, $k_{catS}^* = k_2^*$ and from Equations (5.11) and (5.12),

$$K_{MS}^* = \frac{k_{-1}^* + k_2^*}{k_1^*} = K_S^* \left(1 + \frac{k_{catS}^*}{k_{-1}^*}\right) \qquad (8.6)$$

Because k_{catS}^* is strongly salt-dependent (Section 8.4) and K_{MS}^* and K_S^* are not, the last equality on the right-hand side suggests that $k_{-1}^* \gg k_{catS}^*$, so that $K_{MS}^* = K_S^*$ for the hydrolysis of micelles of short-chain phospholipids by PLA2. In addition, $k_{-1}^* \gg k_{catS}^*$ is consistent with the observations that the values of K_S^* are within a factor of two of K_{MS}^* and that K_{MS}^* does not have the oxy/thio effect. Values of K_{MS}^* are less than 1 (Table 8.1), so that $k_1^* > k_{-1}^* \gg k_{catS}^*$ should hold and the chemical step must be the slow step for interfacial turnover. This conclusion differs from the results for DMPM vesicles where $K_{MS}^* \gg K_S^*$ and substrate release from E*S is slow, $k_{-1}^* \ll k_{catS}^*$. This analysis suggests that the chain length of the substrate may have a significant influence on the magnitude of the association and dissociation rates, k_1^* and k_{-1}^*, although the equilibrium constant $K_S^* = k_{-1}^*/k_1^*$ is fairly independent of chain length.

8.4 Analysis of the Interfacial Anionic Charge Preference: k_{catS}^* Activation

A kinetic model for the origin of the enhanced activity of pancreatic PLA2 at the anionic interface is shown in Figure 8.7. Results outlined below show that the overall effect is based on an enhanced binding of the enzyme to the anionic interface as well as a direct effect on k_{catS}^* (Berg *et al.*, 1997; Yu *et al.*, 2000b). The basis for the anionic charge preference under the kinetic conditions is a particularly difficult problem to resolve. To appreciate the complexity of the underlying problem, consider the fact that the processive turnover in the scooting mode is observed with the anionic substrate interface (Chapters 4 and 5) but not at the zwitterionic interfaces. Also,

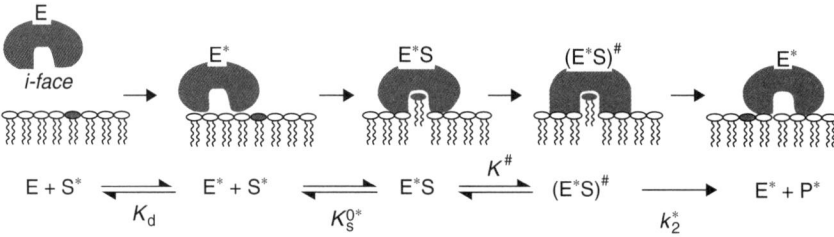

Figure 8.7 The two sequential state models for the k_{catS}^* activation of pancreatic PLA2 by the interfacial anionic charge. The enzyme in the aqueous phase (E) binds to the interface as a prelude to the catalytic turnover at the interface through the species marked with an asterisk (E*, E*S). In this minimal modification to the interfacial kinetic paradigm, the ternary Michaelis complex exists in two forms, with $K^\# = $ [E*S]/[E*S]$^\#$. Only (E*S)$^\#$ undergoes the chemical change. Thus E*S predominates in the wild-type enzyme (WT) at the zwitterionic interface; it is converted by charge compensation of certain cationic residues to the activated (E*S)$^\#$ form at the anionic interface, or by added NaCl, or by the addition of lysine to methionine substitutions

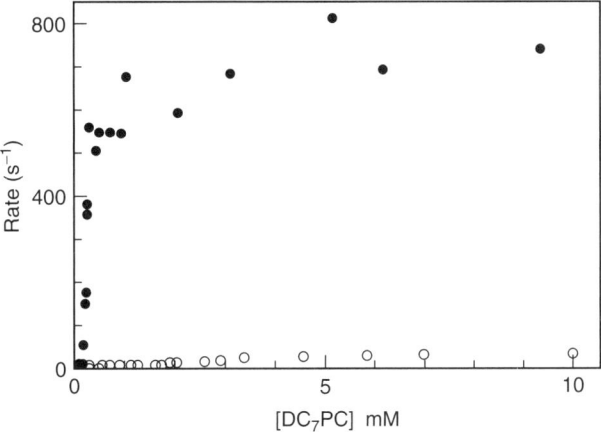

Figure 8.8 The DC$_7$PC concentration dependence of the rate of hydrolysis by pig pancreatic PLA2 in the presence of 0.1 M NaCl (open circles) or in the presence of 4 M NaCl (filled circles). Adapted with permission from Berg *et al.* (1997) and Rogers *et al.* (1998). Copyright (1997 and 1998) American Chemical Society

the rate of hydrolysis at the zwitterionic interface is enhanced in the presence of anionic additives, including the products of hydrolysis. Binding of the enzyme to the interface is a necessary precondition for the interfacial turnover. After binding of the enzyme to the zwitterionic interface, it is not possible to rule out an activating effect of the locally accumulated product on the interfacial turnover as well as the delay in the reaction progress at the monolayer and bilayer of phosphatidylcholines (Chapter 10).

Kinetic studies with micelles of short chain phospholipids have provided insights into the basis for anionic charge preference. The observed rate at the zwitterionic micellar substrates is lower than that with the anionic substrates (Table 8.1). Also, as shown in Figure 8.8, the maximum rate for the hydrolysis of zwitterionic micelles increases in the presence of 4 M NaCl. In terms of the interfacial turnover in the quasi-scooting mode (Figure 8.2), these results raise the possibility that in 4 M NaCl the zwitterionic phospholipid interface becomes anionic (compare the results in Figures 8.6 and 8.8), for which evidence is developed in the next section.

8.4.1 Possible consequences of product accumulation in zwitterionic micelles

The effect of accumulated products at the enzyme-containing zwitterionic interface during the steady state turnover may be significant (Theory Box 8.1). For example, even if the partitioning equilibrium of the product is low, the possibility remains that at least some product remains at the interface during the steady state turnover. This raises uncertainty about the true rate of hydrolysis on zwitterionic micelles in the total absence of the anionic charge at the interface. For example, the rate of hydrolysis of zwitterionic phosphatidylcholine in the absence of NaCl is small but significant (Figure 8.8), which could be due to the anionic charge induced by product accumulation. This assertion is consistent with the results that k^{app}_{cat} in the absence of NaCl increases with the chain length of the substrate because more hydrophobic products are more likely to accumulate in the steady state. Together these results suggest that the potential kinetic problems from product accumulation during the steady state hydrolysis of DC$_7$PC micelles are not insignificant. Even though the observed rate with zwitterionic micelles in the absence of an additive may be several hundred-fold higher than the monomer rate, 15 versus 0.1 s^{-1}, it is still likely that there is enough charge accumulation in the enzyme-containing micelle for the $K^{\#}$ activation (Figure 8.7).

8.4.2 Activation by added NaCl

The activating effect of accumulated product is likely to be relatively insignificant in the presence of 4 M NaCl, which induces an anionic charge at the zwitterionic micellar interface. Coupled with a rapid-substrate replenishment assumption, the following observations provide a basis for developing a model (Figure 8.7) for the NaCl-induced activation of pancreatic PLA2:

(a) Direct binding studies show that the binding of PLA2 to the anionic interface, in the E* form without the occupancy of the active site, is significantly higher than to the zwitterionic interface. Also, the affinity for zwitterionic micellar or bilayer interfaces increases 25-fold in the presence of 4 M NaCl.

(b) The maximum rate increases and $K_{\text{M}}^{\text{app}}$ decreases with increasing NaCl concentration for the hydrolysis of zwitterionic micelles. Virtually all the effect of NaCl on $K_{\text{M}}^{\text{app}}$ is due to a change in K_{d} (cf. Equation 8.4). Independent measurements show that K_{MS}^{*} does not change significantly with added NaCl.

(c) The onset for the increase in the observed rate above the CMC tracks the micelle concentration at all [NaCl], and the presence of salt has no noticeable effect on the apparently negligible monomer rate.

(d) The chemical step for the hydrolysis of the anionic and zwitterionic substrates remains rate-limiting with or without added NaCl.

(e) Virtually all the NaCl-dependent increase in $k_{\text{cat}}^{\text{app}}$ is attributed to a change in k_{catS}^{*} because K_{MS}^{*} does not change (cf. Equation 8.3).

(f) $k_{\text{cat}}^{\text{app}}$ is maximum for the anionic dinonanoylphospholipid or the corresponding zwitterionic phospholipid in the presence of added NaCl. Both the shorter and longer chain substrates have a lower $k_{\text{cat}}^{\text{app}}$.

Based on these observations, the activating effect of added NaCl on the hydrolysis of zwitterionic micelles is attributed to the enhanced affinity of the enzyme for the interface and to an increase in k_{catS}^{*}. As developed in the next section, the results with site-directed mutants show that virtually all the salt effect is due to the charge compensation of only three of the 12 cationic residues on bovine pancreatic PLA2 (Table 8.2). The results are consistent with the k_{catS}^{*} activation model (Figure 8.7), according to which the charge compensation of certain cationic residues on the

Table 8.2 Effect of CMC and added NaCl on secreted PLA2 catalyzed hydrolysis of DC_7PC micelles

Type PLA2 (K or R changed in bovine PLA2[a]	Effects	
	CMC	NaCl
IB Pancreas of		
Bovine	Yes	Yes
Pig (12A, 57N)	Yes	Yes
Iso-pig (12T)	Yes	Yes
Horse (10Q, 12T, 43A, 53T, 57E, 113P)	Yes	Yes
Sheep (108E,113N)	Yes	Yes
IA Venom nmDEI (10H, 53D, 56E, δ62–66, 120N)	Yes	No
IIA Human (10I, 12I, 43F, 57D, 108C, 113A, 115A, 121N)	Yes	Yes
IIB Venom appPLA (53 G, δ62–66, 108D, 113S, 120P, 121L)	Yes	Yes
III Bee venom PLA2 (not homologous)	Yes	No

[a]Sequence information (Verheij *et al.*, 1981). Bovine PLA2 has Arg or Lys at 10, 12, 43, 53, 56, 57, 62, 108, 113, 115, 120, 121.

pancreatic PLA2 changes the inactive E*S form to a catalytically more efficient $(E^*S)^{\#}$ form.

8.5 The Structural Basis for the Anionic Interface Preference

The anionic charge preference of pancreatic PLA2 is relevant for the function in its physiological environment consisting of codispersed phospholipids and anionic bile salts (Figure 1.7). Other secreted PLA2s also show enhanced binding to anionic interfaces (Jain *et al.*, 1991b), but the rate enhancement by increased anionic charge at the interface appears to be less pronounced for some of these PLA2s (Table 8.2). Based on the positions of the cationic residues in these enzymes it can be deduced that, of the 12 cationic residues present in the bovine pancreatic PLA2, substitution of only three at the positions 53, 56 and 120 could account for virtually all of its anionic charge preference (Yu *et al.*, 2000b). For example, as shown in Figure 8.9, the K53,56,120M mutant is 15-fold more active than the wild type (WT). Note that the CMC effect is virtually the same as for the WT and the monomer rate is also virtually zero. The rate of hydrolysis of zwitterionic substrate in the presence of 4 M NaCl, or the rate with anionic substrates, is virtually the same (within a factor of two) with WT and the triple mutant. K_{MS}^* for the triple mutant is moderately lower, and virtually all the rate increase from the lysine to methionine substitution is due to a change in k_{catS}^*. As discussed below, the anionic charge effect is due to charge compensation of the three well-separated residues away from the catalytic site and possibly the i-face.

8.5.1 Additivity of the incremental effect on the primary parameters

The thermodynamics of the incremental change due to a particular substitution, irrespective of the mutations in other positions, provides a rigorous basis for

---

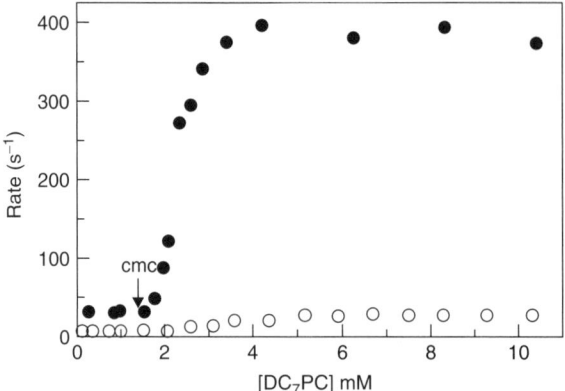

Figure 8.9 The DC$_7$PC concentration dependence of the rate of hydrolysis in stirred 1 mM NaCl by bovine pancreatic PLA2 (open circles) and its K53,56,120M mutant (filled circles). In the presence of 4 M NaCl the observed rate increases modestly for the triple mutant (not shown), whereas the effect on the wild-type bovine PLA2 is comparable to that shown for the pig enzyme in Figure 8.8. Reprinted with permission from Yu *et al.* (2000b). Copyright (2000) American Chemical Society

assigning the functional change to the structural effect of the substitution. Incremental activation effects of the lysine-to-methionine substitution in each of the four positions are shown by results summarized in Table 8.3. The key arguments for assigning the structural origin of the effect on k^*_{catS} of Lys-to-Met substitutions follow. These substitutions will lead to charge neutralization but can also induce steric changes in the enzyme, for example, due to side-chain packing. Salt addition, on the other hand, seems to lead only to charge neutralization. The following results are noteworthy:

(a) K^*_{MS} and K^*_I are virtually independent of the charge neutralization from salt addition. This suggests that the charge neutralization by itself does not influence the enzyme conformation, at least not in the E* and E*S forms, leaving the substrate affinity unchanged.

Table 8.3 Effects of mutation on rate activation and stability of Michaelis complex (Equation 5 to 7 in Theory Box 8.2)

Mutant	K_{charge}	$\Delta_m G^*_S$ kcal/mole	$\Delta_m (G^\# - G^*_S)$ kcal/mole
#1 WT	1	0	0
#2 K53M	2.7	−0.5	0.0
#3 K56M	4.4	−1.9	0.1
#4 K53, 56M	6.5	−2.5	0.4
#5 K120, 121A	1.2	−1.1	0.7
#6 K53, 56, 120M	6.0	−2.9	0.2
#7 K53, 56, 121M	7.0	−2.5	0.3
#8 K53, 120, 121M	2.7	−2.2	0.6
#9 K56, 120, 121M	4.0	−2.7	0.4
#10 K53, 56, 120, 121M	7.6	−2.9	0.3

(b)   The mutations have some effect on K^*_{MS} and K^*_I.

(c) k^*_{catS} increases with increasing charge neutralization, either through salt addition or residue substitution. All mutants reach roughly the same maximum k^*_{catS} when sufficient salt is added.

As a consequence of observations (a) and (b), the effects of the mutations can be subdivided into a charge compensation effect and a charge-independent conformational effect. The salt independence of K^*_{MS} and K^*_I together with the strong salt dependence of k^*_{catS} also suggests that $K^*_{MS} = K^*_S$ (cf. Equation 8.6). Thus K^*_{MS} can be used to calculate the stability of the E*S complex. Observation (c) shows that the steric changes induced by the mutations have no effect on the catalytic rate but only charge matters.

In terms of the two-state model (Figure 8.7), it follows that charge neutralization (compensation) affects $K^\#$, but not K^*_S, while the steric changes due to the mutations affect K^*_S, but not $K^\#$ and k^*_2. While it is possible to assign all charge effects directly to changes in k^*_2, the whole point of the model is to account for the activation effects through a special state, $(E^*S)^\#$, which serves as an activated state on the path to the transition state complex. Thus, $K^\#$ carries all the effects of charge compensation and $K^\# \ll 1$ must hold to account for the salt independence of K^*_{MS} (cf. Equation 3 in Theory Box 8.2).

At high salt, charge neutralization is already maximal and all changes in the standard free energy levels refer only to the binding properties of the two states. Furthermore, the mutations leave k^*_{catS} fairly invariant at high salt, suggesting that substrate binding effects are roughly the same in E*S and $(E^*S)^\#$. In the standard free energy diagram (Figure 8.10), this is seen by the approximate equality

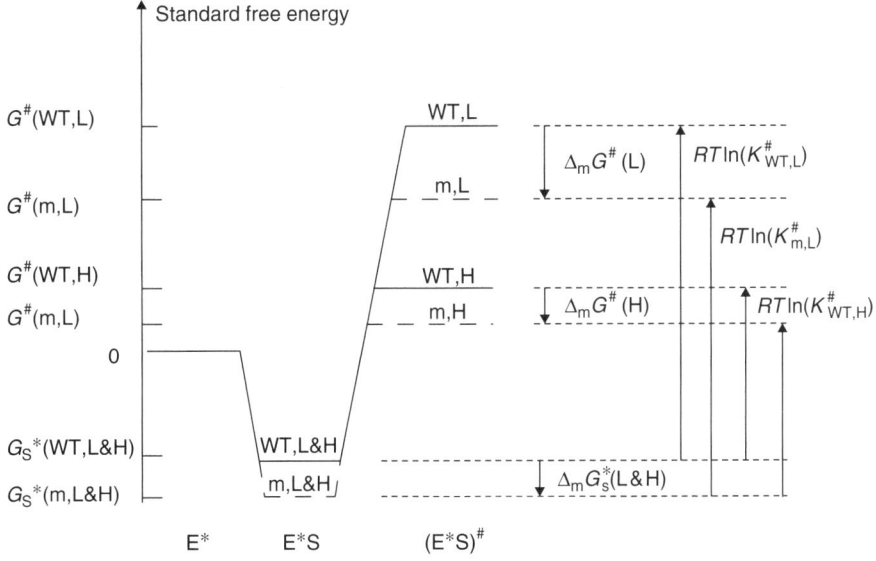

Figure 8.10   Standard free energy diagram for the activation $E^* - E^*S - (E^*S)^\#$ for WT and mutant (m) at high (H) or low (L) salt. Data show that the stability of E*S is fairly independent of salt (L&H), but influenced by mutation (m). $RT \ln(K^\#)$ is the activation free energy from E*S to $(E^*S)^\#$

between $\Delta_m G_S^*$ and $\Delta_m G^\#(H)$ for each mutation. However, there are some systematic deviations such that $\Delta_m G^\#$ is larger (by between 0 and 0.7 kcal/mol; Table 8.3) than $\Delta_m G_S^*$ in almost all cases; i.e. the mutations do not stabilize $(E^*S)^\#$ quite as much as E^*S. This suggests that the mutations bring the unbound enzyme E^* somewhat closer to the substrate-binding conformation than to the active conformation $E^\#$. It is as though some transition-state stabilization effect in the WT is relaxed in the mutants; however, this effect is small. Also note that virtually all substitutions always further stabilize the E^*S state, although the effect on the free energy is not additive.

The effect of charge neutralization can be calculated from the changes in $K^\#$ or, in this case equivalently, from $k_{catS}^* \approx k_2^* K^\#$. When mutations are added at high salt, only the steric differences contribute to the small change in k_{catS}^*. Consequently, by dividing out the binding effects calculated at high salt (H) from the results at low salt (L), we can estimate the charge neutralization effect at low salt for each mutant m:

$$K_{charge}(m) = \frac{k_{catS}^*(m, L)}{k_{catS}^*(WT, L)} \frac{k_{catS}^*(WT, H)}{k_{catS}^*(m, H)} \tag{8.7}$$

These results applied to the Lys-to-Met substitutions lead to the following conclusion. In Table 8.3, the mutation at 53 (i.e. #2 and #8) has $K_{charge} = 2.7$, those at 56 (i.e. #3 and #9) have $K_{charge} = 4.0$–4.4 and those with both 53 and 56 (i.e. #4, #6, #7 and #10) have $K_{charge} = 6.0$–7.6. These numbers correspond to the factor with which k_{catS}^* increases due to the charge neutralization alone after the effect on binding has been factored out. The presence or absence of mutations at positions 120 and 121 show no systematic effect on K_{charge}.

The clear separation of the substitution effects on K_{charge} supports the assumptions on which the calculation was based. Although the model cannot be proven in this way, it is consistent with the kinetic data. The analysis suggests that Lys-53 and Lys-56 have strong charge compensation effects, while 120 and 121 do not. However, 120 and 121 do contribute to conformational changes in the protein that strengthen substrate binding and decrease k_{catS}^*. The main effect of removing the charged residues at 120 and 121 may be to strengthen the interface binding and lower K_S^* of the enzyme at the zwitterionic interface. The stabilization of the active complex, $(E^*S)^\#$, caused by the charge neutralization must affect all active states up to and including the transition state; otherwise it would not have a full impact on k_{catS}^*. Thus charge neutralization leads to transition-state stabilization.

THEORY BOX 8.2 Two-state model for k_{catS}^* activation in the interface

Starting from the two-step sequential scheme in Figure 7.12, assuming that all enzyme is in the interface and that all steps prior to the chemical step are equilibrated, the flux per enzyme is

$$j = \frac{k_2^* \dfrac{X_S^*}{K_S^{*0}} K^\#}{1 + K^{\#0} + \dfrac{X_S^*}{K_S^{*0}}(1 + K^\#)} = \frac{\dfrac{k_2^*}{1 + 1/K^\#} X_S^*}{X_S^* + K_S^{*0} \dfrac{1 + K^{\#0}}{1 + K^\#}} \tag{1}$$

Thus, the interfacial MM parameters can be identified as

$$k_{catS}^* = \frac{k_2^*}{1 + 1/K^\#} \xrightarrow[K^\# \ll 1]{} k_2^* K^\# \tag{2}$$

$$K_{MS}^* = K_S^* = K_S^{*0} \frac{1 + K^{\#0}}{1 + K^\#} \xrightarrow[K^\# \ll 1]{} K_S^{*0}\left(1 + K^{\#0}\right) \tag{3}$$

If the active state, $(E^*S)^\#$, is unfavorable, the limits as given by the arrows in Equations (2) and (3) are reached. K_{MS}^* corresponds to the effective dissociation constant for substrate in the interface, K_S^*. When both $K^\# \ll 1$ and $K^{\#0} \ll 1$, also $K_S^{*0} = K_S^*$ and the activation scheme in Figure 7.12 reduces to that of Figure 8.7. As argued in the text, this is the reasonable limit for PLA2 and its mutants. In this way all differences are assigned to allosteric changes in the equilibria prior to the catalytic step. However, $(E^*S)^\#$ appears as an activated state, and it would be equally possible to assign some or all differences in k_{catS}^* directly to changes in k_2^*.

If the standard free energy of E^* is set equal to zero, the standard free energy of E^*S is

$$G_S^* = RT \ln(K_{MS}^*) \tag{4}$$

and that of $(E^*S)^\#$ is

$$G^\# = RT \ln\left(\frac{K_{MS}^*}{K^\#}\right) = -RT \ln\left(\frac{k_{catS}^*}{k_2^* K_{MS}^*}\right) \tag{5}$$

In the last equality, $k_{catS}^* = k_2^* K^\#$ (Equation 2) was used. The standard free energy changes from mutations (m) at high or low salt can be calculated as

$$\Delta_m G_S^* = G_S^*(m) - G_S^*(WT) = RT \ln\left(\frac{K_{MS}^*(m)}{K_{MS}^*(WT)}\right) \tag{6}$$

for the E^*S complex and

$$\Delta_m G^\# = G^\#(m) - G^\#(WT) = -RT \ln\left(\frac{k_{catS}^*(m)}{k_{catS}^*(WT)} \frac{K_{MS}^*(WT)}{K_{MS}^*(m)}\right) \tag{7}$$

for the $(E^*S)^\#$ complex.

8.6 Modeling the i-face of PLA2

The interfacial kinetic paradigm is about the coupling between the interfacial binding of the enzyme and the catalytic turnover events at the interface. The phenomenological basis for the interfacial turnover is provided by the scooting mode reaction progress with high processivity. The detailed balance condition provides clear evidence that the binding to the interface and the occupancy of the active site are coupled for the K_S^* activation. Also k_{catS}^* activation by the anionic charge at the interface is attributable to charge compensation of selected residues. Having established the functional basis of PLA2, we present results that have provided insights into the identity of the i-face of PLA2 and its relationship to the catalytic site.

8.6.1 Conceptualization of the i-face versus the catalytic site

From the initial X-ray structure of pancreatic PLA2 it was noted that the catalytic His-48 is accessible through a slot on a face of the protein (Dijkstra *et al.*, 1981). In this interface recognition region the active site slot and the surrounding rim contained hydrophobic residues, which were surrounded by a collar of cationic residues. The topography of the recognition region was such that its interactions with the interface required penetration of the protein well beyond the glycerol backbone region of the interface to access the substrate. This means energetically unfavorable disruption of the interface organization and accommodation of the acyl chains in contact with the protein surface. Based on the following observations a slightly tilted but remarkably flat face of PLA2 was suggested as the i-face that makes the high affinity contact with the interface (Ramirez and Jain, 1991):

(a) The scooting mode kinetics showed high affinity binding of virtually all secreted PLA2 with anionic vesicles. The magnitude of the interaction energy for the binding to the anionic interface suggested that although charged residues at the collar may participate in the long-range electrostatic interactions, the E to E* binding is dominated by close-range specific and hydrophobic interactions along the i-face.

(b) Fluorescence properties of Trp-3 and the membrane-localized dansyl probe showed that the contact region between the interface and the bound enzyme is excluded from the bulk aqueous phase and possibly desolvated. The chemical modification, spectroscopic and quenching results also suggest that certain amino acid residues in E* and E*L forms of the enzyme become inaccessible from the aqueous phase.

(c) The gel–fluid transition properties of the bilayer in contact with the bound enzyme suggest that the i-face creates its own microenvironment without disrupting the bilayer organization. This conclusion is also consistent with the observation that the gel–fluid or bilayer–micelle phase change has little effect on the intrinsic rate of catalytic turnover.

(d) The binding of PLA2 to interfaces is rapid, and does not depend on the bilayer or micellar state of the interface.

(e) The binding stoichiometry suggested that about 40 amphiphiles at the interface support the binding of one enzyme molecule. Considering the area per phospholipid molecule to be about 45 Å^2, the area of the contact surface would be about 1800 Å^2, in good agreement with the 1700 Å^2 area of the assigned i-face with about 1400 Å^2 in the hydrophobic close contact.

(f) The dependence of the E to E* equilibrium at the zwitterionic interface to the presence of the anionic additives or the products of hydrolysis showed a maximum at 12 mole % additive. The stoichiometry of 1:8 suggests that about five anionic sites may be present at the i-face that makes contact with a total of 40 phospholipid molecules.

These results provided a functional basis for formulating the interactions of the i-face of PLA2 (Ramirez and Jain, 1991; Zhou and Schulten, 1996; Pan *et al.*, 2001). In the proposal for the interfacial recognition site (Dijkstra *et al.*, 1981), the hydrophobic face was significantly distorted and uneven. To make a close contact to within 5 Å with the interface, such a surface must significantly distort the organization of the polar groups at the interface, which may be energetically unfavorable. In contrast, a remarkably flat nonpolar surface of the i-face can make close contacts to within 5 Å with the interface. Another significant difference is that residues 31, 53 and 56 are at the edge of the i-face, and thus they remain accessible from the bulk aqueous phase.

The high affinity binding of PLA2 to the interface implies significant hydrogen-bonding interactions and the hydrophobic drive along the i-face would desolvate the close-contact region by substituting the H-bonding and dipolar interactions of water molecules at the i-face with the functional groups at the substrate interface. Possibly, the dehydrated microinterface along the i-face facilitates the access of the substrate from the interface to the active site by precluding unfavorable contacts of the acyl chains of the substrate with the bulk aqueous phase. This model of the i-face has been a useful guide for a decade, and recently it has been found to be identical to the contact face of the crystallographic dimer of pig pancreatic PLA2 in containing five coplanar divalent anions (Figures 8.11 and 8.12).

8.6.2 The anion-assisted dimer of PLA2 has five coplanar anions

The sulfate or phosphate anion-assisted dimer structure of pig pancreatic PLA2 has provided significant insights into the atomic level interactions along the i-face (Pan *et al.*, 2001). The nature of the interactions along the i-face and the interface may complement the interactions of the subunit contact face in the crystallographic dimer. As shown in Figure 8.11, only one of the subunits of the dimer has one MJ33 molecule. The contact face of the dimer is virtually free of bound water molecules. It has five anions in the same plane and the hexadecyl chain of MJ33 passes through the center of the plane. This would suggest that the active site is accessible through the slot surrounded by hydrophobic residues. The five anion binding sites in the dimer bind sulfate or phosphate through the same ligands provided by the opposing faces of the dimer. Each of the anions is coordinated with from two to six ligands of the protein and also several water molecules are in close contact. All of the anion binding ligands are within 2.5 to 3.5 Å. Remarkably, only three of the five anions have the cationic ligands from the side chains of Arg-6 or Lys-10. Otherwise the anion binding is stabilized largely through the oxygen and nitrogen functional groups from the back bone and side chains of the protein. While the cationic residues on the protein participate in the binding of the anions, it is quite likely that the fixed anionic charges at the interface may come closer than 4 Å contact along the i-face via coordination to unionized potentially polarizable and hydrogen-bonding groups.

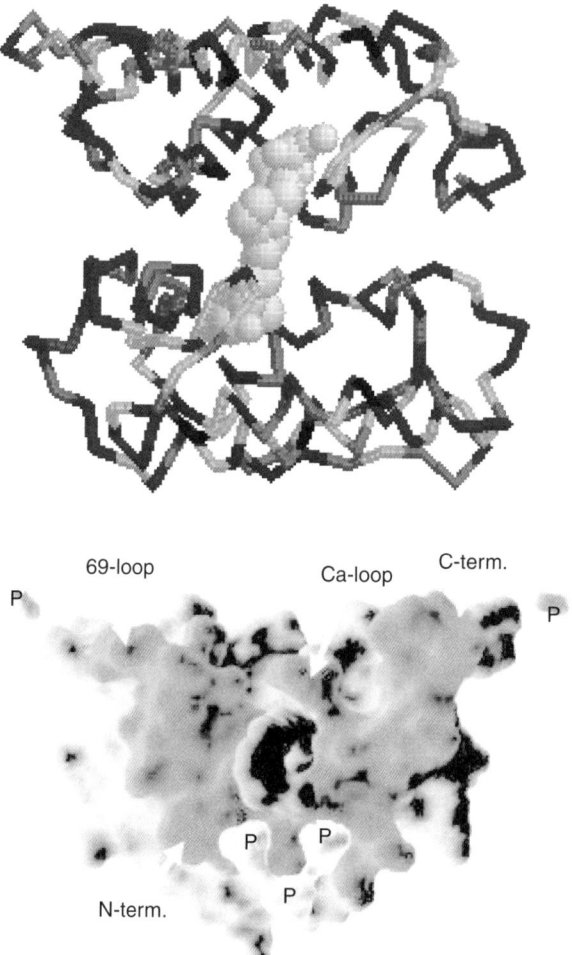

Figure 8.11   (Top) The backbone structure of the anion-assisted dimer of pig pancreatic PLA2 with a shared MJ33 molecule shown with the van der Waals sphere (PDB 1FXF and 1FXP). The contact face of the dimer is virtually identical to the putative i-face of PLA2 (Ramirez and Jain, 1991). (Bottom) The view of the subunit–subunit contact surface from a perspective orthogonal to the plane of the bound anions. The five bound phosphate anions are shown with P. In the center of the footprint is the gray alkyl tail of the MJ33 inhibitor. The darker shaded region represents the surface formed with the residues that are within 5 Å of the other subunit. Reprinted with permission from Pan *et al.* (2001). Copyright (2001) American Chemical Society

8.7   Site-Directed Mutagenesis to Discern Interactions Along the i-face

Scores of residues on PLA2 have been substituted by site-directed mutagenesis to evaluate their role in the binding of the enzyme to the interface as well as the active site events (Yu *et al.*, 1999b; Berg *et al.*, 2001). These studies confirm the independent identities of the active site and the i-face. Typically, substitutions at the i-face change

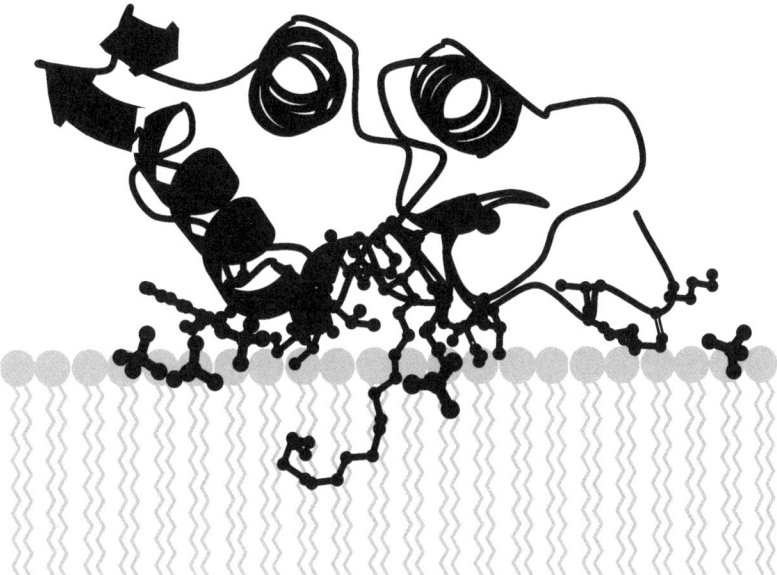

Figure 8.12   Conceptualization of the residues of the i-face of pig pancreatic sPLA2 that putatively interact with the phospholipid bilayer interface. The position of the five anions along the interface is shown along with the extension of the alkyl chain of MJ33. The figure was constructed by Professor Bahnson based on the PDB 1FXF structure from Pan *et al.* (2001)

the interface binding affinity by less than 10-fold, which is consistent with a model where the i-face makes multiple contacts with only modest contributions from each contact (Figure 8.12). If the interface binding occurred though a recognition site or pocket for a single amphiphile at the interface, considerably larger changes would be expected.

8.7.1   The Trp-3 environment

The lone tryptophan in pig and bovine pancreatic PLA2, located on the i-face near the active site slot, is a useful probe for the changes associated with the binding of the enzyme to the interface. Such studies show that in the E* and E*L forms Trp-3 becomes inaccessible to quenchers from the aqueous phase (Jain and Maliwal, 1993), and at least a part of the i-face environment is desolvated (Jain and Vaz, 1987). Substitution of Trp-3 by less hydrophobic residues lowers K_d by a factor of 2 to 20. It appears that Trp residues in membrane proteins tend to be in the interface binding region (Yau *et al.*, 1998).

8.7.2   Energetics of the substitutions at the *N* terminus

Results in Figure 8.13 show that the incremental energy change in the binding of the enzyme to the interface correlates with the change in the area due to the residue substitution. The volume changes associated with the residue substitution do not

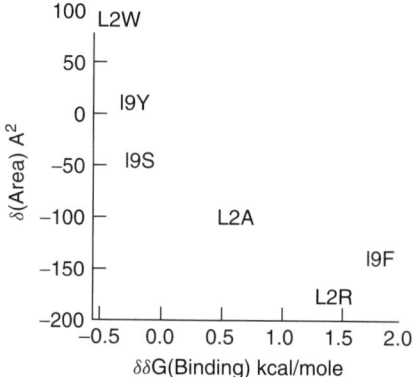

Figure 8.13 The relationship between the incremental free energy for the binding of PLA2 mutants to the interface as a function of the change in the hydrophobic area due to the substitution. Reprinted with permission from Yu *et al.* (1999b). Copyright (1999) American Chemical Society

correlate well with the changes in K_d (not shown). These results are consistent with the energetics of the area-wise contact; i.e. the hydrophobic effect is proportional to the area of the residue that is exposed to the bulk solvent in the E form of the enzyme. Such incremental changes (Table 7.3) due to the hydrophobic effect and related to the area are significantly smaller than those expected for a change in the environment of the whole side chain of an amino acid residue. The advantage of the correlation based on the residue substitution is that one does not require an *a priori* knowledge of the environment in which the residue may be partitioned.

8.7.3 Charge compensation versus surface electrostatics

The charge compensation as well as the charge reversal mutants of the cationic residues show a modest change in K_d (Berg *et al.*, 2001). Of course, the affinity is lower for the charge-reversal mutants and higher for the isosteric lysine-to-methionine substitutions. Since the electrostatic interactions are weak, long range and nonspecific, it is usually not possible to assign a specific role for the ionic residues at the i-face. The problem is further compounded by the fact that the binding of the anionic groups at the interface may not be through the cationic groups on the protein, as mentioned in the preceding section (also Chakrabarti, 1993).

8.7.4 Conformational reciprocity

Even though the secondary structures of PLA2 do not show a significant change, the changes in one region appear to bring about changes in the other regions, as seen in the catalytic function and the fluorescence emission from Trp-3. In the X-ray crystallographic structure, the side chains of residues 3, 53 and 56 and in the 69 loop are noticeably perturbed by the K120,121A mutation at the other end of the molecule. Similarly, in the K53,56M mutant, residues 3 and 120 and of the 69 loop are perturbed (Yu *et al.*, 2000b). Such a structural reciprocity of the changes in the side chains induced by the substitutions at the far-removed residues suggests that the side-chain conformer of the residues on the i-face are possibly generated through domain motions. Considering the role of residues 53 and 56 in the k^*_{catS} activation, the role of Trp-3 in the substrate binding, the role of Tyr-69 in the chemical step and a role of Lys-120 in the binding to the interfacial anion site, the reciprocity of the structural changes in these three well-separated regions suggests the possibility of long-range coupling between the domain motions.

8.7.5 Multiple conformational changes

The 14 kDa secreted PLA2s contain five to eight disulfide bridges, yet these molecules are remarkably plastic in the active site and flexible in the i-face regions of the structure. For example, the catalytic active site is able to accommodate structurally diverse active-site-directed inhibitors. The coordination number of calcium appears to be 7 in most PLA2s. However, the 26 to 34 calcium loop, which provides three backbone carbonyl oxygen coordinating ligands is flexible. The loop is located at the edge of the i-face, but it is also connected to residues 19 to 22 which make hydrophobic contacts with the interface. In effect, such contacts could change the orientation of the calcium loop and possibly influence k^*_{catS}.

On the other side of the same surface but at the edge of the i-face, a part of the 60 to 74 loop also makes contact with the interface. The regions of the i-face that could come in the hydrophobic and H-bonding close contact with the interface include the N-terminus residues including Trp-3, the 18 to 22 segment, the 62 to 69 loop and the C-terminus segment including residue 120 of bovine pancreatic PLA2. The solution structure of PLA2 is more flexible and the degree of flexibility decreases on the binding to the interface and on the occupancy of the active site. As a possible basis for the interfacial activation, note that residues 53 and 56, and the calcium loop are not directly part of the i-face. However, the charge compensation at 53 and 56 could influence the 69 loop, where Tyr-69 is known to control the rate-limiting chemical step.

8.8 Summary: Residues Involved in Charge Compensation

The kinetic basis of the enhanced activity of pancreatic PLA2 at the micellar interface under the rapid-replenishment conditions is resolved. The monomer rates with short chain phosphatidylcholines are virtually negligible. The micellar rates with these substrates show a 25- to 60-fold activation in the presence of 4 M NaCl. Since the products of hydrolysis can have a steady state level in the micellar interface that is well above the equilibrium level, the magnitude of the micellar rates in the total absence of anionic charge in the interface remains uncertain.

The observed rate of hydrolysis of DC$_n$PC micelles increases in the presence of 4 M NaCl; the rates are comparable to those obtained with micelles of the corresponding anionic phospholipids. The reaction progress under these conditions occurs in the quasi-scooting mode with rapid substrate replenishment. The effect of the steady state accumulation of the product under these conditions is expected to be relatively modest in comparison to the effect of the charge already present at the interface. The anionic charge at the interface promotes not only the binding of the enzyme to the interface but the analysis shows that most of this activating effect is due to k^*_{catS} activation by the charge neutralization or compensation on Lys-53 and Lys-56.

Functional resolution of the binding of PLA2 to the interface and the catalytic and activation events provide the rationale for structural identification of the active site (Chapter 5) and its functional coupling to the i-face events. A working model for the i-face is that the close contact along the i-face involves several residues with scores of phospholipid molecules. Additional details emerge from the observation that the putative i-face is virtually identical to the subunit contact face in the crystallographic dimer of PLA2 cocrystallized with MJ33 and divalent anions. In this model (Figure 8.12), to access the active site the substrate is dislodged from the bilayer through the dehydrated i-face contact region with the interface. Operationally, such close contact of the i-face also allows for high interfacial processivity. A degree of flexibility along the remarkably flat i-face probably accommodates a variety of phospholipid head groups that may be present at the interface. In effect, the allosteric coupling of the interface interactions with the catalytic events (Figure 8.7) requires multiples states of the bound enzyme.

9 Inhibition: Specific or Nonspecific

It is said, with half-a-dozen, constants, you can fit an elephant.
Today I shall powder my elephant's ears and paint his posterior red,
I'll turn all his toe nails with suitable shears and place a toupee on his head,
There will be a warm smile on my elephants face
As we are welcomed to Pachyderm's ball.
Yes it is a fit, yet we do not know, if it is fit for an elephant.

Collective Wisdom, based on a nursery rhyme

In enzymology, the term 'inhibition' of an enzyme-catalyzed reaction is applied only for the lower rates attributable to a specific change in the catalytic parameters resulting from a direct interaction of a solute with the enzyme to form an EI or E*I complex. Nonpolar and amphiphilic solutes can modulate one or more steps in Figure 3.1. Observed rates may also change, not only with changes in the intrinsic catalytic properties of the enzyme but also with the substrate accessibility and replenishment associated with changes in the properties of the interface. In such cases, the observed rate may increase or decrease, or both, as seen in a biphasic solute concentration dependence. It is possible to distinguish between the following modes for a solute-induced reduction in the rate of an interfacial enzyme (Table 9.1):

(a) *Specific* inhibitors interfere with the catalytic turnover events by binding to a specific site at the enzyme, such as the catalytic site or an allosteric site. For example, a *competitive* inhibitor competes for the substrate or cofactor binding site. The potency of a competitive inhibitor is related to the stability of E*S versus E*I at the interface, which results in an increase in the apparent K_{MS}^* without an effect on k_{catS}^*. Other modes of specific inhibition mechanisms according to the standard solution kinetic classification include: the inhibitor binds at a site other than the substrate binding site on E*S (*uncompetitive*) to impair both K_{MS}^* and k_{catS}^*, or binds to both E* and E*S (*noncompetitive*) to impair only k_{catS}^*. By the same convention, modulation of the *allosteric* site may also influence one or both turnover cycle parameters. Note that the kinetic criteria used for distinguishing the inhibition mechanism for the soluble enzymes cannot be applied for the interfacial enzymes because the significance

Table 9.1 Solutes that reduce the observed rate of hydrolysis by secreted PLA2

Site	Impaired	Solutes
Active site	S binding	In Figure 9.1
Cofactor site	Ca^{2+} binding	Zn^{2+}, Cu^{2+}, Cd^{2+}
Catalytic residue	k_{catS}^*	Alkylation of His-48
i-face	N-terminus NH_2	Gossypol
	multiple Lys residues	Manoalides, scalaradial
Cationic residues	Aggregation with E	Heparins and glycoconjugates
	Bind E	Proteins from venoms and inflammatory tissues
Interface	Higher K_d	Cationic (amphiphiles, proteins)
	?	Nonpolar solutes

of the concentration variables is different. The interfacial inhibitors dispersed in a two-dimensional solution are characterized by concentrations X_S^*, X_I^*, etc. A specific inhibitor could also bind the enzyme and impair its ability to bind substrate by blocking the enzyme–interface binding. Such an inhibitor would be apparently competitive, since it blocks substrate accessibility, but in reference to the bulk concentration of interfacial substrate, $[S_0^*]$, rather than X_S^*.

(b) *Nonspecific* inhibition results from a decrease in the interfacial turnover processivity or the substrate replenishment rate. Such inhibitors could affect the enzyme conformation without binding at a specific site. Also a complex of E with another macromolecule (or a substrate-free interface), not necessarily along the i-face or at the active site, could reduce the fraction of the enzyme at the interface. The effects on the substrate accessibility or the replenishment can sometimes be dissected only through independent controls. The distinction between specific and nonspecific inhibitors is not always clear; for the interfacial enzymes, it is more to the point to distinguish between inhibitors that affect the E to E* step and those that impair the E* to E*S step.

(c) *Covalent modifiers* include the reagents for the catalytic residues, substrate binding site, the i-face or the allosteric site. The same considerations also apply for the evaluation of the effect of site-directed mutagenesis.

9.1 Specific Inhibitors

Specific inhibitors with a high affinity for the enzyme active site are in demand for a variety of purposes. They have been used for mechanistic and structural studies, in bioassays to block specific metabolic steps, and as drugs to compete with a metabolite. In Chapters 5 and 6 we considered the kinetic consequences of the active-site-directed inhibitors of PLA2. Such inhibitors (Figure 9.1) have provided rigorous kinetic tests for the specific inhibition of the interfacial turnover cycle, and they have also been useful for the characterization of the active site of the enzyme by biochemical, biophysical and crystallographic methods. The interface binding properties of PLA2 with an occupied active site have also provided insights into the detailed balance condition that constitutes the basis for identifying K_L^* activation. Additional considerations are developed below.

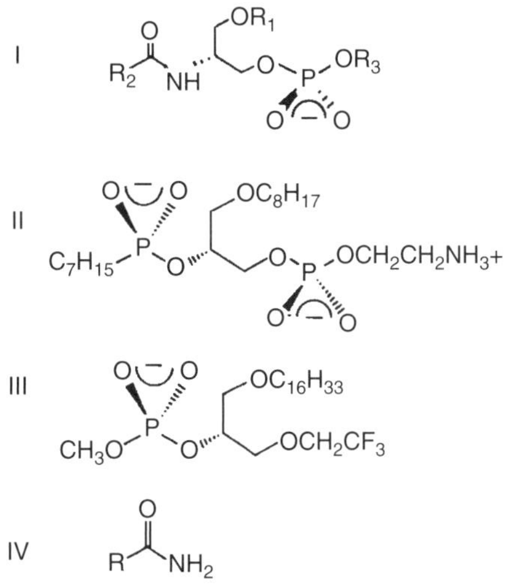

Figure 9.1 Competitive inhibitors of secreted PLA2 that interact with the calcium cofactor and His-48 in the catalytic site. Structural variations on the *sn*-2-amides (Bonsen *et al.*, 1972; de Haas *et al.*, 1990, 1993, 1995), *sn*-2-phosphonates II (Lin and Gelb, 1993), the *sn*-2-phosphates of MJ series III (Jain *et al.*, 1991d) and the fatty acid amides IV (Jain *et al.*, 1992a; Seshadri *et al.*, 1994)

9.1.1 Concerns for the interfacial inhibitor design

The specificity and high affinity of the active-site-directed inhibitors (Figure 9.1) comes from the multiple contacts they make with the enzyme (Figure 9.2). The active-site-directed mimics, especially the analogs of the rate-limiting transition state, are among the most specific and potent inhibitors. Although a natural metabolite is likely to be recognized to varying degrees by other enzymes involved in its metabolism, the underlying chemistry, and therefore the intermediates and transition state, is different. In other words, with the same ground state conformation of the substrate, the conformers for the interactions along the reaction path are expected to be quite different. In this sense the mimics of the intermediates, especially those of the rate-limiting transition state, are likely to be significantly different. By paying attention to such differences, it has been possible to obtain inhibitors that show a million-fold difference in their specificity towards the evolutionarily divergent relatives.

High affinity, or a low K_I^* value in mole fraction units, for an interfacial inhibitor ensures a low mole fraction, $X_I^*(50)$, for 50% inhibition. As developed later (Section 9.6), the mole fraction of an inhibitor in the microenvironment of the target enzyme depends not only on $[S_T]$ and $[I_T]$ in the reaction mixture but also on the value of K_I' for the partitioning of the inhibitor. For designing the inhibitor screens, it is critical to reduce potential nonspecific effects of a solute on the substrate interface. For example, the effects resulting from changes in the processivity, due to a shift in the E to E* equilibrium, are minimized in scooting mode assays (Chapter 5). For the interpretation of bioassay results, the accessibility and X_I^* and X_S^* in the microenvironment of the enzyme are not readily established because of the presence of the membrane and compartments (Figure 1.1). The problem in cellular and tissue

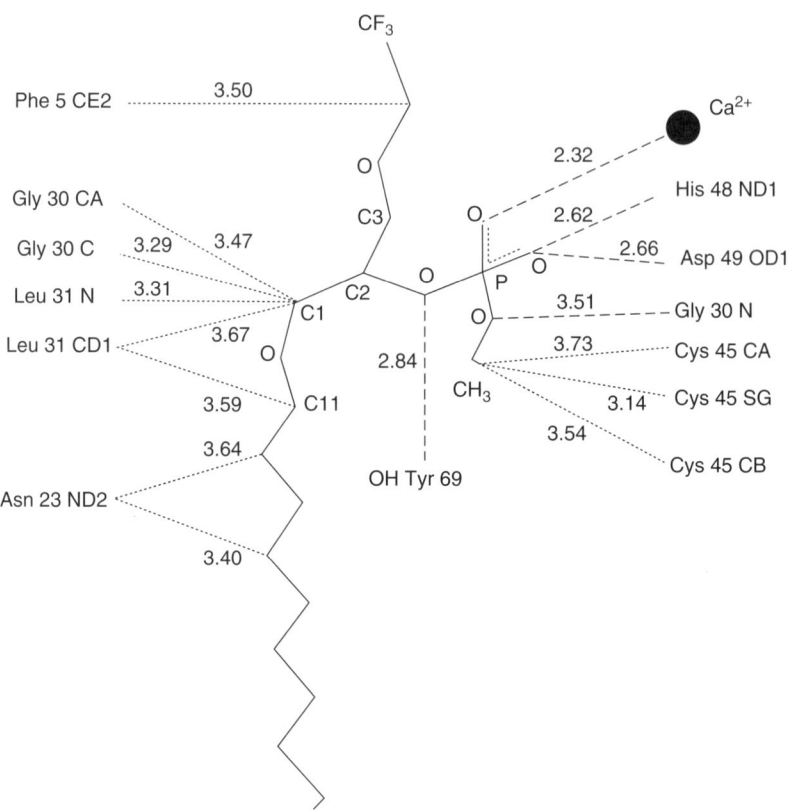

Figure 9.2 Contacts of the head group of MJ33 with the active site residues of bovine pancreatic PLA2. Reprinted with permission from Sekar *et al.* (1997a). Copyright (1997) American Chemical Society

assays is further compounded by the fact that evolutionary relatives of the selected target are sometimes expressed in specialized cellular compartments.

9.1.2 Active-site-directed competitive inhibitors of secreted PLA2

In the catalytic mechanism of secreted PLA2, the *sn*-2-carbonyl of the ester is polarized by coordination to the calcium cofactor (Figure 5.10). Both calcium and His-48 remain in contact with the *sn*-2-tetrahedral mimic MJ33 (Figures 9.2 and 5.9). Both of these structures exhibit multiple-contact sites, but the differences may be mechanistically significant. For example, a water molecule (w6) seen in the anion-assisted dimer structure (Figure 5.9) is not seen in the structure obtained with crystals of the monomer form of EI without the anions (Figure 9.2). Such differences may be related to the activation of the enzyme at the anionic interface (Figure 8.7).

The *sn*-2-amide analogs of phospholipids (I in Figure 9.1) are also specific inhibitors of secreted PLA2. They retain the ability to bind to δNH of His-48 and the calcium cofactor in the active site. As a result of such interactions, the affinity of the active-site-directed substrate mimics, relative to that for the substrate, is higher by about 100 to 5000-fold ($= K_{MS}^*/K_I^*$) for the various secreted PLA2s. For

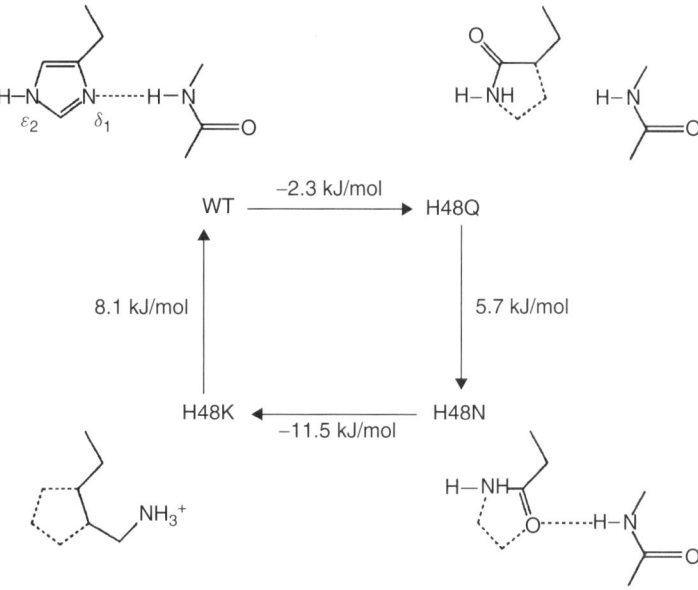

Figure 9.3 The thermodynamic box for the binding (K_I^*) of a competitive inhibitor I (Figure 9.1) to pig pancreatic PLA2 and its three His-48 substitution mutants. From Janssen *et al.* (1999), reproduced by permission of Oxford University Press

the best cases, this ratio corresponds to as little as one inhibitor molecule per vesicle at $X_I^*(50) < 0.0005$.

9.1.3 Detailed balance condition for the active site binding and stability

The hydrogen bonding of *sn*-2-amide NH from the inhibitor to δNH of His-48 is suggested by the crystallographic structure. It is supported by the results in Figure 9.3. The possibility of H bonding of the *sn*-2-amide NH is significant with a suitable conformation. For example, the H bonding is possible in H48N with the indicated conformer of the side chain amide of Asn. Such a conformer is unlikely with the side chain of Gln or Lys substitution mutants of PLA2. The fact that a similar bond may be formed in the chemical step is also supported by the turnover number results, where H48N shows 0.1 % activity, whereas the activities of H48Q and H48K is about 0.01 % relative to the WT PLA2. A similar thermodynamic relationship has also been observed in the stability measurements. Here the results are complementary, i.e. the conformational stability is in the order WT > H48Q > H48N > H48K. These results are attributed to the contribution of an H-bond between εN of His-48 and the carboxylate of Asp-99. In this case H bonding with the amide of Gln is preferred. For further discussion see Berg *et al.*, 2001.

9.1.4 Sulfated glycoconjugates bind to the E form of PLA2

Polyanionic glycoconjugates, heparin and the heparan-conjugated peptidoglycans bind to the E form of PLA2, and thus lower the fraction of the enzyme present

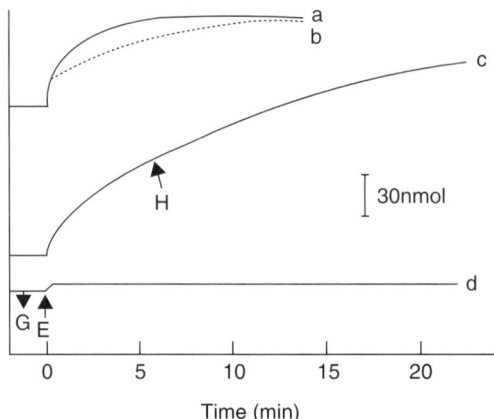

Figure 9.4   Reaction progress for the hydrolysis of DMPM vesicles by 14 pmol of pig pancreatic (curves a and b) or 36 p mol of V3W mutant of human synovial IIA PLA2 (curves c and d). The reaction mixture for curves b and d also contained 2.5 µg of heparan-conjugated proteoglycan I from human aorta. Note that the conjugate added (at H) after the initiation of the reaction had no effect on the reaction progress (curve c). Reprinted with permission from Yu *et al.* (1997b). Copyright (1997) American Chemical Society

at the interface (Yu *et al.*, 1997b). The order-of-addition effects on the scooting mode reaction progress distinguish between certain possibilities if the complex formation is tight. As shown in Figure 9.4, virtually no inhibition is observed with the glycoconjugates added after initiating the scooting mode reaction progress by pig or synovial PLA2. As also shown in this figure, the synovial PLA2 forms a tight complex with the conjugate, and the complex in the aqueous phase does not dissociate if DMPM vesicles are added to the complex. On the other hand, the complex with pig pancreatic PLA2 (curve b) is fully active on DMPM vesicles, and the enzyme does not dissociate from the vesicles in the presence of the glycoconjugates. A reduction in the observed rate due to the formation of a tight complex may also be attributed to the PLA2 inhibitory proteins found in certain tissues.

9.1.5   Covalent modifiers of PLA2

Covalent modifiers of the catalytic site residues His-48 and Asp-49 inactivate PLA2. A remarkable selectivity toward a particular residue can be induced by structural fine-tuning of such reagents. However, the covalent modifiers are generally not useful in tissue or cellular assay conditions, except as agents of chemical warfare. *In vitro*, the modifiers of specific functional groups, such as the catalytic His-48 of PLA2 (Chapter 6), have been useful to quantify the occupancy of the active site. A similar strategy has also been useful with gossypol, which modifies only the NH_2 group of Ala-1 of PLA2. The half-time for modification of the E form (without the cofactor Ca^{2+}) is 55 minutes and the modified enzyme retains 5 % of the total activity at the interface (Yu *et al.*, 1997a). In contrast, the half-time for the modification of the ECa and ECaI forms by gossypol is only 5 minutes, whereas the half-time for the modification of the E*, E*Ca and E*CaI forms is over 1000 minutes. These results show a clear difference between the reactivity and/or accessibility of the free amino group of Ala-1 in the free and interface-bound forms of PLA2.

9.2 Kinetic Effects of Nonspecific Inhibitors

Some of the kinetic consequences of amphiphilic and nonpolar solutes added to the assay mixtures are examined in this section. Such solutes show little structural specificity, and the only common structural requirement for such interface modifiers is that they partition into the interface. Thus it is likely that such solutes can influence the binding of the enzyme to the interface, and also influence the turnover by binding to the catalytic site or to other sites that may influence the catalytic turnover.

9.2.1 Substrate dilution and phase change

As an example of the solute-induced nonspecific effects, consider the kinetic effects induced by a neutral diluent (Figure 9.5). When partitioned into the interface, at low mole ratios it changes in the surface density of the substrate. Decreasing turnover rates below $X_S^* < 0.7$ are a direct consequence of the lower surface density of the substrate as predicted by Equation (5.2). The fit gives $K_{MS}^* = 0.37$ mole fraction, which is consistent with the value obtained independently by other methods (Table 5.1). The observed rate decreases abruptly to less than 20 % at >40 mole % diluent in the reaction mixture. This is accompanied by disruption of the vesicles as the fraction of the accessible substrate increases from about 65 % (outer layer of small sonicated vesicles) to 100 %. Note that the diluent forms micelles. Therefore, at about a 0.3 mole ratio of the two lipids, an abrupt change is expected in the organization

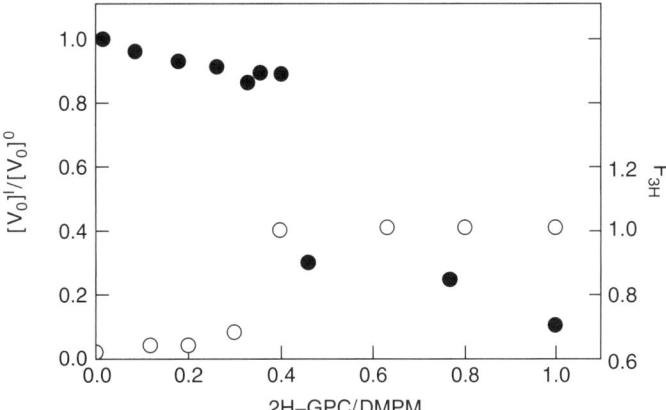

Figure 9.5 The relative rate (filled circles) or the extent of hydrolysis (open circles) of DMPM vesicles by pig pancreatic PLA2 as a function of the mole ratio of 2-hexadecyl-*rac*-3-glycerophosphocholine (2H-GPC). Note that in both cases an abrupt change is seen at 0.3 mole ratio of 2H-GPC, beyond which the vesicles are disrupted and possibly mixed micelles are formed. Reprinted with permission from Jain *et al.* (1991a). Copyright (1991) American Chemical Society

of bilayer to mixed micelles. As discussed in Chapter 11, such a decrease in the observed rate is attributed to the fact that the processive turnover becomes limited by the substrate replenishment rate in mixed micelles.

9.2.2   Nonspecific inhibition of PLA2 by alkanols

In general, the origins of the effect of *nonspecific inhibitors* are varied and not yet fully established in any particular case. In this section we describe one particularly well-characterized, yet unresolved, example to illustrate the complexity of the problem. For example, consider the concentration dependence of the effects of alkanols on the apparent initial rate for the hydrolysis of phosphatidylcholine vesicles by bee venom PLA2. Results in Figure 9.6 show that the apparent initial rate of hydrolysis displays a biphasic dependence on the concentration of isomeric octanols, where the observed rate is increased more than 50-fold at the peak concentration followed by a several-fold decrease at a higher concentration. The biphasic alkanol concentration dependence is seen with virtually all secreted PLA2s. Also, alkanols, anesthetics and many weakly amphiphilic nonpolar solutes exhibit such anomalous effects with lipolytic and membrane-associated enzymes. In all such cases, as a first approximation, at least two opposing alkanol concentration-dependent processes are at work: an increase in the rate at low concentrations and a decrease at higher alkanol concentrations. Independent controls show that alkanols partition into vesicles and that the bilayer organization is retained even at

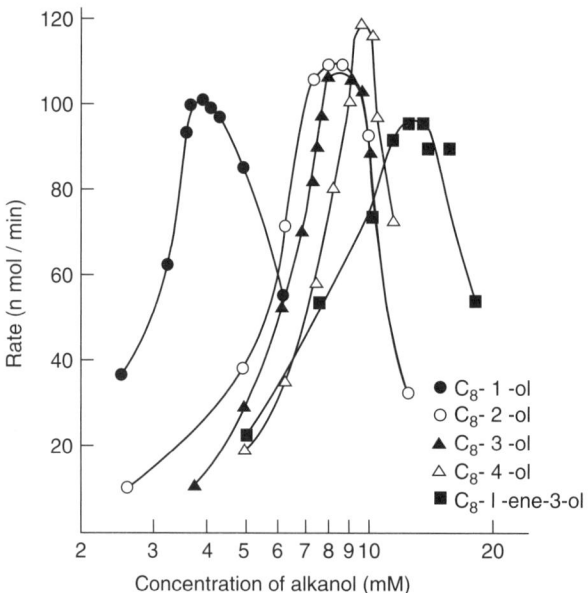

Figure 9.6   Effect of isomeric octanols on the apparent initial rate of hydrolysis of osmotically shocked multilamellar vesicles of 0.5 mM egg phosphatidylcholine by 0.05 μg of bee venom PLA2 in 5 ml of reaction mixture. The maximum alkanol concentration used in these studies is below the solubility limit. From Upreti *et al.* (1980). Reproduced by permission of Springer-Verlag New York, Inc

the inhibitory concentrations. Therefore, as a basis for the effect on the observed rate, it is likely that the solutes partition into and possibly induce changes in the bilayer interface.

As developed later in Chapter 10, the time course of hydrolysis of zwitterionic phosphatidylcholine vesicles by PLA2 is complex with a very low initial rate. The initial rate also increases in the presence of the products of hydrolysis with a maximum initial rate at $X_P^* = 0.12$ mole fraction. The presence of products increases the binding of the enzyme to the interface. As shown in Figure 9.7, the effect of hexanol is still biphasic at low X_P^* in the substrate vesicles, although the rate increase is modest. In the presence of products the maximum rate is observed at lower hexanol concentrations, and no activation is seen at $X_P^* = 0.3$ (Figure 9.7A). These results suggest that at $X_P^* > 0.3$ it is possible to investigate the apparent inhibitory effect of alkanols.

The results in Figure 9.8 show that the inhibitory potency of hexanol also depends on the total substrate concentration. These plots, analogous to the Dixon plot, would for a solution enzyme suggest an apparently noncompetitive mechanism of inhibition by hexanol. As summarized in Figure 9.7B, a 50 % decrease in the peak rate at a constant vesicle concentration is observed at the same hexanol concentration, i.e. the inhibitory effect of hexanol does not change with X_P^*. Also, the K_i values, obtained from the Dixon plots, do not change significantly with X_P^*. An

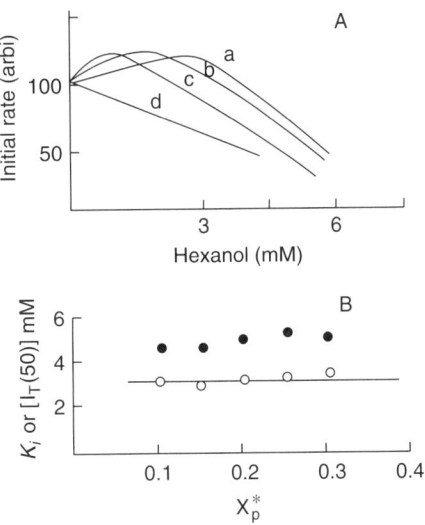

Figure 9.7 Panel A: normalized apparent initial rate of hydrolysis of 1,2-dimyristoylphosphatidylcholine (DMPC) containing 0.1, 0.15, 0.2 and 0.3 mole fraction of 1:1 palmitic acid + 1-palmitoyl-lyso-phosphatidylcholine by 0.5 µg of pig pancreatic PLA2 at 30 °C. Panel B: the hexanol concentration for (filled symbols) 50 % inhibition of the peak rate from the peak concentration, $[I_T(50)]$, or the corresponding K_i (unfilled symbols) obtained from the plot in Figure 9.8 do not shift as a function of the mole fraction of the products. Note that the $[I_T(50)]$ concentration at lower X_P^* is from the concentration at which the peak rate is observed. This is based on the assumption that the activating effect of the products and alkanol is additive. Reprinted from Jain and Jahagirdar (1985b). Copyright (1985) with permission from Elsevier Science

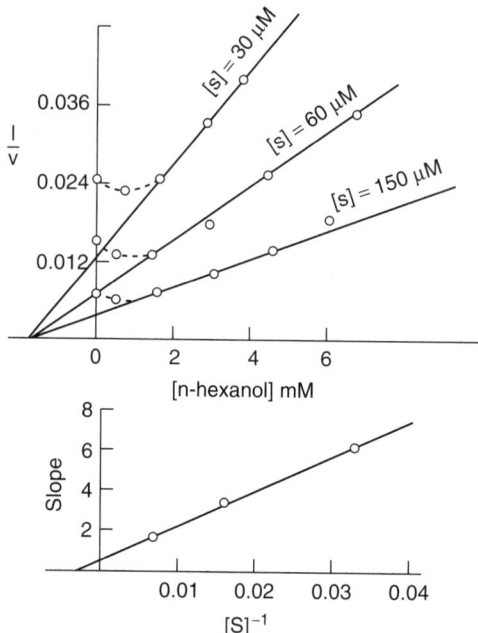

Figure 9.8 (Upper panel) Dixon plot for the effect of n-hexanol concentration on the reciprocal of the initial rate of hydrolysis of DMPC covesicles with 0.18 mole fraction of products of hydrolysis. (Lower panel) The slope of the plots in the upper panel changes with the total DMPC concentration. In the lower plot the slope $= K_m/(K_i V_{max})$ and the y intercept $= 1/(K_i V_{max})$ when interpreted as Dixon plots of a solution enzyme (Figure III-19 in Segel, 1975). These results provide K_i with V_{max} obtained at saturating substrate vesicle concentrations in the absence of hexanol. Reprinted from Jain and Jahagirdar (1985b). Copyright (1985) with permission from Elsevier Science

alternative interpretation of the Dixon plot, in the context of the interfacial turnover path, is developed below in Section 9.6, Equation (9.35).

The biphasic concentration dependence of the initial rate is seen with homologous alkanols. Both the $[I_T(50)]$ and K_i values change with the chain length of the alcohol. As summarized in Figure 9.9, the maximum inhibitory concentration is seen with C_9–C_{11} alkanols, and the higher alkanols are less potent.

9.3 Kinetic Effects of the Interface-Based Competitive Inhibitors

Within the paradigm of Scheme I (Figure 3.1), several kinetic paths are conceivable depending on the ability of E, S, P and I to exchange between the interface and the aqueous phase. To arrive at an analytically tractable kinetic model, consider the case where a neutral diluent, A, forms the interface, and substrate and inhibitor are present only in the interface at mole fractions X_S^* and X_I^*, respectively (see Figure 9.10 where the diluent is not shown). The extra state E*I leads to an extra term X_I^*/K_I^* in the denominator of the flux expression given in Equation (5.23):

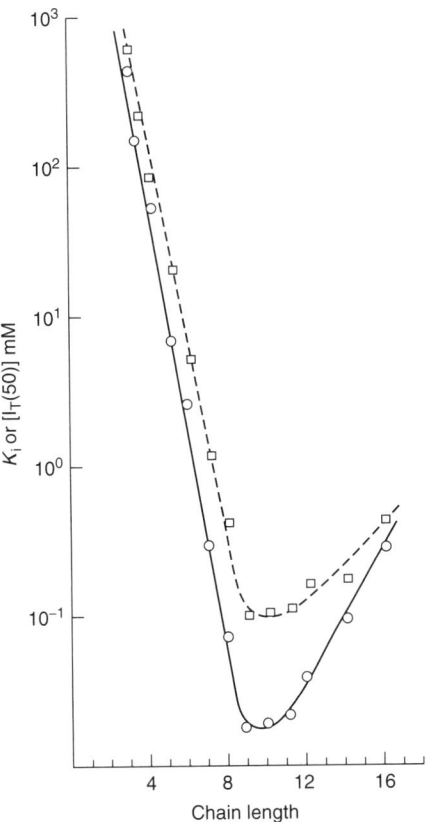

Figure 9.9 The alkanol concentration (squares) for 50% inhibition, $[I_T(50)]$, or the apparent K_i obtained from the Dixon plot (circles) is plotted as a function of the chain length. Reprinted from Jain and Jahagirdar (1985b). Copyright (1985) with permission from Elsevier Science

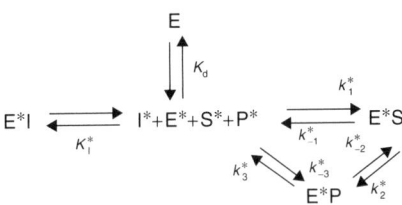

Figure 9.10 Steps of the kinetic path for the interfacial turnover when all the inhibitor, substrate and product remain bound to the interface and only the enzyme exchanges between the interface and the aqueous phase

$$ j = \frac{\dfrac{k^*_{catS}}{K^*_{MS}}X^*_S - \dfrac{k^*_{catP}}{K^*_{MP}}X^*_P}{1 + \dfrac{X^*_I}{K^*_I} + \dfrac{X^*_S}{K^*_{MS}} + \dfrac{X^*_P}{K^*_{MP}} + \dfrac{K_d}{[M^*]}} \tag{9.1} $$

For the initial rate, $X^*_P = 0$ and the concentration of the interface is $[M^*] = [A_T] + [S_T] + [I_T]$. Furthermore, when all A, S and I are in the interface, $X^*_S = [S_T]/([A_T] +$

$[S_T] + [I_T])$ and $X_I^* = [I_T]/([A_T] + [S_T] + [I_T])$. Thus, the initial rate is

$$j_0 = \frac{\dfrac{k_{catS}^*}{1 + K_{MS}^*}[S_T]}{[S_T] + \dfrac{K_d + [A_T]}{1 + 1/K_{MS}^*} + [I_T]\dfrac{1 + 1/K_I^*}{1 + 1/K_{MS}^*}} = \frac{k_{cat}^{app}[S_T]}{[S_T] + K_M^{app}} \tag{9.2}$$

This initial rate can be sustained only when substrate replenishment is not limiting. Here it is assumed that the inhibitor contributes to the surface area of the accessible interface, and thereby to some extent may increase interface binding and activity. For an efficient inhibitor, where $X_I^* \ll 1$, this surface expansion will be very small. If the total substrate concentration $[S_T]$ is manipulated while keeping $[I_T]$ constant, the initial rate (Equation 9.2) will change according to the MM equation (cf. Equation 5.24) if the apparent MM parameters are identified as

$$k_{cat}^{app} = j_0^{max} = \frac{k_{catS}^*}{1 + K_{MS}^*} \tag{9.3}$$

$$K_M^{app}([I_T]) = \frac{K_d + [A_T]}{1 + 1/K_{MS}^*} + [I_T]\frac{1 + 1/K_I^*}{1 + 1/K_{MS}^*} = K_M^{app}(0) + [I_T]B \tag{9.4}$$

where

$$B = \frac{1 + 1/K_I^*}{1 + 1/K_{MS}^*} \tag{9.5}$$

Thus B is a measure of the intrinsic strength of the interfacial inhibitor relative to the substrate. The presence of an interfacial competitive inhibitor will affect only the apparent K_M value, not k_{cat}^{app}, just as is the case for the usual solution kinetics (Equation III-9 in Segel, 1975). The values of k_{cat}^{app} and K_M^{app} can be found from the initial rate studied as a function of $[S_T]$, e.g. through a Lineweaver–Burk plot (Figure 3.3) where $1/j_0$ is plotted versus $1/[S_T]$ at constant $[I_T]$. The slope in such a plot gives K_M^{app}/k_{cat}^{app} while the y intercept gives $1/k_{cat}^{app}$. Then plotting K_M^{app} versus $[I_T]$ gives the slope B (Equation 9.5) from which K_I^* can be deduced if K_{MS}^* is independently known. In the common solution kinetics, this slope is K_m/K_i (Equation III:9 in Segel, 1975).

9.3.1 The inhibitor concentration dependence of the apparent rate

Alternatively, one can plot k_{cat}^{app}/j_0 versus $[I_T]$ (Dixon plot, Figure 9.11A):

$$\frac{k_{cat}^{app}}{j_0} = 1 + \frac{K_M^{app}([I_T])}{[S_T]} = 1 + \frac{K_M^{app}(0)}{[S_T]} + [I_T]\frac{B}{[S_T]} \tag{9.6a}$$

The slope is

$$Slope_I = \frac{1 + 1/K_I^*}{1 + 1/K_{MS}^*}\frac{1}{[S_T]} = \frac{B}{[S_T]} \tag{9.6b}$$

Plotting these slopes versus $1/[S_T]$ gives a straight line that extrapolates to the origin (Figure 9.11B). One can also consider the local inhibitor concentration in terms of

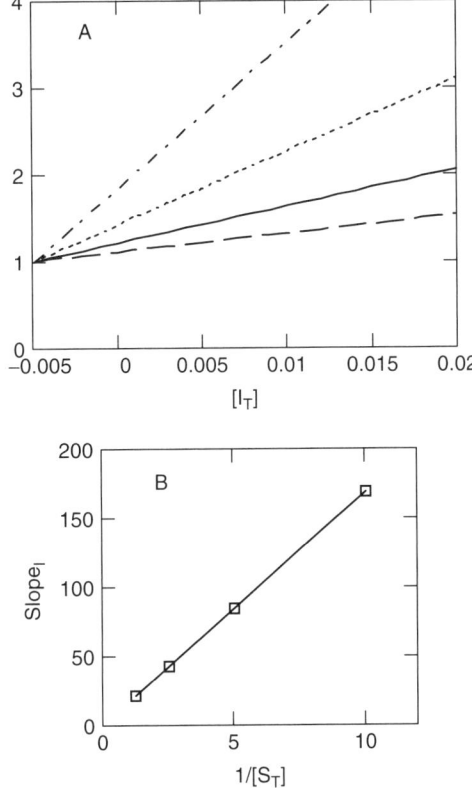

Figure 9.11 Dixon plot from Equations (9.6) and (9.6b). Data used: $K_{MS}^* = 0.2$, $K_I^* = 0.01$, $K_d + [A_T] = 0.5$. Panel A shows the inverse flux versus $[I_T]$ at substrate concentrations (from upper to lower) $[S_T] = 0.1, 0.2, 0.4, 0.8$. The extrapolated lines for the different substrate concentrations meet at $[I_T] = -K_M^{app}(0)/B$; in common solution kinetics, this point is at $[I] = -K_i$ (Figure III-6 in Segel, 1975). Panel B shows the slopes of the lines in panel A plotted versus $1/[S_T]$; the slope of this line equals B. For a competitive inhibitor this line extrapolates to the origin according to Equation (9.6b)

the mole fraction X_I^*:

$$\frac{k_{cat}^{app}}{j_0} = 1 + \frac{K_M^{app}([I_T])}{[S_T]} = 1 + \frac{K_M^{app}(0)}{[S_T]} + B\frac{X_I^*}{1 - X_I^*} \tag{9.7}$$

In this representation, the slopes of the inverse flux versus X_I^* are independent of $[S_T]$ and equal to B (if $X_I^* \ll 1$).

From Equation (9.2) one finds that the initial rate is reduced by 50 % at inhibitor concentration $[I_T(50)]$, given by

$$[I_T(50)] = \frac{1 + \dfrac{1}{K_{MS}^*}}{1 + \dfrac{1}{K_I^*}}[S_T] + \frac{K_d + [A_T]}{1 + \dfrac{1}{K_I^*}} = \frac{1}{B}\left([S_T] + K_M^{app}(0)\right) \tag{9.8}$$

Thus plotting $[I_T(50)]$ versus $1/[S_T]$ gives the slope $1/B$ and the x intercept $-K_M^{app}(0)$, i.e. the same information as the Dixon plot (Figures 9.8 and 9.11). In the scooting limit where $K_d \to 0$ and in the absence of a neutral diluent, where $K_M^{app}(0) = 0$, Equation (9.8) gives

$$\frac{X_I^*(50)}{1 - X_I^*(50)} = \frac{1}{B} = \frac{1 + 1/K_{MS}^*}{1 + 1/K_I^*} \tag{9.9}$$

for the mole fraction $X_I^*(50)$ in the interface where the initial rate is reduced by 50 %.

Equations (9.6) to (9.8) describe the inhibitory effect of a competitive inhibitor in the interface. These predictions of the model do not correlate with the inhibitory effect of alkanols (e.g. Figure 9.8). One possible reason is that the alkanols are not competitive inhibitors or that they are not completely partitioned in the interface. The effect of incomplete partitioning is considered in Section 9.6. Note that even if the inhibitor does not bind the enzyme, i.e. if $K_I^* = $ infinity, the inhibitor has an effect both due to its dilution of the substrate concentration in the interface as well as to an increased interface area. In this limit, the 'inhibitor' is a neutral diluent (see Chapters 3 and 6). Due to this dilution effect, the interfacial inhibitor gives somewhat different relations for the initial rate in terms of the local concentrations than the normal solution case (Equation III-7 in Segel, 1975).

The scheme in Figure 9.10 assumes that E* can dissociate while both E*I and E*S are strongly interface bound. This is possible only if the ligand (I or S) helps to anchor the enzyme in the interface. As discussed in Chapter 8, such anchoring is not expected for enzymes that, like PLA2, engulf their ligands in the active site, leaving little room for a residual ligand–interface interaction. In such a case, the model above is applicable primarily in the scooting limit where E* is also tightly bound to the interface (i.e. when $K_d \to 0$).

9.3.2 The Z-factor for inhibitor potency

De Haas and coworkers (Ransac *et al.*, 1990; de Haas *et al.*, 1993, 1995) have defined the inhibitory power, Z, which is determined by the mole fraction $\alpha_I = [I_T]/([S_T] + [I_T])$ required to reduce enzyme activity by 50 % when the total $[S_T] + [I_T]$ is kept constant in the bile–salt dispersed phospholipid interface, as

$$\alpha_I(50) = \frac{1}{2 + Z} \tag{9.10}$$

From Equation (9.2) this gives

$$Z = \frac{B - 1}{1 + \dfrac{K_M^{app}(0)}{[S_T] + [I_T]}} = \frac{\dfrac{1}{K_I^*} - \dfrac{1}{K_{MS}^*}}{\left(1 + \dfrac{1}{K_{MS}^*}\right)\left(1 + \dfrac{K_M^{app}(0)}{[S_T] + [I_T]}\right)} \tag{9.11}$$

This shows that the inhibitory power Z depends on the total concentrations of both I + S and diluent A (implicit), except when all enzymes are in the interface, in which case $Z = B - 1$. However, relative values of Z for different inhibitors should reflect relative values of their affinity ($1/K_I^*$) for the enzyme, as long as rapid exchange or replenishment is assured, so that the initial rate as given by Equation (9.2) can be sustained. The assay for the determination of Z uses the mixed-micellar codispersions with bile salts as a diluent in a 1:1 mole ratio with $[S_T]$. For a variety of reasons, primarily the difficulty of ensuring substrate replenishment as developed in Chapter 11, a more detailed analysis of the Z-factors in terms of the primary parameters is not possible under these conditions.

9.4 Influence of Cofactor on the Scooting Kinetics

When a cofactor C is required for binding I or S at the catalytic site, we can reinterpret the reaction scheme above such that E^* denotes the cofactor-bound (activated) enzyme in the interface. Then there will also be a cofactor-free enzyme state, E_0^*, which will be present at concentration (Equation 6.20)

$$[E_0^*]/[E^*] = K_C^*/[C] \tag{9.12}$$

Here $[C]$ is the concentration of cofactor in aqueous solution and K_C^* the corresponding dissociation constant for binding C to E^* (see Figure 6.4). If cofactor preferentially accumulates in the interface and binds the enzyme from there, the relation remains valid due to the detailed balance condition. However, if such accumulation is significant, the aqueous concentration $[C]$ may be smaller than the total concentration $[C_T]$ (see Equation 6.21). The extra enzyme state leads to an extra term in the denominator of Equation (9.1). In the limit where all enzyme is in the interface ($K_d \to 0$) this gives, for the initial rate ($X_P^* = 0$),

$$j_0 = \frac{\frac{k_{\text{catS}}^*}{K_{\text{MS}}^*} X_S^*}{1 + \frac{X_S^*}{K_{\text{MS}}^*} + \frac{X_I^*}{K_I^*} + \frac{K_C^*}{[C]}} \tag{9.13}$$

The activation by cofactor can be expressed as the ratio of the flux at saturating C ($[C] \gg K_C^*$) and the flux at smaller $[C]$:

$$\frac{j_0\,(C_{\text{sat}})}{j_0\,([C])} = 1 + \frac{K_C^*}{[C]} \frac{1}{1 + \frac{X_S^*}{K_{\text{MS}}^*} + \frac{X_I^*}{K_I^*}} \tag{9.14}$$

Thus cofactor activation takes place with an effective dissociation constant

$$K_C^{\text{eff}} = \frac{K_C^*}{1 + \frac{X_S^*}{K_{\text{MS}}^*} + \frac{X_I^*}{K_I^*}} \tag{9.15}$$

A comparison with Equation (6.16) shows that under kinetic conditions the effect of substrate binding is determined by K_{MS}^* rather than the equilibrium constant K_S^*. In the absence of inhibitor, the cofactor activation in terms of Equations (9.14) and (9.15) can be used to determine K_{MS}^*, as discussed in Chapter 5 (Equation 5.17).

Similarly, these equations can also be used to determine K_I^* from the kinetic results (Yu *et al.*, 1993, 1997a).

Except in situations where the influence of cofactor is studied, as above, it has been assumed that cofactor is present in saturating amounts in the aqueous phase so that all enzyme states are cofactor-bound. In this case, cofactor can be suppressed in the reaction scheme. For instance, PLA2 has Ca^{2+} as an obligatory cofactor, which has been suppressed in the general reaction scheme in Figure 3.1 as well as in virtually all the discussions in the subsequent chapters.

9.5 Effects of the Interface-Based Inhibitor on the Integrated Rate Equation in the Scooting Mode

The integrated rate equation for the scooting kinetics, as given in the form of Equation (5.8), holds also in the presence of inhibitor. With scooting kinetics in the presence of a diluent A, the initial rate j_0 is given by Equation (9.1) (with $K_d \to 0$, $X_P^* = 0$ and $X_S^* = X_S^{*0}$) as

$$j_0\left(X_I^*\right) = \frac{k_{catS}^*}{K_{MS}^*} \frac{X_S^{*0}}{1 + \dfrac{X_I^*}{K_I^*} + \dfrac{X_S^{*0}}{K_{MS}^*}} \qquad (9.16)$$

where X_S^{*0} is the initial mole fraction of substrate in the interface. In the limit where the equilibrium is far to the product side, $X_P^{*eq} = X_S^{*0}$ and one finds the relaxation rate

$$k_i\left(X_I^*\right) = \frac{n_E}{N_S^0} \frac{k_{catS}^*}{K_{MS}^*} \frac{X_S^{*0}}{1 + \dfrac{X_I^*}{K_I^*} + \dfrac{X_S^{*0}}{K_{MP}^*}} \qquad (9.17)$$

Thus the reaction progress is determined by the integrated rate equation (Equation 5.8), where the effective rates j_0 and k_i are replaced by the expressions from Equations (9.16) and (9.17).

When there is no exchange of enzyme or substrate and product between aggregates (i.e. in the scooting mode), Equation (9.17) expresses the relaxation rate for product accumulation in an aggregate with n_E enzyme molecules and N_S^0 substrate molecules initially. If there is a population distribution range of these numbers n_E and N_S^0, the overall rates will be given only after an appropriate averaging, as discussed in Chapter 5. When the enzyme-to-aggregate ratio is small, so that there is at most one enzyme per aggregate and the aggregate sizes are fairly uniform, all enzymes will see the same environment and the overall relaxation rate will be as given by Equation (9.17). If there is a fast exchange or mixing between aggregates, each enzyme will see the same environment and the overall rate would also be given by Equation (9.17); in this case, n_E/N_S^0 should be interpreted as the initial ratio of interface-bound enzymes to substrates (see Equation 5.21).

Studying the reaction progress as discussed in Chapter 5 with different X_I^* present allows identification of the two rate parameters j_0 and k_i if n_E/N_S^0 is known. The initial rate is affected by the presence of the inhibitor as given by Equation (9.16). The ratio of the relaxation rates in the absence and presence of inhibitor is

$$\frac{k_i(0)}{k_i(X_I^*)} = 1 + [I_T] \frac{1 + \dfrac{1}{K_I^*}}{[A_T] + \left[S_T^0\right]\left(1 + \dfrac{1}{K_{MP}^*}\right)} \tag{9.18}$$

The relaxation rate is reduced by 50 % when the inhibitor concentration is

$$[I_T(50)] = \left[S_T^0\right] \frac{1 + \dfrac{1}{K_{MP}^*}}{1 + \dfrac{1}{K_I^*}} + [A_T] \frac{1}{1 + \dfrac{1}{K_I^*}} \tag{9.19}$$

$[I_T(50)]$ also corresponds to the inverse of the slope of the rate ratio in Equation (9.18) plotted versus $[I_T]$.

In the absence of diluent ($[A_T] = 0$), one finds from Equation (9.17) that the relaxation rate k_i is reduced by 50 % when the inhibitor mole fraction is $X_I^* = n_I^*(50)$, given by

$$\frac{n_I^*(50)}{1 - n_I^*(50)} = \frac{1 + \dfrac{1}{K_{MP}^*}}{1 + \dfrac{1}{K_I^*}} \tag{9.20}$$

The nice symmetry can be noted between this expression and that for the initial rates in Equation (9.9) in the limit when there is no enzyme in solution. While the inhibitor influence on the initial rate depends on the substrate K_{MS}^*, the influence on the relaxation rate depends on the product K_{MP}^*. Taken together, Equations (9.9) and (9.20) give the relationship

$$\frac{n_I^*(50)/\left[1 - n_I^*(50)\right]}{X_I^*(50)/\left[1 - X_I^*(50)\right]} = \frac{1 + 1/K_{MP}^*}{1 + 1/K_{MS}^*} \tag{9.21}$$

This relation is valid for an interface-based competitive inhibitor in the limit when all enzyme, substrate and inhibitor molecules are in the interface and see the same environment. The ratio in Equation (9.21) is the same as the ratio of the rates j_0 and k_i without inhibitor, as given in Equation (5.10). The validity of these relations has been demonstrated for pig pancreatic (Berg *et al.*, 1991; Jain *et al.*, 1991a), human synovial (Bayburt *et al.*, 1992) and bee venom (Yu *et al.*, 1997b) PLA2.

9.6 Partitioning of the Inhibitor and Substrate between the Interface and the Aqueous Phase

When inhibitor and substrate are in solution as well as in the interface, the reaction scheme is as shown in Figure 9.12. Here it has been assumed that both inhibitor

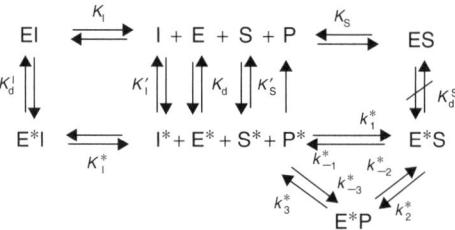

Figure 9.12 The kinetic path with exchangeable substrate, enzyme and inhibitor. The association–dissociation steps between ES and E*S are assumed to be slow, to prevent interference in the flux through the catalytic loop. This could keep ES to E*S out of equilibrium although the equilibrium constant K_d^S is well defined

and substrate can bind the enzyme in solution, that the EI and ES complexes can bind the interface (with dissociation constants K_d^I and K_d^S, respectively) and that the ES state is catalytically incompetent. The ES state introduces a branch that bridges part of the interfacial catalytic loop. As a consequence, there could be steady state fluxes also through the closed reaction loops outside the catalytic one. To avoid these complications, it will be assumed that the step E*S $\rightarrow$ ES is slow, such that all enzyme states, except E*S, will be equilibrated relative to E*. Furthermore, it will be assumed that substrate replenishment is sufficiently fast so that partitioning of S and I can be considered as equilibrated. A three-component partitioning becomes messy to describe in general (Theory Box 9.1) and therefore only the situation without diluent A is considered below. Then the initial rate is determined for $X_P^* = 0$ and $X_S^* = 1 - X_I^*$.

Partitioning I and S into the interface gives (Equation 3.10)

$$[I] = K_I' X_I^* \tag{9.22a}$$

$$[S] = K_S' \left(1 - X_I^*\right) \tag{9.22b}$$

Furthermore, the *detailed balance conditions* must hold (Equations 6.6b and 6.8):

$$K_I K_d^I = K_d K_I^* K_I' \tag{9.23a}$$

$$K_S K_d^S = K_d K_S^* K_S' \tag{9.23b}$$

The total concentration of inhibitor in the system is

$$[I_T] = [I] + [I^*] = X_I^* (K_I' + [M^*]) \tag{9.24a}$$

and the total substrate concentration is

$$[S_T] = [S] + [S_0^*] = \left(1 - X_I^*\right) (K_S' + [M^*]) \tag{9.24b}$$

The total interface concentration is

$$[M^*] = [S_0^*] / \left(1 - X_I^*\right) \tag{9.25}$$

Eliminating $[S_0^*]$ from Equations (9.24a) and (9.24b) gives a quadratic equation from which X_I^* can be calculated as a function of $[S_T]$ and $[I_T]$ (cf. Equation 3.15):

$$X_I^* = \frac{1}{2} \left\{ 1 + \frac{[S_T] + [I_T]}{K_I' - K_S'} \pm \sqrt{\left(1 + \frac{[S_T] + [I_T]}{K_I' - K_S'}\right)^2 - \frac{4\,[I_T]}{K_I' - K_S'}} \right\} \tag{9.26}$$

Due to the new enzyme states EI and ES, the flux relation (Equation 9.1) will get two new terms in the denominator:

$$
j = \frac{\dfrac{k_{catS}^*}{K_{MS}^*} X_S^* - \dfrac{k_{catP}^*}{K_{MP}^*} X_P^*}{1 + \dfrac{K_d}{[M^*]}\left(1 + \dfrac{[I]}{K_I} + \dfrac{[S]}{K_S}\right) + \dfrac{X_I^*}{K_I^*} + \dfrac{X_S^*}{K_{MS}^*} + \dfrac{X_P^*}{K_{MP}^*}}
\tag{9.27}
$$

This is the general flux relation, which holds irrespective of the assumptions about the partitioning equilibria, which for the ideal interfacial mixture of I and S considered here are given by Equations (9.22) to (9.26). In this case, changes in $[S_T]$ will lead to changes in X_I^* as well. Unless X_I^* can be controlled separately and kept constant, the initial rate will have higher-order terms in $[S_0^*]$ and will no longer conform exactly to the MM relation, $j_0 = k_{cat}^{app}[S_T]/(K_M^{app} + [S_T])$, except in special limits. This shows up as a nonlinear Lineweaver–Burk plot (Figure 9.13), as also seen experimentally under appropriate conditions. Thus, k_{cat}^{app} and K_M^{app} cannot be properly defined in this situation, except in the absence of inhibitor, where

$$
k_{cat}^{app}(0) = j_0^{max} = \frac{k_{catS}^*}{1 + K_{MS}^*}
\tag{9.28}
$$

$$
K_M^{app}(0) = \frac{K_d\left(1 + K_S'/K_S\right)}{1 + 1/K_{MS}^*}
\tag{9.29}
$$

Rewriting Equation (9.27) with $X_P^* = 0$, $X_S^* = 1 - X_I^*$ and $[M^*] = [S_T]/(1 - X_I^*) - K_S'$ from Equations (9.24b) and (9.25) gives, for the inverse of the initial rate as a

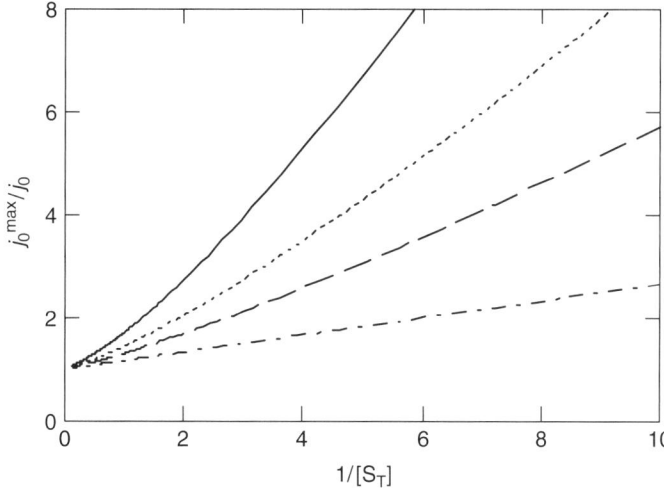

Figure 9.13 Lineweaver–Burk plot of the inverse flux versus inverse substrate concentration from Equations (9.27) and (9.28). Only the curve (dash–dot) for $[I_T] = 0$ is properly linear. Data used: $K_{MS}^* = 0.2$, $K_I^* = 0.01$, $K_d = K_d^I = 1$, $K_I' = 0.2$, $K_S' = 0$ and $[I_T] = 0.2, 0.1, 0.05, 0$ (from upper to lower curves)

function of X_I^* and $[S_T]$,

$$\frac{j_0^{max}}{j_0} = 1 + B\frac{X_I^*}{1 - X_I^*} + \frac{K_M^{app}(0)}{[S_T] - K_S'\left(1 - X_I^*\right)}\left(1 + CX_I^*\right) \tag{9.30}$$

where $j_0^{max} = k_{catS}^*/(1 + K_{MS}^*)$ is the maximal rate with saturating bulk concentration of substrate and in the absence of inhibitor (Equation 9.28), $B = (1 + 1/K_I^*)/(1 + 1/K_{MS}^*)$ from Equation (9.5) and

$$C = \frac{\dfrac{K_I'}{K_I} - \dfrac{K_S'}{K_S}}{1 + \dfrac{K_S'}{K_S}} \tag{9.31}$$

In the limit when both substrate and inhibitor are all in the interface ($K_S' = K_I' = 0$), Equation (9.30) reduces to Equation (9.7). It may be most convenient to measure the initial rate at constant substrate concentration, varying $[I_T]$. Then calculate X_I^* from Equation (9.26) and plot the flux ratio from Equation (9.30) versus X_I^*; this procedure requires that K_S' and K_I' are known independently. For a strong inhibitor, $X_I^* \ll 1$, this gives a straight line with (initial) slope

$$\text{Slope}_{X_I^*} \approx B + \frac{K_M^{app}(0)}{[S_T] - K_S'}\left(C - \frac{K_S'}{[S_T] - K_S'}\right) \tag{9.32}$$

The intrinsic inhibitor strength B can be found by plotting this slope versus $1/([S_T] - K_S')$ and extrapolating to the y axis; some curvature, particularly for large values of K_S', may appear also in this plot. Although I is a competitive inhibitor in the Scheme of Figure 9.12, the results in Figure 9.14 look like those of a noncompetitive one (cf. Figure III-19 in Segel, 1975). This is because the dependence on the inhibitor concentration is considered from the local X_I^*, while that of substrate comes through the interface concentration $[S^*] = [S_T] - K_S'$, and in terms of the enzyme–interface binding the inhibitor is noncompetitive; i.e. it binds both E and E*. Figure 9.14 is a variation on the Dixon plot, which is useful for interfacial enzymes when X_I^* is a known variable (see also Figure 9.16).

Alternatively, one could measure the amount of inhibitor required for a 50% reduction in enzyme activity, $[I_T(50)]$, and calculate the corresponding $X_I^*(50)$ from Equation (9.26). Taking the ratio of the initial rate in the absence ($X_I^* = 0$) and in the presence of inhibitor, one finds from Equation (9.30) that

$$\frac{j_0(0)}{j_0\left(X_I^*\right)} = \frac{1 + B\dfrac{X_I^*}{1 - X_I^*} + \dfrac{K_M^{app}(0)\left(1 + CX_I^*\right)}{[S_T] - K_S'\left(1 - X_I^*\right)}}{1 + \dfrac{K_M^{app}(0)}{[S_T] - K_S'}}$$

$$\approx 1 + X_I^*\frac{B + \dfrac{K_M^{app}(0)}{[S_T] - K_S'}\left(C - \dfrac{K_S'}{[S_T] - K_S'}\right)}{1 + \dfrac{K_M^{app}(0)}{[S_T] - K_S'}} \tag{9.33}$$

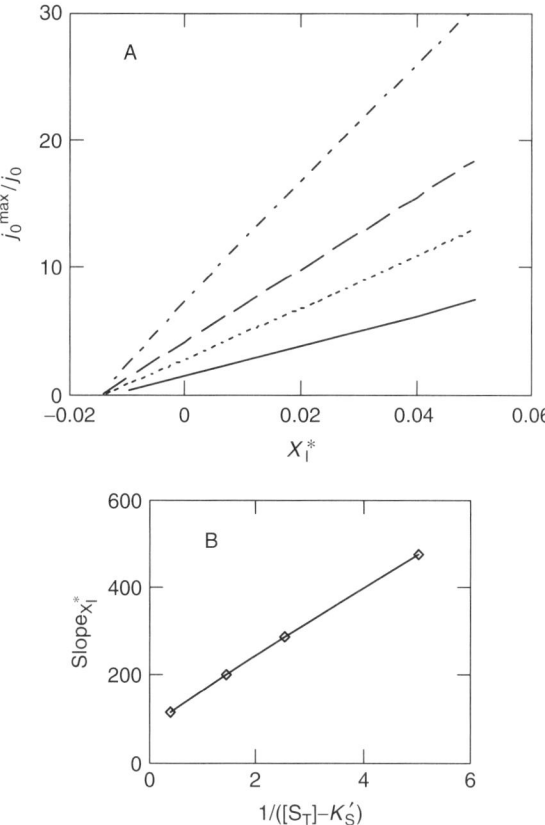

Figure 9.14 Plots for the inverse flux versus the mole fraction X_I^* from Equation (9.30). Data used: $K_{MS}^* = 0.2$, $K_S^* = 0.15$, $K_S' = 0.3$, $K_S = 0.015$, $K_I^* = 0.002$, $K_I' = 0.2$, $K_I = 0.0004$ and $K_d = K_d^I = K_d^S = 1$. Panel A shows the inverse flux versus X_I^* at different $[S_T]$; X_I^* is calculated from Equation (9.26) with $[S_T] = 0.5, 0.7, 1, 3$ (from upper to lower curves). Panel B shows the plot of the initial slopes in panel A versus $1/([S_T] - K_S')$; the extrapolated y intercept of this plot gives B

The approximation is valid to first order in X_I^* and is useful for strong inhibitors where $X_I^* \ll 1$. From Equation (9.33) and neglecting terms of order $(X_I^*)^2$, one finds

$$\frac{X_I^*(50)}{1 - X_I^*(50)} \approx \frac{1 + \dfrac{K_M^{app}(0)}{[S_T] - K_S'}}{B + \dfrac{K_M^{app}(0)}{[S_T] - K_S'}\left(C - 2\dfrac{K_S'}{[S_T] - K_S'}\right) - \dfrac{K_S'}{[S_T] - K_S'}} \tag{9.34}$$

When $K_M^{app}(0)$ and K_S' have been determined separately, B and C can be deduced from a best fit, as discussed for Figure 9.15.

9.6.1   Effect of the chain length on the inhibitor potency

A comparison of the inhibitory effect of the homologous solutes on the hydrolysis of DC$_7$PC micelles by pancreatic PLA2 has shown that the chain length has an

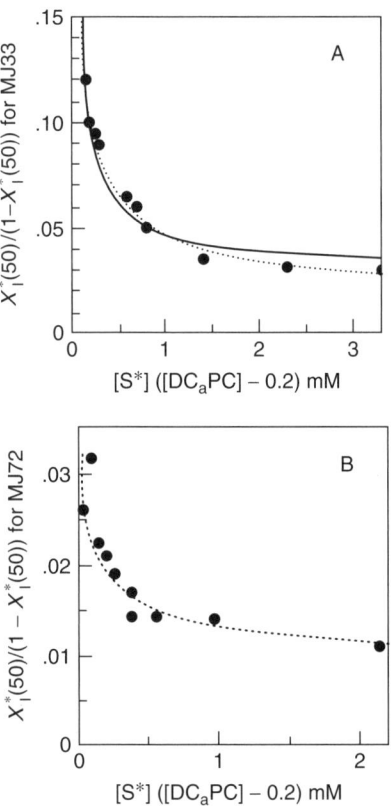

Figure 9.15 Effect of homologous inhibitors on the hydrolysis of DC_7PC micellar dispersions by pig pancreatic PLA2. Panel A shows MJ33 with hexadecyl chain (Figure 9.1), and Panel B shows MJ72 with octyl chain. Reprinted with permission from Berg *et al.* (1997). Copyright (1997) American Chemical Society

insignificant effect on the binding of a competitive inhibitor to the active site. Equation (9.34) has been used to evaluate the results in Figure 9.15 (Berg *et al.*, 1997):

(a) The [S*] dependence of the inhibition by MJ33 (Figure 9.1) is fitted with K_M^{app} and the fit parameters $B = 40$ and $C = 6$. With the CMC of 0.008 mM, MJ33 is virtually completely partitioned into the interface. At high $[S_T]$ the B term in the denominator dominates and C dominates at low $[S_T]$. The B dominated relationship assumes the form of Equation (9.9) for the inhibition in the scooting mode (Berg *et al.*, 1991), i.e. $X_I^*(50)$ does not depend on $[S_T]$ for DC_8PC, as also shown for anionic DC_8PM micelles (Rogers *et al.*, 1996) or $DC_{14}PM$ vesicles (Jain *et al.*, 1991a, 1991b). These conditions provide a basis for the calculation of K_{MS}^* (Chapter 5), using $K_I^* = 0.0014$ mole fraction for MJ33.

(b) The dependence of $X_I^*(50)$ on $[S_T] - K_S'$ for MJ72, an octyl chain homolog with $K_I' = 0.3$ mM in deoxy-LPC (Jain *et al.*, 1993b), gave $B = 106$ and $C = 42$. Since K_{MS}^* remains constant, the value of B (Equation 9.34) gives K_I^* for MJ72 as 0.0005 mole fraction.

Compared to MJ33, the value of C for MJ72 is several-fold higher, although the uncertainty in these values may approach 50 %. According to Equation (9.24), with the same substrate, a change in C must be due to the K'_I/K_I term, calculated with $K'_S = 0.2$ mM and $K_S = 0.065$ mM (from Equation 9.29 based on the measured $K_M^{app}(0)$). As expected, the values for the ratio are higher for the short chain inhibitors, and the absolute values of the K'_I/K_I ratio are different from those obtained from the components determined independently. A basic assumption in the analysis is that the mimics partition into all interfaces in the same way. A departure could account for some of the variation in the calculated parameter values depending on the nature of the interface.

THEORY BOX 9.1 Partitioning equilibria for a three-component ideal interfacial mixture

Consider the partitioning equilibrium of three components, A, S and I. The basic relations for an ideal interfacial mixture are (cf. Equations (9.24))

$$[A_T] = X_A^* (K'_A + [M^*]) \tag{1a}$$

$$[S_T] = X_S^* (K'_S + [M^*]) \tag{1b}$$

$$[I_T] = X_I^* (K'_I + [M^*]) \tag{1c}$$

Together with the condition $1 = X_A^* + X_S^* + X_I^*$, this constitutes four equations in the four unknowns X_A^*, X_S^*, X_I^* and $[M^*]$. To describe the enzyme flux (Equation 9.27) requires knowledge of X_S^*, X_I^* and $[M^*]$ as functions of the known concentrations, $[A_T]$, $[S_T]$ and $[I_T]$. Solving for X_S^* or X_I^* leads to a third-degree equation. While it may be possible to handle this numerically, it is more convenient to use $[A_T]$, $[S_T]$ and X_I^* as independent variables. This gives

$$X_S^* = \frac{1}{2}\left\{ 1 - X_I^* + \frac{[S_T] + [A_T]}{K'_S - K'_A} \pm \sqrt{\left(1 - X_I^* + \frac{[S_T] + [A_T]}{K'_S - K'_A}\right)^2 - \frac{4(1 - X_I^*)[S_T]}{K'_S - K'_A}} \right\}$$

$$\tag{2}$$

$$[I_T] = X_I^* \left(\frac{[S_T]}{X_S^*} - K'_S + K'_I \right) \tag{3}$$

$$[M^*] = \frac{[S_T]}{X_S^*} - K'_S \tag{4}$$

Clearly, X_I^* is not under direct experimental control, but by varying X_I^* in the relations above, one can relate values of $[I_T]$ with values for X_S^* and $[M^*]$. Together with the enzyme flux relation (Equation 9.27), Equations (2) to (4) make it possible to create Dixon plots, by calculating $1/j_0$ and $[I_T]$ for different values of X_I^*, keeping $[A_T]$ and $[S_T]$ at given values, and then plotting $1/j_0$ versus $[I_T]$ (Figure 9.16). While this can give apparent straight lines (Figure 9.16A) for many parameter values, the slopes do not necessarily fall on a straight line (Figure 9.16B).

Without independent determination of most of the parameters involved, the interpretation of experimental curves is not feasible. If partitioning parameters are known so that X_I^* can be calculated, the inverse flux can be plotted versus X_I^* and the corresponding slopes used (Figure 9.16C). In order to keep the partitioning equilibrium as undisturbed as possible, it may be better to vary S and A in parallel at constant ratio $[S_T]/[A_T]$ (Figure 9.16D).

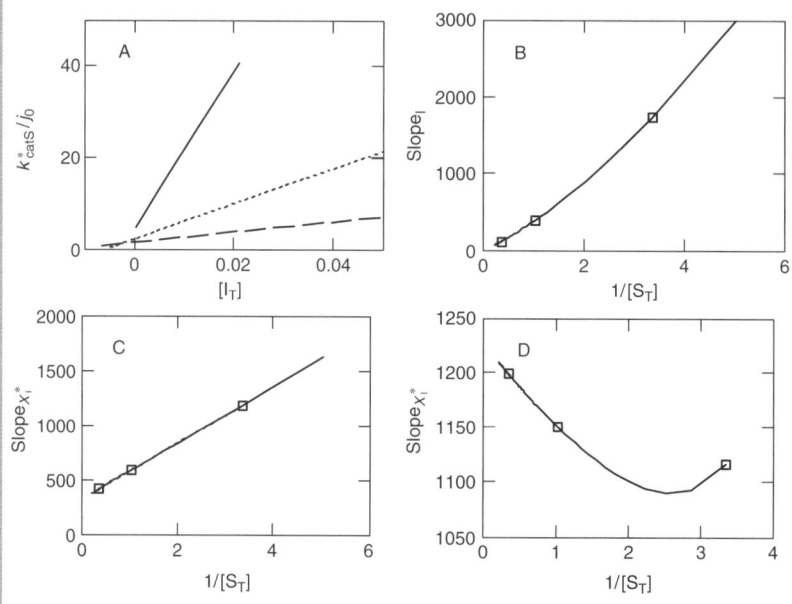

Figure 9.16 Dixon plot for the competitive inhibition in a three-component mixture. Data used: $[A_T] = 1$, $K_A' = 1$, $K_{MS}^* = 0.3$, $K_S^* = 0.1$, $K_S' = 0.5$, $K_I^* = 0.001$, $K_I' = 0.2$, $K_d = 1$ and $K_d^I = K_d^S = 0.1$. Panel A: $[S_T] = 0.3$, 1, 3 (from upper to lower curves). Panel B: slopes from the apparently straight lines in panel A. Panel C: same data as in panels A and B but with the slopes from k_{catS}^*/j_0 versus X_I^* plotted versus the inverse substrate concentration; the extrapolated y intercept is $K_{MS}^*(1 + 1/K_I^*)$. Panel D: same as panel C except that $[S_T]$ and $[A_T]$ are varied at constant $[A_T]/[S_T] = 3$; the shape of this plot varies depending on the parameters, but the slope$_{X_I^*}$ is approximately equal to $K_{MS}^*(1 + 1/K_I^*)(1 + [A_T]/[S_T])$

The representation in Equations (2) to (4) is not useful for a Lineweaver–Burk plot, $1/j_0$ versus $1/[S_T]$ at constant $[A_T]$ and $[I_T]$ (cf. Figure 9.13), since Equation (3) does not allow a way of holding $[I_T]$ constant when $[S_T]$ and X_S^* vary.

9.6.2 Solution-based inhibitor

In the limit where there is little inhibitor in the interface, $K_I' \gg [S_0^*]$, $X_I^* \ll 1$ and $[I] \approx [I_T]$, the term X_I^*/K_I^* in the flux relation (Equation 9.27) will be replaced by $K_d[I_T]/K_I K_d^I$ from Equations (9.22) and (9.23). Assuming, furthermore, that all the substrate is in the interface, $K_S' \ll [S_0^*]$ and $[S_0^*] = [S_T]$, the initial rate from Equation (9.27) with $X_S^* = 1$ and $X_P^* = 0$ can be expressed as

$$j_0 = \frac{\dfrac{k_{catS}^*}{K_{MS}^*}}{1 + \dfrac{K_d}{[S_T]}\left(1 + \dfrac{[I_T]}{K_I}\right) + \dfrac{[I_T]\,K_d}{K_I K_d^I} + \dfrac{1}{K_{MS}^*}} = \frac{k_{cat}^{app}\,[S_T]}{[S_T] + K_M^{app}} \tag{9.35}$$

When the bulk substrate concentration, $[S_0^*]$, is changed at constant inhibitor concentration, $[I_T]$, the initial rate can be written on the general MM form if the apparent MM parameters are identified as

$$k_{cat}^{app}\,([I_T]) = \frac{k_{catS}^*}{1 + K_{MS}^*}\;\frac{1}{1 + \dfrac{[I_T]}{K_I}\,\dfrac{K_d K_{MS}^*}{K_d^I\left(1 + K_{MS}^*\right)}} \tag{9.36}$$

$$K_M^{app}\,([I_T]) = \frac{K_d K_{MS}^*}{1 + K_{MS}^*}\;\frac{1 + \dfrac{[I_T]}{K_I}}{1 + \dfrac{[I_T]}{K_I}\,\dfrac{K_d K_{MS}^*}{K_d^I\left(1 + K_{MS}^*\right)}} \tag{9.37}$$

Thus, the solution-based inhibitor influences the apparent k_{cat} and K_M values differently and their ratio is

$$\frac{k_{cat}^{app}\,([I_T])}{K_M^{app}\,([I_T])} = \frac{k_{catS}^*}{K_d K_{MS}^*\left(1 + \dfrac{[I_T]}{K_I}\right)} \tag{9.38}$$

The inhibitor concentration $[I_T(50)] = K_I$ reduces the value of the ratio k_{cat}^{app}/K_M^{app} by 50 %; i.e. for a solution-based inhibitor, the inhibitor dissociation constant, K_I, can be determined by studying the behavior of the ratio k_{cat}^{app}/K_M^{app} as a function of $[I]$. Taking the ratio of the maximum rates, k_{cat}^{app}, without and with inhibitor gives

$$\frac{k_{cat}^{app}\,(0)}{k_{cat}^{app}\,([I_T])} = 1 + \frac{1}{1 + 1/K_{MS}^*}\,\frac{K_d}{K_d^I}\,\frac{[I_T]}{K_I} \tag{9.39}$$

The inverse of the rate can be expressed as

$$\frac{k_{cat}^{app}\,(0)}{j_0} = 1 + \frac{K_M^{app}\,(0)}{[S_T]}\left\{1 + \frac{[I_T]}{K_I}\left(1 + \frac{[S_T]}{K_d^I}\right)\right\} \tag{9.40}$$

The plot of inverse flux versus $[I_T]$ (Dixon plot) gives straight lines (Figure 9.17A) with slope

$$\text{Slope}_I = \frac{K_M^{app}\,(0)}{K_I}\left(\frac{1}{[S_T]} + \frac{1}{K_d^I}\right) \tag{9.41}$$

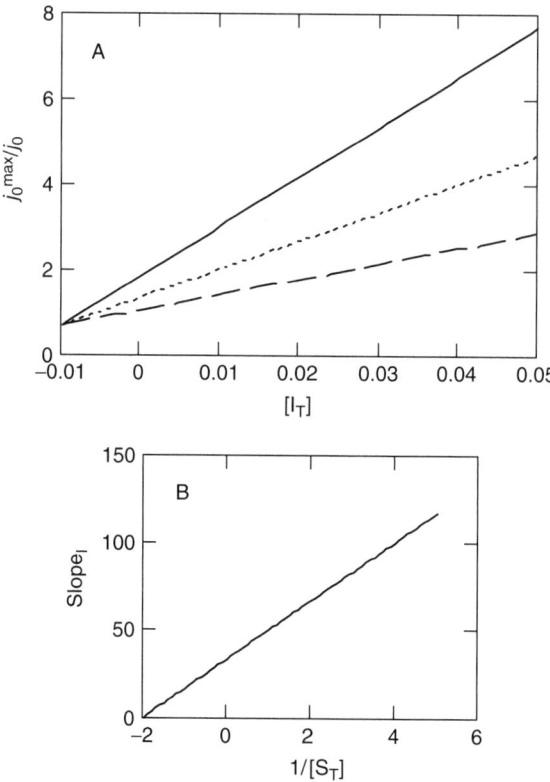

Figure 9.17 Dixon plot for the solution-based inhibitor. Data used: $K_{MS}^* = 0.5$, $K_I = 0.01$, $K_d = K_d^I = 0.5$. Panel A shows the inverse flux $k_{cat}^{app}(0)/j_0$ versus $[I_T]$ from Equation (9.40) at $[S_T] = 0.2, 0.5, 5$ (from upper to lower curves). Panel B shows the slopes of the curves in panel A plotted versus $1/[S_T]$ as described by Equation (9.41). This line has the x intercept at $1/[S_T] = -1/K_d^I$

As shown in Figure 9.17B, a plot of these slopes versus $1/[S_T]$ gives the extrapolated x intercept $1/[S_T] = -1/K_d^I$ and slope $K_M^{app}(0)/K_I$. This is in contrast to the interface-based competitive inhibitor where the x intercept is at $1/[S_T] = 0$ (Equation 9.7 and Figure 9.11) but similar to the results in Figure 9.8.

Note that these results are independent of how the inhibitor accesses the enzyme. The inhibitor can bind to the enzyme through the free enzyme, $E + I$, or by binding from the aqueous solution to the interface-bound enzyme, $E^* + I$. Both pathways give the same state E^*I. By detailed balance, the dissociation constant for $E^*I \rightarrow E^* + I$ will be $K_I K_d^I/K_d$, whether or not this pathway exists under the rapid exchange condition.

In the limit where $K_d^I = $ infinity, i.e. when the inhibitor works by blocking enzyme binding to the interface, k_{cat}^{app} is independent of inhibitor concentration. In this respect, the inhibitor works like a classical competitive inhibitor whose effect can be competed out by increasing the bulk concentration of the interface. Thus the presence of such an inhibitor has no effect on k_{cat}^{app} while K_M^{app} increases by the factor $(1 + [I_T]/K_I)$, just like the effect of a competitive inhibitor in solution kinetics (cf. Equation III-9 in Segel, 1975). It should be noted that this effect is independent of whether the inhibitor binds the catalytic site or the i-face of the enzyme; in either case, substrate access for the catalytic turnover is blocked.

9.6.3 Apparent noncompetitive behavior

In the limit where the inhibitor binding does not change the enzyme affinity for the interface very much, if $K_d^I \approx K_d/(1 + 1/K_{MS}^*)$, K_M^{app} of Equation (9.37) becomes independent of $[I_T]$; i.e. in this limit the inhibitor behaves like a traditional noncompetitive inhibitor (Equation III:28 in Segel, 1975). The relationship, which does not have to hold exactly, implies that the E^*I complex is bound to the interface as strongly as $E^* + E^*S$. Thus in this limit, even if the inhibitor does block the catalytic site, it appears to be noncompetitive by conventional criteria. This counterintuitive result is due to the fact that variation in bulk substrate concentration does not influence the concentration of substrate that the enzyme 'sees' at the interface for the formation of E^*S; it only changes the fraction of enzymes that are interface bound. Clearly, in this situation, increasing the bulk substrate concentration cannot compete out the inhibitor, which therefore appears noncompetitive. With the results at hand, the apparent noncompetitive behavior for the inhibition by alkanols (Figure 9.8) could be attributed to such an effect.

9.7 Summary: Multiple Pathways for Reduction of the Observed Rate

The search for PLA2 inhibitors has been motivated not only to understand prototypical interfacial inhibition kinetics and the catalytic mechanism, but also because these enzymes are implicated in a variety of physiopharmacological processes ranging from tissue necrosis to inflammation. The active-site-directed inhibitors of PLA2 have been useful for the identification of the catalytic site and for examination of the interfacial kinetic mechanisms. Many of these cases are now on firm quantitative footing for the secreted PLA2 and the results have provided insights into relationships between a range of underlying variables.

As discussed in Chapter 5, the scooting mode reaction progress offers the best conditions for the assay of specific inhibitors and also for the interpretation of the kinetic changes induced by specific inhibitors. Also low mole fractions of hydrophobic solutes, including the nonspecific inhibitors, have little or no effect on the scooting mode reaction progress unless the vesicle organization is disrupted. Under less stringent assay conditions, including those in the cellular and tissue environments, the nonspecific kinetic effect of a solute is most likely to manifest if the enzyme is not tightly bound to the substrate interface. Thus, a change in the kinetic path for the turnover cycle or the processivity leads to dramatic effects on the reaction progress or the solute concentration dependence.

Based on the interfacial kinetic paradigm the observed rate can decrease by a variety of perturbations. The reaction progress in the scooting mode can adequately sort out most of the nonspecific effects. As developed in Chapter 5, scooting mode kinetics provides a basis for resolving the relationship of the inhibitory potency with K_I^*, K_{MS}^* and K_P^*. The detailed balance condition (Chapters 6 and 8) shows that the binding of the inhibitor to the active site of PLA2 increases on the binding of the enzyme to the interface. The general treatment developed in this chapter provides a basis for identifying the interfacial competitive inhibitors, the cofactor dependence for the inhibitor binding and the effect of the partitioning of the inhibitor. Several interesting situations that develop under the limiting conditions of the fast exchange may be useful for diagnostic purposes. Although such considerations should be generally useful to account for the behavior of interfacial inhibitors, the number of underlying parameters certainly precludes the possibility of analysis in terms of the primary inhibition parameters.

Further Reading

Segel I.H. (1975). *Enzyme Kinetics: Behavior and analysis of Rapid Equilibrium and steady-state Enzyme systems*, Wiley, New York, 957 pp.

The Delay to the Steady State
in the Reaction Progress

. . . Are you the smoke from a fire that never burned?
Derek Walcott

Anomalous consequences of the properties of the zwitterionic substrate interface on the observed reaction progress have been extensively characterized. For example, a delay to the onset of the steady state phase with a maximum rate in the reaction progress of PLA2 is observed for the phosphatidylcholine vesicles and monolayers at the air–water interface. The delay to the steady state in the reaction progress at the monolayer interface (Figure 10.1) was first interpreted to suggest that the rate constant for the binding (penetration) of the enzyme to the interface is intrinsically slow (Zographi *et al.*, 1971; Verger *et al.*, 1973; Panaiotov *et al.*, 1997; Ransac *et al.*, 1997). Thus the reciprocal of the delay was attributed as the penetrating power of the enzyme as influenced by the quality of the interface.

In this chapter we analyze salient features of the delay phenomenon. Since virtually no delay is seen with the reaction progress in the scooting mode, the origin of the delay is in the pre-steady state step for the initial equilibration of the enzyme to the monolayer. Analytical strategies outlined in this chapter provide a basis for characterizing and understanding the origins of the pre-steady state complexities in terms of the information already established from the interpretation of the scooting and quasi-scooting mode reaction progress. Recall that unless the rate-limiting step is identified and the kinetic path is defined, it is virtually impossible to obtain the primary parameters for the elementary steps. For such reasons, from the reaction progress alone it is difficult, if not impossible, to distinguish the interfacial versus matrix catalytic path (Chapter 3). As developed in Theory Box 5.3, exchange of the enzyme results in lower processivity due to a lower residence time, and thus the reaction progress becomes virtually uninterpretable if the enzyme, substrate or product exchange between the coexisting interfaces is rate-limiting. Also, anionic charge at the interface lowers K_d (Chapter 6) and induces k^*_{catS} activation (Chapter 8). Coupled with such factors that determine the microscopic steady state environment in an ensemble of dispersed interfaces, the origin of the long delays to the steady state is shown to be due to a product-induced shift in the E to E* equilibrium,

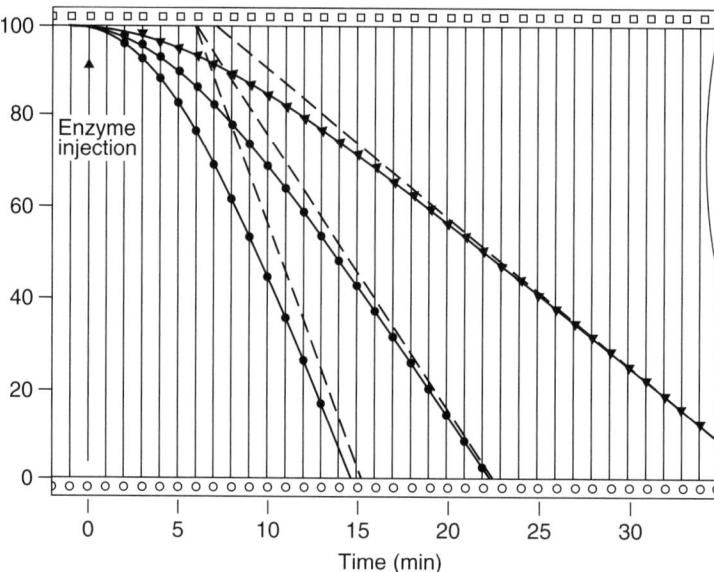

Figure 10.1 Reaction progress for the hydrolysis of DC$_9$PC monolayer at 5 mN on the injection of (from left) 11.8, 5.9 and 2.9 μg of pig pancreatic PLA2. The ordinate is the normalized area of the remaining substrate. From Verger *et al.* (1973). Reproduced by permission of the American Society for Biochemistry & Molecular Biology

but not in the underlying rate constant for the E to E* step. In fact, the delay is not a primary event to be described by a single rate constant; it is the cumulative consequence of an ensemble of the catalytic events that lead to the accumulation of product. The microscopic steady state is thus established by the binding of the PLA2 to the modified interface, followed by the activation of the bound enzyme by the anionic charge developed on the zwitterionic interface.

10.1 Effects of the Accumulated Products in Zwitterionic Bilayers

The delay to the steady state in the reaction progress for the hydrolysis of phosphatidylcholine monolayer and vesicles depends on the presence of the additives and shows a complex behavior with changing environmental conditions. The course of the reaction progress depends not only on the source and concentration of PLA2, or the structure of the substrate, or the presence of additives or the curvature of vesicles, but also on the reaction temperature. For example, as shown in Figure 10.2, the delay is completely abolished in the presence of the products of hydrolysis. A similar effect is seen with other anionic amphiphiles, including a codispersed second substrate. These results show that the rapid rate of reaction after a delay is due to the accumulation of the products of hydrolysis in the substrate vesicles. As discussed in Chapters 4 and 5, the products of hydrolysis of long chain phospholipids by PLA2 do not leave the interface. Thus results in Figure 10.2 clearly

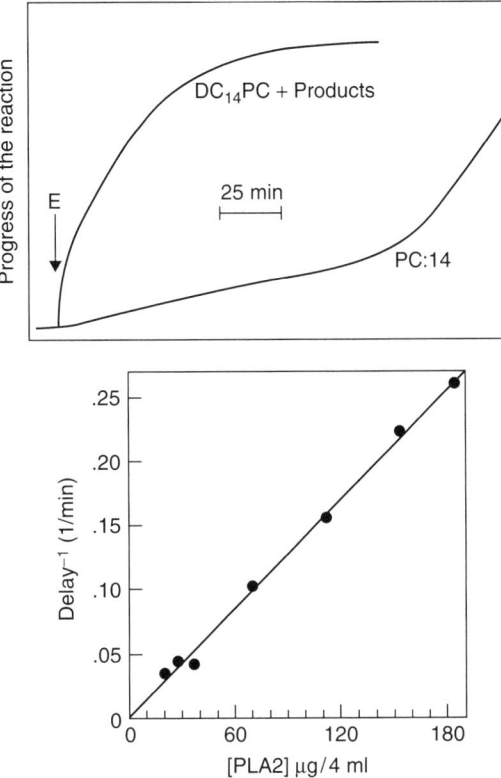

Figure 10.2 (Top) Reaction progress curves for the hydrolysis of $DC_{14}PC$ vesicles alone or of vesicles of the ternary codispersion of $DC_{14}PC$ + myristic acid + 1-myristoyl-lysophosphatidyl-choline (100:15:15 mole ratio) at 22 °C. From Apitz-castro *et al.* (1982). (Bottom) Dependence of 1/delay for the hydrolysis of 0.4 mM $DC_{16}PC$-luv at 38 °C as a function of the pig pancreatic [PLA2]. The conditions for determining the delay are as described by Romero *et al.* (1987)

show that the long delay to a rapid observed rate is due to the accumulation of the products of hydrolysis of long chain phosphatidylcholine vesicles.

Based on the evidence in the preceding chapters, there are at least four different effects of the accumulated product on the reaction progress: substrate dilution in the interface, product inhibition, enhanced binding of the enzyme to the interface due to a shift in the E to E^* equilibrium and the k^*_{catS} activation due to the charge compensation on the enzyme due to the build-up of the anionic charge at the interface. Such effects together open up possibilities for numerous kinetic paths as the residence time of the enzyme, as well as the equilibrium between the E^*S and $(E^*S)^{\#}$ forms (Figure 8.7), changes with the reaction progress. Although these effects depend on the mole fraction of the product in the interface, the dependence on X_P^* is not linear in most cases.

Interestingly, the delay time shown in Figure 10.2 (bottom) is inversely proportional to the enzyme concentration; or, put in another way, the delay time multiplied by enzyme concentration is an invariant. This is exactly as expected if the activation is determined by the extent of reaction progress (X_P^*) while the rate of change,

(Equation 10.2) is directly proportional to enzyme concentration (see also Theory Box 10.1). It is not expected if the delay is caused by an intrinsically slow interface binding of the enzyme.

10.1.1 Effect of the gel–fluid phase transition temperature

One of the most dramatic effects of the interface on the apparent rate of hydrolysis of phosphatidylcholine vesicles is shown in Figure 10.3. In this single time-point assay, the amount of product formed in 10 minutes shows a maximum at the gel–fluid phase transition temperature of the phosphatidylcholine substrate vesicles. Neither the delay nor the biphasic apparent zero-order or first-order rate is seen for the processive turnover in the scooting mode at the anionic interface (Figure 5.12). Thus the anomalous reaction progress with zwitterionic vesicles suggests that the thermotropic changes in the bilayer may also change the E to E* equilibrium and k^*_{catS} due to the accumulation and distribution of the products in the vesicles during the reaction progress.

The binding of pig pancreatic PLA2 to zwitterionic interfaces is rather poor, and the phase transition does not have a significant effect on the E to E* equilibrium on the ether analogs of the substrate alone. However, in the presence of the products of hydrolysis the binding of the pancreatic PLA2 increases in the gel and fluid coexistence region. The laterally phase-separated product, or the phase boundaries, in the bilayer could be the preferred site of the binding of PLA2 to the interface and the k^*_{catS} activation. This possibility is consistent with independent observations showing that the underlying processes are nonlinear.

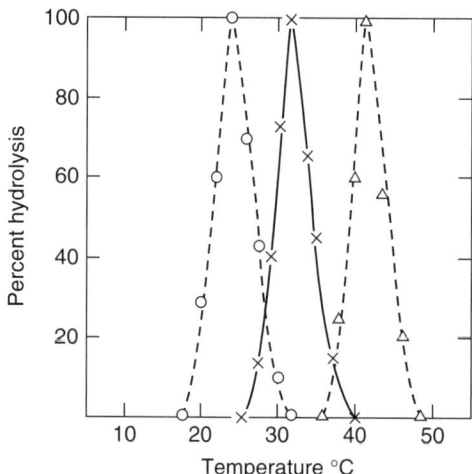

Figure 10.3 Normalized amount of the product formed after 10 minutes of incubation of $DC_{14}PC$ (circles), $DC_{16}PC$ (triangles), and their equimolar mixture with pig pancreatic PLA2 (crosses) at indicated temperatures. The maximum rate is observed at the gel–fluid phase transition temperature of the dispersion. Reprinted from Op den Kamp *et al.* (1975). Copyright (1975) with permission from Elsevier Science

10.1.2 Origin of the anomalous effect of the gel–fluid phase transition

The results in Figure 10.4 show that the magnitude of the delay is minimum at the gel–fluid transition temperature of the bilayer. The maximum rate at the end of the delay does not depend significantly on the phase transition temperature. However, with a suitably chosen single time point (Figure 10.3), a dramatic anomalous increase in the apparent rate would be seen at the transition temperature, where the delay to the steady state would be minimum. In fact, the delay also shows a strong dependence on the phase properties of the interface, as influenced by additives to the bilayer, as well as on the structural features of the phospholipid. The key features of this phenomenon for the pig pancreatic PLA2 are summarized below to provide a basis for the model in the next section:

(a) The delay decreases with the enzyme concentration (Figure 10.2).

(b) The delay is not seen with phosphatidylcholine vesicles containing about 0.15 mole fraction of the products of hydrolysis near the phase transition temperature (Figure 10.2). The steady state rate after the delay, or the initial rate in the presence of the products, does not show a significant dependence on the thermotropic phase properties of the zwitterionic vesicles.

(c) The products of hydrolysis lower the apparent K_d of the pig pancreatic PLA2 for the zwitterionic bilayer or micellar interface from 3 mM to about 0.1 mM. As developed in Chapter 8, the build-up of the negative charge at the interface has an allosteric activating effect on k^*_{catS} of the pancreatic PLA2, which is

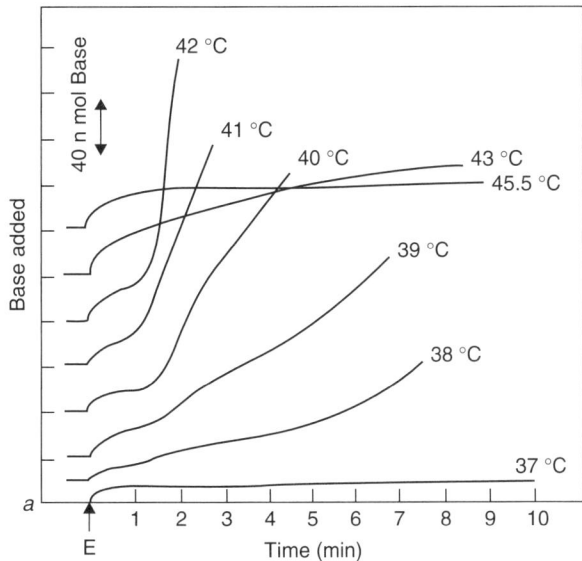

Figure 10.4 Reaction progress curves for the hydrolysis of multilamellar vesicles of $DC_{16}PC$ by bee venom PLA2 at the indicated temperatures. The gel–fluid phase transition temperature range for these dispersions is 40 to 42 °C. From Upreti and Jain (1980). Reproduced by permission of Springer-Verlag New York, Inc

difficult to resolve under the conditions of the reaction progress where the products accumulate in the vesicles.

(d) The rate of binding of PLA2 to the micellar or bilayer interface is rapid. The binding relaxation in the stopped-flow measurements consists of a second-order vesicle concentration-dependent diffusion-limited process coupled with a first-order process with a rate constant of about $4\,\text{s}^{-1}$ (Jain *et al.*, 1988).

(e) The fact that the long delay is seen under certain conditions but not others suggests that the underlying process is not a single event.

(f) The micelle–bilayer organization of the interface, or the gel–fluid coexistence in the anionic bilayer, has little effect on the turnover rate.

Together, these observations suggest that the delay to the onset of the rapid steady state phase of reaction depends on the changes in the interface during the pre-steady state. Such changes shift the equilibrium toward the PLA2 bound to the interface, which is followed by the activation of the bound enzyme by the charge compensation (Chapter 8).

10.2 Model for the Delay Due to the Product Accumulation during the Reaction Progress on Phosphatidylcholine Vesicles

Consider the situation where the enzyme binds only weakly to the interface, as depicted in Sections 5.6 and 5.7. The momentary flux per enzyme at some time when the mole fraction of the product is X_P^* is given by Equation (5.23). If the substrate and product are the only constituents of the interface and if the back reaction P→S is blocked, this gives the flux as

$$j = \frac{\dfrac{k_{catS}^*}{K_{MS}^*}\left(1 - X_P^*\right)}{1 + \dfrac{1 - X_P^*}{K_{MS}^*} + \dfrac{X_P^*}{K_{MP}^*} + \dfrac{K_d}{[M^*]}} \tag{10.1}$$

As before, $[M^*]$ equals the initial concentration of substrate accessible in the outer layer of the vesicles, $[S_0^*]$. For PLA2 we know that both the interfacial binding and the catalytic rate depend on the anionic charge from the amount of accumulated product in the interface. However, it is not known exactly how the affinity or catalytic rate depends on the mole fraction of product in the interface. In principle, such information could be found from a series of initial-rate measurements at well-defined X_P^*, which show an abrupt increase in the rate at $X_P^* > 0.08$ for $DC_{14}PC$ vesicles. Thus both K_d and k_{catS}^* in this expression must be considered as functions of X_P^*. When all product stays in the interface, $[P^*] = X_P^*[S_0^*]$. Thus the rate of change in X_P^* is

$$\frac{dX_P^*}{dt} = j\frac{[E_0]}{[S_0^*]} \tag{10.2}$$

from which the reaction progress, $X_P^*(t)$, could be calculated if the explicit X_P^*-dependence of j in Equation (10.1) were known. The experimental results discussed above suggest that this dependence will also be different at different stages in the gel–fluid phase transition range, which also changes with the mole fraction of the additives. At the transition temperature, the delay is the shortest, suggesting that the product-induced activation is the most efficient under these conditions. To see what the effect of this activation can be, simple linear expressions are applied in Theory Box 10.1. Even this simple description contains too many parameters for any quantitative (statistically uncorrelated) fit with the data.

The product-induced activation introduces a strong nonlinearity in the reaction progress. As can be seen in Figure 10.5, e.g. in the difference between the dashed and dash–dot curves, the effect is very sensitive to the details of this activation. In a single time-point experiment like that of Figure 10.3, it will make a big difference

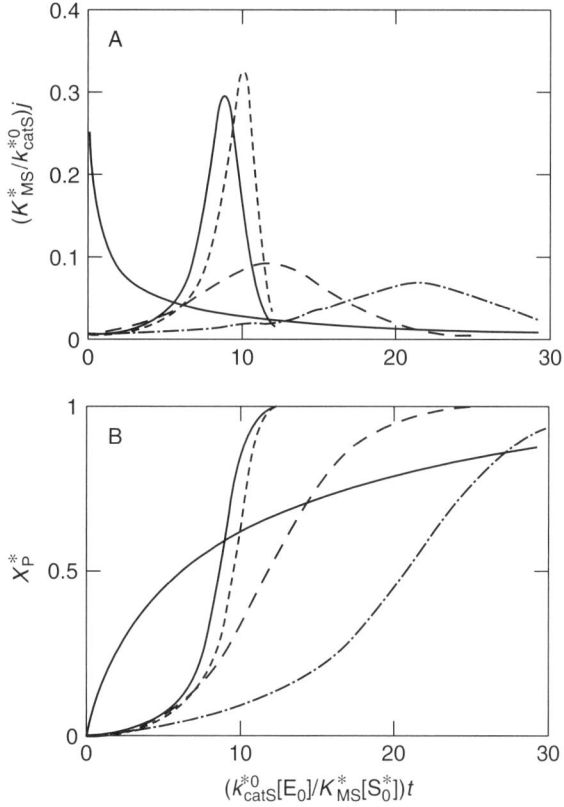

Figure 10.5 Expected time delay at zwitterionic vesicles from Equation (5) in Theory Box 10.1: solid (hyperbolic, no activation), $a = 0, b = 0, K_{MS}^* = 0.5, K_{MP}^* = 0.05, K_d^0/[M^*] = 1$; solid (sigmoid), $a = 50, \ b = 10, \ K_{MS}^* = 0.5, \ K_{MP}^* = 0.05, \ K_d^0/[M^*] = 200$; (dotted), $a = 500, \ b = 10, \ K_{MS}^* = 0.05,$ $K_{MP}^* = 0.5, \ K_d^0/[M^*] = 200$; (dashed), $a = 75, \ b = 0, \ K_{MS}^* = 0.5, \ K_{MP}^* = 0.05, \ K_d^0/[M^*] = 200$; (dash–dot), $a = 10, b = 10, K_{MS}^* = 0.5, K_{MP}^* = 0.05, K_d^0/[M^*] = 200$. Time on the x axis is scaled as $(k_{catS}^{*0}[E_0]/K_{MS}^*[S_0^*])t$. Panel A: flux per enzyme scaled as jK_{MS}^*/k_{catS}^{*0} as a function of time; panel B: X_P^* as a function of time

if this time is before or after the peak (Figure 10.5A) where most of the enzyme reactions take place. In terms of Figure 10.3, the activation is largest at the gel–fluid transition where all of the reactions take place before the 10 min time-point.

The product accumulation in phosphatidylcholine vesicles accounts, at least qualitatively, for the key features of the reaction progress. The underlying model is based on the properties established in earlier chapters. Additional assumptions required to account for the complexity resulting from the phase transition appear to be about the effect of the phase state of the bilayer on the distribution of the products, rather than a direct effect of the interface on the primary parameters for the turnover cycle at the interface.

THEORY BOX 10.1 Model for the time delay at zwitterionic vesicles

Although the mixing of the products of hydrolysis in phosphatidylcholine vesicles is not ideal, the simplest assumption is a linear form for activity and affinity:

$$k_{catS}^* = k_{catS}^{*0}(1 + aX_P^*) \tag{1}$$

$$K_d = \frac{K_d^0}{1 + bX_P^*} \tag{2}$$

where a and b are measures of the strength of activation. These forms are not likely to hold except as first-order expansions for small X_P^*, but to see the effects it will be assumed that they are valid for all X_P^*. Furthermore, if we consider the rapid-exchange limit (quasi-scooting), all vesicles will have the same mole fraction of products and all bound enzymes will 'see' the same environment. Then, from Equations (10.1) and (10.2), the rate of product formation is

$$\frac{dX_P^*}{dt} = j\frac{[E_0]}{[S_0^*]} = \frac{k_{catS}^{*0}[E_0]}{K_{MS}^*[S_0^*]} \frac{(1 + aX_P^*)(1 - X_P^*)}{1 + \frac{1 - X_P^*}{K_{MS}^*} + \frac{X_P^*}{K_{MP}^*} + \frac{K_d^0}{[M^*](1 + bX_P^*)}} \tag{3}$$

This differential equation can be written in the form

$$\frac{dX_P^*}{dt} = \frac{k_{catS}^{*0}[E_0]}{K_{MS}^*[S_0^*]} \frac{1}{\frac{A}{1 + aX_P^*} + \frac{B}{1 + bX_P^*} + \frac{C}{1 - X_P^*}} \tag{4}$$

This is a separable equation that can be integrated to give

$$\frac{k_{catS}^{*0}[E_0]}{K_{MS}^*[S_0^*]}t = \frac{A}{a}\ln(1 + aX_P^*) + \frac{B}{b}\ln(1 + bX_P^*) - C\ln(1 - X_P^*) \tag{5}$$

The parameters are

$$A = \frac{a - 1/K_{MP}^*}{a + 1} + \frac{1}{K_{MS}^*} + \frac{K_d^0}{[M^*]}\frac{a^2(1 + b)}{(a - b)(1 + a + b + ab)} \tag{6}$$

$$B = -\frac{K_d^0}{[M^*]} \frac{b^2(1+a)}{(a-b)(1+a+b+ab)} \tag{7}$$

$$C = \frac{1+1/K_{MP}^*}{a+1} + \frac{K_d^0}{[M^*]} \frac{1}{1+a+b+ab} \tag{8}$$

These expressions satisfy

$$A + B + C = 1 + \frac{1}{K_{MS}^*} + \frac{K_d^0}{[M^*]} \tag{9}$$

$$C - \frac{A}{a} - \frac{B}{b} = \frac{1}{a}\left(\frac{1}{K_{MP}^*} - \frac{1}{K_{MS}^*}\right) \tag{10}$$

The concentration of interface molecules $[M^*]$ is equal to the initial substrate concentration $[S_0^*]$. Together, Equations (5) to (8) describe the time-dependent reaction progress. In the limit when a and $b \to 0$, there is no activation and the result reduces to that of Equation (5.20) combined with Equations (5.27) and (5.28). Equation (5) shows that the time variable t and $[E_0]$ only occur together, so that changing enzyme concentration will only change the time scale, leaving the shape of the reaction progress curve invariant (Figure 10.5).

In the scooting limit where all enzyme remains bound without exchange, the reaction progress on a vesicle with n_E substrates bound would be given by Equation (5), where $K_d^0 = 0$ and $[E_0]/[S_0^*]$ is replaced by n_E/N_S^0 (cf. Equation 5.7). In this limit, the value of b is irrelevant and only k_{catS}^* activation contributes to the time delay.

When product activation has a more complicated form than the linear assumption in Equations (1) and (2), the resulting reaction progress can probably not be solved analytically. However, if the form of the activation as a function of X_P^* is known, the resulting equation corresponding to Equation (3) can be integrated numerically.

10.3 Effect of the Accumulated Products on the Delay in the Monolayer Reaction Progress

The elegant simplicity of the two-dimensional phase at the air–water interface has attracted considerable attention for the study of interfacial kinetic and equilibrium binding behavior of lipolytic enzymes in general and of PLA2 in particular. The principal features of the monolayer trough are shown in Figure 10.6. The lipolytic reaction progress is monitored as a change in the surface area or the surface pressure when at least some of the products leave the monolayer. In its simplest form the rectangular Teflon trough is filled with the aqueous subphase. As a known amount of the substrate amphiphile is spread at the surface, the surface tension of the aqueous phase changes. The change, defined as the monolayer pressure at the air–water interface, is monitored through a platinum plate attached to a balance. The surface pressure changes with the amount of an amphiphile partitioned into a given

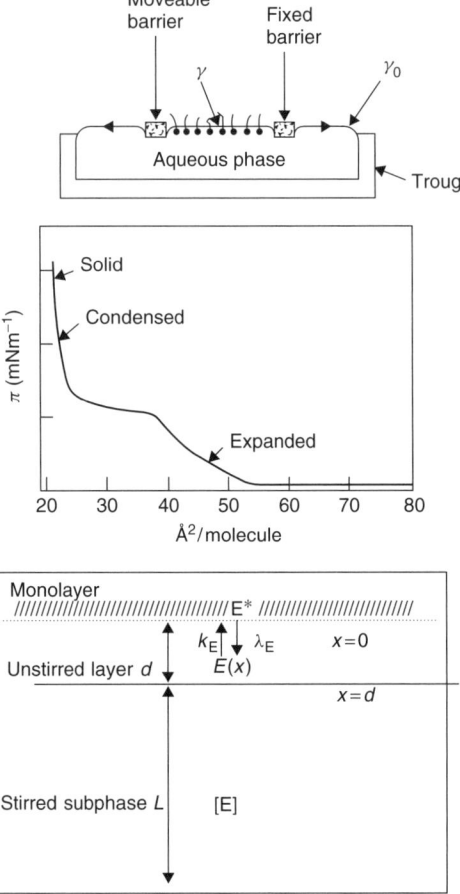

Figure 10.6 (Top) A schematic diagram for the monolayer trough in which the amphiphile monolayer at the air–water interface can be compressed systematically by changing the position of the movable barrier. The surface pressure changes with the area per molecule of the amphiphile at the interface (Chapter 2). (Middle) The dependence of the surface pressure with the area per acyl chain shows a complex dependence because the average molecular orientation changes with the two-dimensional phase states and depends on the degree of compression. The expanded, condensed and solid phases are comparable to the three-dimensional liquid, liquid crystalline and crystalline phases. The organization of the amphiphile in a particular phase state also depends on the structure of the amphiphile as well as the temperature. The laterally disordered compressible expanded phase is often used for monitoring the reaction progress with interfacial enzymes. (Bottom) Cross-section of the vicinity of the monolayer interface where the interfacial turnover occurs. It emphasizes the microscopic features in the vicinity of the head groups of the amphiphile. Note that the stirred bulk aqueous subphase is separated from the head group region (the horizontal dotted line) by an unstirred layer of the aqueous phase

area of the interface. Often the amphiphile is chosen such that the substrate forms an essentially 'insoluble' monolayer; the product does not form the monolayer. Thus after adding enzyme to the subphase a change in the pressure sensed by the plate (first-order conditions) or the area (zero-order conditions) swept by the movable barrier to maintain a constant surface pressure provides a measure of the change in the number of amphiphiles at the interface as a function of time (Figure 10.7).

Essentially all the anomalous characteristics of the reaction progress in vesicles of long chain phospholipids can be produced with the reaction progress for the hydrolysis of the monolayer of medium chain phospholipids. For example, in a first-order reaction progress where the surface pressure is allowed to change (Figure 10.7), a delay is seen which changes with the surface pressure. The steady state of the reaction progress by pig pancreatic phospholipase A2 (PLA2) on a phosphatidylcholine monolayer is preceded by a period of pre-steady state delay. The delay has been attributed to the penetrating power of PLA2 controlled by the intrinsic kinetic constants k_E and λ_E (see Figure 10.6 and Theory Box 10.2). According to this formulation, the delay to the steady state is attributed to the slow

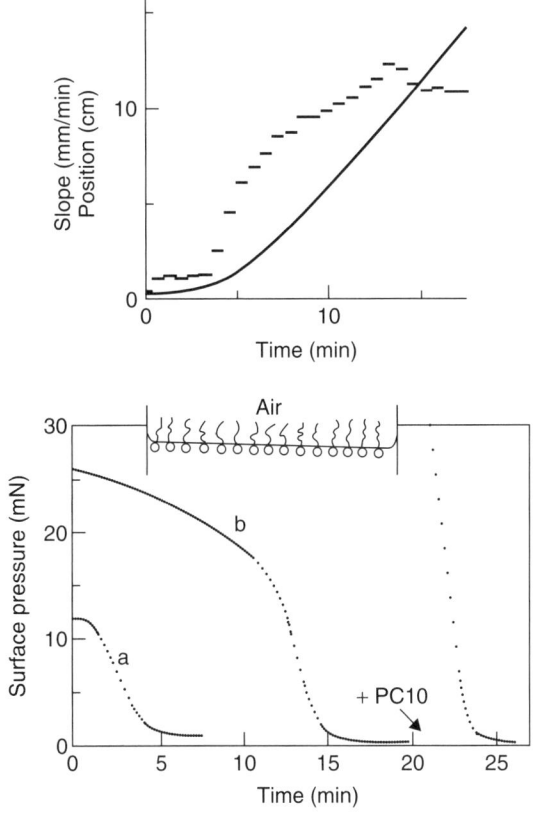

Figure 10.7   The reaction progress for the hydrolysis of didecanoylphosphatidylcholine (DC$_{10}$PC) monolayer at the air–water interface by 5.6 µg of pig pancreatic PLA2. (Top) The reaction progress monitored under the zero-order conditions as the change in the surface area (continuous line). The rate at any time-point, the derivative of the reaction progress, is shown as the dashed line. (Bottom) Reaction progress monitored under the first-order conditions as the change in the surface pressure (7 cm diameter). (a) The reaction at 12.5 mN/m or (b) 26 mN/m was initiated by the addition of the enzyme in the 40 mL aqueous subphase. At the end of the reaction progress in (b) at 21 min (arrow), the same amount of the substrate was spread on the surface. Reprinted with permission from Cajal *et al.* (2000a). Copyright (2000) American Chemical Society

binding to the interface as an intrinsic kinetic property of the enzyme. According to this interpretation the delay depends on the 'penetrating power' of the enzyme as well as the 'quality of the interface' such as the surface pressure (Verger *et al.*, 1973; Panaiotov *et al.*, 1997). Critical tests show that this model is inconsistent with the experimental results.

Key observations described below provide a basis for a model for the monolayer reaction progress. However, intrinsic features of the monolayer assay make it impossible to analyze the reaction progress in terms of the primary events of the turnover cycle at the interface:

(a)   If the lag is a measure of the intrinsic rate constant for the E to E* step, the onset of the steady state would be a first-order process. The onset of the delay obtained as the derivative of the zero-order reaction progress is sigmoidal (Figure 10.7, top).

(b)   A first-order process for the binding of the enzyme to the interface leading to the onset of the steady state should be independent of the amount of enzyme in the subphase. As shown in Figure 10.8, the enzyme concentration dependence of the delay to the steady state is significant at the 20 mN surface pressure. These results suggest that the delay at high pressures has an enzyme concentration-dependent component as well as an enzyme concentration-independent component. The enzyme concentration-dependent delay is not seen at lower pressures. This implies that the delay at low surface pressure is not caused by the enzyme-dependent product accumulation. At high surface pressure the delay is inversely proportional to the enzyme concentration (Figure 10.8) just as it was for the delay on zwitterionic vesicles (Figure 10.2). This is consistent with the proposition that the activation is caused by the extent of the reaction progress.

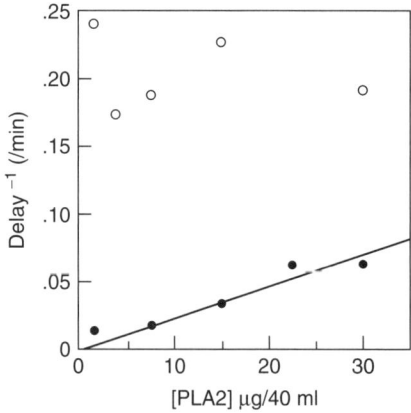

Figure 10.8   The pig pancreatic PLA2 concentration dependence of 1/delay to the steady state in the reaction progress curve for the hydrolysis of $DC_{10}PC$ monolayer at 20 mN/m (filled circles) or 15 mN/m (unfilled circles) in a zero-order trough at pH 8. The steady state rate of hydrolysis at the end of the delay shows a linear dependence on the enzyme concentration. From Berg, Cajal and Jain (unpublished results)

(c)   As shown in Figure 10.7 (bottom), the lag period is not seen in the presence of the products of hydrolysis. As also shown below, the product release from the monolayer is not instantaneous, and during the reaction progress a significant amount of the product of hydrolysis remains in the monolayer.

(d)   The slope of the steady state phase of the reaction progress exhibits a linear dependence on the amount of enzyme added to the aqueous subphase. Irrespective of the enzyme concentration or the surface pressure, the maximum measured apparent rate is less than 1 s^{-1} for the hydrolysis of $DC_{10}PC$ monolayer, compared to a rate of 500 s^{-1} with the dispersions, possibly as large cylindrical micelles, of the same substrate in the aqueous phase. Also, the measured rates for the hydrolysis of homologous substrates in the micellar or bilayer interfaces do not differ significantly. If there is no direct effect of the monolayer interface on the intrinsic catalytic parameters of the enzyme, lower monolayer rates suggest that only a small fraction of the total enzyme in the reaction compartment is involved in the turnover at the monolayer interface. This is not an unreasonable expectation because the ratio of the aqueous phase to the interface under the monolayer assay conditions is thousands of times higher than in the assays with the bilayer or micellar substrates.

(e)   As outlined below, the most intractable problem for interpretation of monolayer kinetics comes from the effect of the unstirred layer during the pre-steady state diffusion of the enzyme and during the steady state reaction progress for the diffusion of the product away from the monolayer.

10.3.1   Trough geometry and unstirred layer

Based on the results summarized above, key assumptions emerge for a model to analyze the course of PLA2 catalyzed hydrolysis at the air–water monolayer. Virtually all major features of the reaction progress on the monolayer interface can be predicted on the basis of a quantitative model that takes into consideration the geometry of the monolayer trough (Figure 10.6). In this model, the catalytic cycle events at the monolayer interface are identical to those at any other interface. However, the hydrolysis at the monolayer interface requires consideration of the trough geometry that controls the accumulation in and diffusion through the unstirred aqueous layer between the monolayer and the bulk aqueous phase.

10.3.2   Diffusion through the unstirred layer

As shown in Figure 10.6, the monolayer with a total surface area A is spread on top at the air–water interface. The solution volume is V and the depth of the trough is $L = V/A$. If the monolayer does not cover the whole surface of the trough, L should still be defined as V/A in the calculations below. The solution is stirred but there is an unstirred layer of thickness d just below the monolayer. Two factors contribute to the pre-steady state delay to the rapid steady state phase of the reaction progress:

(a) First, the equilibration of the monolayer bound (E^*) and free enzyme in the stirred aqueous subphase (E) requires diffusion through the unstirred aqueous layer adjacent to the monolayer. This causes a delay that is independent of the enzyme concentration, as seen at low surface pressures (Figure 10.8). The *diffusion-limited equilibration* of the enzyme and the products through the unstirred layer between the aqueous phase and the monolayer interface occurs on the time scale of several minutes.

(b) Second, if the anionic products of hydrolysis accumulate in the monolayer, the charge build-up would account for the *enzyme concentration-dependent delay* at higher surface pressures (Figure 10.8), typically above 15 mN for PLA2 and $DC_{10}PC$ monolayer. This component of the delay leads to a build-up of anionic surface charge because the products of hydrolysis do not rapidly leave the monolayer interface.

THEORY BOX 10.2 Stationary diffusion to the monolayer through the unstirred layer

Consider the concentration, $E(x)$, of enzyme in solution at distance $x(0 < x < L)$ from the monolayer (Figure 10.6). (To simplify handling of units, use dm for all distances to make them commensurate with the concentration units, M, as (mole/dm^3.) The rate of binding to the monolayer is $k_E E(0)$, which defines the local binding rate constant k_E (M^{-1}dm^{-2}s^{-1}) per unit surface area of the monolayer. The corresponding rate of desorption of bound enzyme is λ_E(s^{-1}). The surface density of bound enzyme is E_i (mole/dm^2). Thus, expressed as the concentration per solution volume, $[E^*] = E_i/L$ (M). Due to the stirring in the bulk aqueous phase, the enzyme concentration is constant, $E(x) = [E]$ (M) for $d < x < L$. In the unstirred layer, the diffusion flux towards the surface satisfies Fick's first law, $J_E = D_E(dE/dx)$. Thus, for a stationary flux, the enzyme concentration in the unstirred layer can be found by integrating Fick's law as

$$E(x) = (J_E/D_E)(x - d) + [E] \qquad \text{for } 0 < x < d \qquad (1)$$

where J_E (M dm/s) is the association flux per unit surface area of the interface and D_E (dm^2/s) is the diffusion constant of the enzyme. At the surface, $x = 0$, the boundary condition is

$$k_E E(0) - \lambda_E E_i = J_E \qquad (2)$$

Setting $x = 0$ in Equation (1) and inserting $E(0) = [E] - J_E d/D_E$ into Equation (2) gives the net association flux of enzymes as

$$J_E = \frac{k_E[E] - \lambda_E L [E^*]}{1 + k_E d/D_E} \qquad (3)$$

At equilibrium, $J_E = 0$ and $[E]/[E^*] = \lambda_E L/k_E = K_d/[M^*]$, where $[M^*]$ as before is the concentration of the interface-forming molecules per volume of the solution. Equation (3) allows us to identify effective rate constants for dissociation and association:

$$k_{\text{off}} = \frac{\lambda_E}{1 + k_E d/D_E} \xrightarrow[k_E, \lambda_E \to \infty]{} \frac{D_E}{Ld} \frac{K_d}{[M^*]} = \frac{D_E a_M N_{Av}}{d} K_d \quad (\text{s}^{-1}) \quad (4a)$$

$$k_{\text{on}} = \frac{k_E/L \, [M^*]}{1 + k_E d/D_E} \xrightarrow[k_E \to \infty]{} \frac{D_E}{Ld} \frac{1}{[M^*]} = \frac{D_E a_M N_{Av}}{d} \quad (\text{M}^{-1}\text{s}^{-1}) \quad (4b)$$

where a_M is the area per molecule in the monolayer and N_{Av} is Avogadro's number. In this way the on-rate has been counted per molecule in the monolayer so that these expressions are directly comparable to those for the spherical aggregates, Equations (3) and (4) in Theory Box 2.1. The limit when k_E and λ_E are very large corresponds to the diffusion-controlled limit. The effect of the unstirred layer is to slow down both association and dissociation.

Starting from an out-of-equilibrium distribution of enzymes, the equilibration time is determined by

$$\tau_E = \frac{1}{k_{\text{on}} [M^*] + k_{\text{off}}} = \frac{Ld}{D_E} \frac{[M^*]}{K_d + [M^*]} \quad (5)$$

The last equality holds in the diffusion-controlled limit. This can be compared to the mean time for an enzyme molecule to diffuse across a distance d (see Theory Box 12.2):

$$\tau_{\text{diff}} = \frac{d^2}{2D_E} \quad (6)$$

The main difference between Equations (5) and (6) is that Equation (5) allows the enzymes to diffuse through the whole aqueous solution while Equation (6) would be the equilibration time if the enzymes were confined to the unstirred layer (i.e. if $L = d$).

The diffusion of product can be described in exactly the same way if we replace E by P everywhere with an appropriate change for the parameter subscripts.

10.3.3 Time for diffusion to the monolayer versus the equilibration time through the unstirred layer

For the enzyme we can estimate $D_E = 10^{-6}$ cm^2/s, $L = 1$ cm and $d = 10^{-3}$ cm, which gives a maximal diffusion-limited time delay of 17 min from Equation (5) in Theory Box 10.2. If binding is not very strong, e.g. if at equilibrium only 10 % of the enzyme is in the interface, the diffusional time delay for enzyme binding is only 10 % of the maximum estimated for very strong binding, i.e. ca. 1.7 min.

Similarly, for the products, $D_P = 2 \times 10^{-6}$ cm^2/s, which gives a maximal diffusion delay of 8 min. If at equilibrium only 20 % of the product partitions into the monolayer, the delay time is ca. 100 s. Due to the uncertainty, mostly in the thickness of the unstirred layer, d, these are only order-of-magnitude estimates. Also, if the rates are not fully diffusion-limited, the delay times would be larger. The delay times for an unstirred layer of thickness 10^{-3} cm can be substantial. In contrast, the mean time for a single enzyme molecule to diffuse through the unstirred layer, as given by Equation (6) in Theory Box 10.2, is very short – smaller than 1 s in general. However, this time is related primarily to the time taken to set up a stationary diffusion flux through the unstirred layer, and it is not a measure of the time delay for the equilibration of enzyme binding to the interface.

10.3.4 Product accumulation and the build-up of the anionic charge at the interface

The product accumulation in zwitterionic monolayers leads to an enhanced binding and turnover of the enzyme in the interface (Cajal *et al.*, 2000a). For example, the products of hydrolysis alone do not form monolayers. On the other hand, as shown in Figure 10.9, depending on the surface pressure of the substrate monolayer a

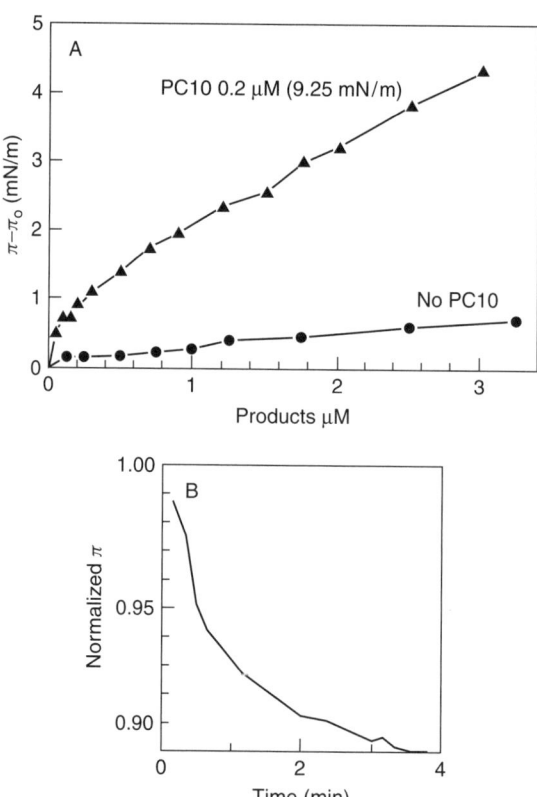

Figure 10.9 Effects of the products (1:1) of PLA2 catalyzed hydrolysis on the properties of $DC_{10}PC$ monolayer. Panel A: the change in equilibrium surface pressure (triangles) of the monolayer of 0.008 μmol $DC_{10}PC$ at 9.25 mN or without any lipid (circles). Panel B: the time course of the change in the surface pressure of $DC_{10}PC$ monolayer on the addition of the product at 9.25 mN. Reprinted with permission from Cajal *et al.* (2000a). Copyright (2000) American Chemical Society

significant amount of the product is retained in the monolayer at equilibrium (top) as well as during the steady state (bottom). Also, the equilibrium partitioning of the product shows a biphasic dependence on the surface pressure of the monolayer, which accounts for a complex dependence of the delay and the turnover on the surface pressure.

When products are present in the monolayer, PLA2 binds better and has a higher activity (k_{catS}^*). Thus some product must accumulate in the monolayer before the enzyme becomes fully active. Initially, enzyme activity is low and product is formed only slowly. As product starts to accumulate in the monolayer, more enzyme is bound and the observed rate increases. The slow initial build-up of product and consequent shift in the E to E* equilibrium leads to a delay before a maximal rate is reached. Since product production depends on the amount of enzyme present, this delay is strongly dependent on enzyme concentration (Figure 10.8). If product desorbs immediately, this delay would be determined primarily by the time it takes to get so many product molecules in solution that they start to partition back into the monolayer as an equilibrium effect. In this limit the delay time could be very long, possibly several hours. Alternatively, if product desorption is slow, product accumulates faster and the delay in enzyme activation will be relatively shorter.

In short, the product effects on the monolayer reaction progress for the hydrolysis of phosphatidylcholines are seen in the equilibrium as well as the steady state conditions (Theory Box 10.3). Also, the anionic charge induced by the accumulated product in the interface increases both the fraction of the total enzyme bound to the interface and the activity per enzyme by k_{catS}^* activation (Chapter 8). The apparent rates thus change with any change in the product accumulation profile.

THEORY BOX 10.3 Delay to steady state in monolayer kinetics

The product flux (molecules per unit surface area) away from the monolayer can be expressed as (cf. Equation 3 in Theory Box 10.2)

$$J_P = k_d (P_i - (k_P/\lambda_P)[P]) \tag{1}$$

where

$$k_d = \frac{\lambda_P}{1 + k_P d/D_P} \xrightarrow{\text{diffusion limit}} \frac{D_P}{Ld} \frac{\lambda_P L}{k_P} = \frac{D_P}{Ld} \frac{K'_P}{[M^*]} \tag{2}$$

is the effective desorption rate (cf. Equation 4a in Theory Box 10.2). Here,

$$K'_P = [P]/X_P^* = [M^*]\lambda_P L/k_P$$

is the dissociation constant for product in the monolayer. This flux gives rise to an increase in the solution concentration of product:

$$\frac{d[P]}{dt} = J_P/L \tag{3}$$

The product density in the monolayer increases due to enzyme activity, by V product molecules per unit surface area and unit time, and decreases by the desorption flux J_P:

$$\frac{dP_i}{dt} = V - J_P \qquad (4)$$

If a product pair in the interface take up roughly the same area as a substrate molecule then the mole fraction of the product in the interface can be introduced as $X_P^* = P_i/S_i^0$, where S_i^0 is the initial surface density of substrate in the monolayer. Also, the product concentration in solution can be normalized to the initial amount of substrate as $Y = [P]L/S_i^0 = [P]/[M^*]$. This gives the rate equations for the two unitless variables as

$$\frac{dX_P^*}{dt} = \frac{V}{S_i^0} - k_d X_P^* + \frac{D_P}{Ld}Y \qquad (5a)$$

$$\frac{dY}{dt} = k_d X_P^* - \frac{D_P}{Ld}Y \qquad (5b)$$

The enzyme activity V depends on X_P^* as well as on the amount of enzyme that is bound. Assuming for simplicity that enzyme activation depends linearly on X_P^* at low mole fractions of the product, while enzyme binding is independent, one can write (cf. Theory Box 10.1)

$$V = \begin{cases} \dfrac{E_i k_{catS}^{*0}}{K_{MS}^*} \dfrac{\left(1 + aX_P^*/X_P^{act}\right)\left(1 - X_P^*\right)}{1 + \left(1 - X_P^*\right)/K_{MS}^* + X_P^*/K_{MP}^*} & \text{for } X_P^* < X_P^{act} \\[4mm] \dfrac{E_i k_{catS}^{*0}}{K_{MS}^*} \dfrac{\left(1 + a\right)\left(1 - X_P^*\right)}{1 + \left(1 - X_P^*\right)/K_{MS}^* + X_P^*/K_{MP}^*} & \text{for } X_P^* > X_P^{act} \end{cases} \qquad (6)$$

Here, the maximum catalytic rate, $k_{catS}^{*0}(1 + a)$, is reached when the product is at mole fraction X_P^{act} or above. When there is no product present the activity, k_{catS}^{*0}, is low. The factor $1 - X_P^*$ accounts for the depletion in the substrate density. The amount of bound enzyme is given from Theory Box 10.2 as

$$\frac{dE_i}{dt} = J_E = k_{on}L[E] - k_{off}E_i = \frac{D_E}{d}\left\{[E_0] - \frac{E_i}{L}\left(1 + \frac{K_d}{[M^*]}\right)\right\} \qquad (7)$$

The last equality holds in the diffusion-controlled limit. If product accumulation shifts the E to E* equilibrium, K_d in Equation (7) must depend on X_P^* (e.g. like Equation (2) in Theory Box 10.1). These rate equations (5) to (7) can easily be integrated numerically.

   To get a feel for the importance of the different parameters, we can first get a rough estimate for the time delay by looking at the initial rates, assuming $Y = 0$, $X_P^* < X_P^{act} \ll 1$ and a negligible time delay for enzyme binding. Introducing for convenience the notation $k_0 = k_{catS}^{*0} E_i^{eq}/[(1 + K_{MS}^*)S_i^0]$ and integrating the rate equation for X_P^*, one finds

$$X_P^*(t) = \frac{e^{\left(ak_0/X_P^{act} - k_d\right)t} - 1}{a/X_P^{act} - k_d/k_0} \qquad (8)$$

> The time to reach $X_P^* = X_P^{act}$, where the enzyme activity is maximal, is
>
> $$\tau \approx \frac{X_P^{act} \ln \left(1 + a - k_d X_P^{act}/k_0\right)}{a k_0 - k_d X_P^{act}} \tag{9}$$
>
> This is very sensitive to the enzyme concentration, E_i (through k_0), as well as to the product desorption rate, k_d. When the desorption rate is too large relative to enzyme catalysis, there is little product build-up in the monolayer until there is so much in solution that product starts to partition back into the monolayer.

10.3.5 Reaction progress from the model

Some results from the model described in Theory Box 10.3 are plotted in Figure 10.10. The monolayer method measures the desorption of product from the interface. Thus Figure 10.10A can be compared with the experimental result displayed in the top panel of Figure 10.7. The analysis clearly shows that the products of hydrolysis partitioned in the $DC_{10}PC$ monolayer qualitatively can account for virtually all the reported features of the reaction progress at the monolayer interface. These predictions of the model are consistent with the experimental results, and thus affirm a unified and consistent basis for the sPLA2 catalyzed turnover at the various organizational forms of phospholipid aggregates. Unfortunately, peculiarities of the monolayer trough preclude complete analysis. A major problem is to account for the large fraction of enzyme that is not in the interface; if that fraction is bound at the walls, it is impossible to know how a redistribution takes place when product accumulation shifts the E to E* equilibrium. At present there is no evidence that there is any effect of the monolayer surface pressure on the intrinsic catalytic parameters of sPLA2 at the interface. The monolayer method is unlikely to provide primary rate and equilibrium parameters for interfacial catalysis because the microscopic steps of the catalytic turnover cycle cannot be unequivocally identified; nor can the steady state condition for the variables be independently established.

10.4 Summary: Activation by the Anionic Charge Induced by the Product Accumulation

Reaction progress for the hydrolysis of phosphatidylcholine vesicle and monolayers exhibits a delay of minutes to hours to the rapid steady state phase of the reaction progress. The magnitude of the delay depends on the experimental conditions such as temperature, composition and the presence of additives. The delay is not seen in the presence of the products of hydrolysis at the zwitterionic monolayer or bilayer interface. Nor is it seen during the scooting mode reaction progress on anionic vesicles, nor during the quasi-scooting mode reaction progress on micelles of zwitterionic or anionic short chain phospholipids. These observations show that during the delay to the steady state the anionic charge builds up due to accumulation of a critical mole fraction of the products of the PLA2-catalyzed hydrolysis in zwitterionic interfaces. The activating effect of the anionic charge

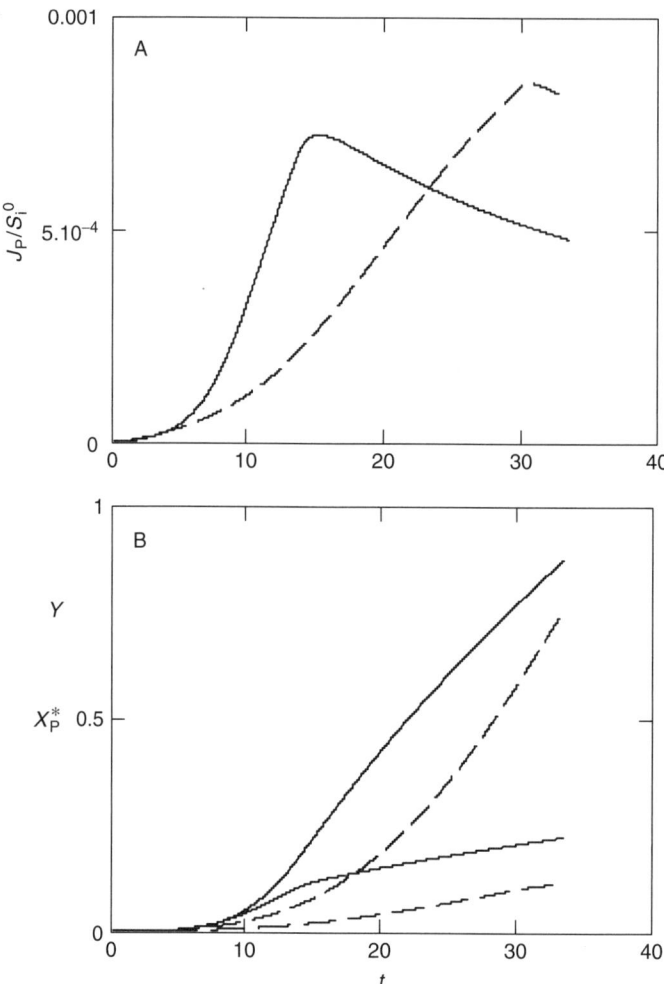

Figure 10.10 Time delay in monolayer kinetics from Equations (5) to (7) in Theory Box 10.3 with k^*_{catS} activation and no product-induced shift in E to E* equilibrium. Data used: $a = 100$, $k^{*0}_{catS}E^{eq}_i/S^0_i = 2.5 \times 10^{-5}$ s^{-1}, $K^*_{MS} = 0.5$, $K^*_{MP} = 0.05$, $X^{act}_P = 0.1$, $(D_E/Ld)(1 + K_d/[M^*]) = 0.008$ s^{-1}, $D_P/Ld = 0.002$ s^{-1}, $k_d = 0.01$ s^{-1} (solid curves) and $k_d = 0.02$ s^{-1} (dashed curves). Panel A: the flux J_P/S^0_i(s^{-1}) as a function of time. Panel B: upper curves, the reaction progress $Y = [P]/[M^*]$ as a function of time (min) and lower curves, mole fraction product in the interface X^*_P as a function of time (min)

is primarily attributed to the E to E* equilibrium and the k^*_{catS} activation of the bound enzyme. In contrast, the intrinsic rate constant for the binding of the enzyme to the interface is less than 1 s. Thus all slower effects are attributed to product accumulation effects or the extraneous factors such as the pre-steady state diffusion through the unstirred layer.

The delay in the onset of the steady state in the reaction progress for the hydrolysis of monolayers at the air–water interface has been a key observation in the debate about the origin of the interfacial activation of pancreatic PLA2. Results and analysis outlined in this chapter clearly show that the delay to the steady state has two components: the pre-steady state diffusion of the enzyme to the monolayer through the unstirred layer and the diffusion of the products from the monolayer to the aqueous phase. Both of these processes occur on the time scale of minutes. Less product partitions in the interface at higher pressures, and therefore the apparent rate decreases, and the delay to the onset of the steady state increases, as both the fraction of the enzyme at the interface and the catalytic rate decreases.

Together, results at hand clearly show that the delay is not a consequence of the 'penetrating power' of the enzyme; i.e. irrespective of the interface the intrinsic rate of binding of the enzyme to the interface is rapid. As a general, unified and consistent basis for the interpretation of interfacial kinetics, we conclude that irrespective of the nature of the interface and the equilibrium for the enzyme, binding to the interface depends on the magnitude of the interfacial anionic charge. Key features of the reaction progress with delay can be modeled with assumptions appropriate for the vesicle or the monolayer interfaces. In addition to the turnover parameters and the k^*_{catS} activation, for such analyses one must also invoke the factors that control the E to E* equilibrium. The number of primary parameters in both cases is simply too large to be determined independently. However, the analysis in this chapter clearly shows that the results are consistent with the assumptions and parameter values established earlier by independent methods.

Considerable effort, at least as judged from the published literature, has been invested in the study of interfacial enzymes at the monolayer interface in general and of saturated phosphatidylcholine vesicles in particular. Although these assays have provided rich and curious phenomenology, it is also clear that the underlying anomalies are dominated by the effects that influence the interface accessibility for the enzyme. The activating effect of the anionic charge is due to accumulated product in bilayers. This may have relevance for the substrate accessibility in the cellular environment; the effect of the phase transition properties on the binding of the enzyme due to an effect on the product distribution may influence the hydrolysis of mixed-lipid bilayers of membranes. While the reaction progress at high surface pressure does exhibit the effect of accumulated product, at low surface pressure these results are dominated by the diffusion through the unstirred layer in contact with the monolayer. Due to the inability to measure the fraction of the enzyme at the interface, coupled with the indications that a significant fraction of the enzyme is associated with the walls of the trough, the usefulness of the monolayer measurements remains to be demonstrated, at least for the quantitative kinetic or equilibrium measurements.

11 Nonidealities of the Dispersed Phases

If you ask me whether there is another world, well, if I thought there were, I would say so. But I don't say so. And I don't deny it. And I don't say there neither is, nor is not another world. And if you ask me about the beings produced by chance; or whether there is any fruit, any result, of good or bad actions; or whether a man who won the truth continues or not after death – to each or any of these questions do I give the same reply.

Sanjaya Belatthiputta (ca. 600 BC)

From the results and discussion of the interfacial kinetic paradigm in the preceding chapters it is clear that the intrinsic interfacial catalytic turnover parameters of a kinetic path are modulated by the variables and allosteric effects controlled by the interface. The key factors controlling the variables include the accessibility of the interface for the binding of the enzyme, the rate of replenishment of the substrate in the environment of the bound enzyme and the interfacial processivity of the successive turnover cycles. In terms of the hierarchy of such effects, the single most important factor that has a dramatic effect on the apparent rate is the fraction of the total enzyme at the interface. The primary variable for K_d is $[M^*]$. Once the enzyme accesses the interface, the observed rate changes with the residence time of the enzyme at the interface, which is, for example, influenced by the salt-induced exchange of the enzyme (Chapter 5) or by the accumulation of the products of hydrolysis (Chapter 10). These analyses show that it is not trivial to analyze the reaction progress corrupted by changing kinetic or equilibrium contributions of the E to E^* step. As developed in this chapter, the situation is intractable for the reaction on small particles where the substrate replenishment rate is not rapid enough on the time scale of the catalytic turnover time (Figure 11.1).

11.1 The Exchange Limit

In this chapter we examine the limits in which small particles provide the interface for the catalytic turnover. Recall that the substrate 'concentration' that the bound enzyme (E^*) 'sees' for the formation of the interfacial Michaelis complex (E^*S) is related to the mole fraction of the substrate in the interface, X_S^*. To define a kinetic

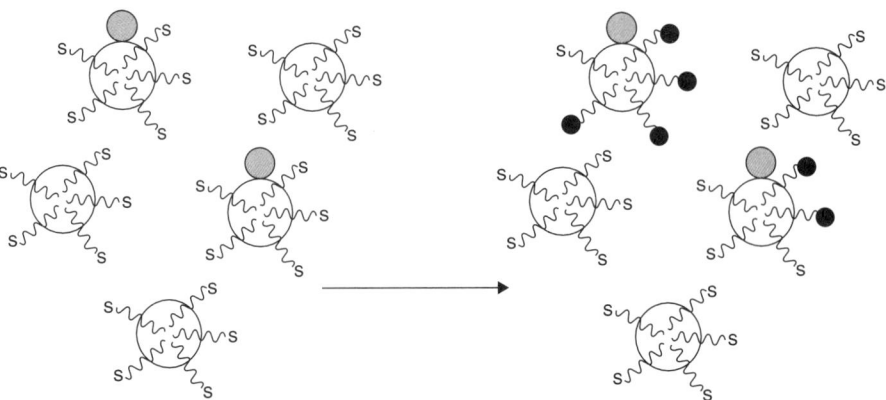

Figure 11.1 The time-dependent changes within a hypothetical ensemble of five small aggregate particles with two tightly bound enzyme (small circles) molecules. The initial condition (left) changes after a few turnover cycles (right), but without the exchange of the substrate, product or enzyme molecules between the aggregates. Note that in order to maintain the bulk-averaged X_S^* and X_P^* on the enzyme-containing particles, both S and P must exchange rapidly on the time scale of the single turnover time. Obviously, without such rapid replenishment the microscopic steady state condition would not remain constant for more than a few turnover cycles

path consider the problems associated with the changes in the microenvironment of a bound enzyme within a few turnover cycles after the enzyme is added to a particle containing a few substrate molecules. For example, in Figure 11.1, in an ensemble of particles containing five substrate molecules each, two enzyme molecules are added to an ensemble of five particles. The probability of the two enzyme molecules being on the same particle is 20 % with five particles, i.e. the random probability that the second enzyme binds to the same particle as the first is one in five. If both the bound enzyme and substrate molecules do not exchange at all, after about three turnover cycles the enzyme-containing particles will have 40 % lower substrate concentration. In effect, if the product does not leave the interface and substrate is not replenished from excess particles, X_S^* will drop to 0.6 and X_P^* would increase to 0.4 within a few milliseconds on the enzyme-containing particles.

Obviously, such a large and rapid change in the microenvironment of the enzyme-containing interface would influence the kinetic path and the observed turnover rate. In other words, the microscopic steady state condition of the variables for the initial rate cannot be maintained unless the substrate replenishment rate on the enzyme-containing particle is rapid on the time scale of the turnover cycle. If not, within a few turnover cycles the observed rate will be limited by events other than the steps of the interfacial turnover cycle. In this somewhat contrived situation what matters is neither the molar excess of the substrate over the enzyme nor the initial interface concentration, but the factors that maintain constancy of the microenvironment at the enzyme-containing interface over the observation window of several minutes during which the reaction progress is monitored.

In this chapter, we set up the problem in certain limits to provide an appreciation of the complexities encountered in dealing with the observed rate information from the kinetic assays with dispersed interfaces. The relationship of the primary events to the observed turnover flux for the substrate replenishment is developed in Theory Box 11.1 in the monomer exchange limit and in Theory Box 11.2 for the fusion–fission limit. Some consequences of the nonidealities at mixed interfaces are developed in Theory Box 11.3, because often the particles of concern here are near-equimolar mixed micelle or more complex microemulsions (Chapter 2) of nonpolar solutes with detergent.

THEORY BOX 11.1 Substrate replenishment by free monomer diffusion between particles

When enzymes are sequestered on the surface of aggregates, not only the amphiphiles but also the enzymes will have a microscopically heterogeneous distribution in solution. Consider the case where n_E enzymes are bound to each spherical aggregate of radius R in solution. The enzymes get their substrate from solution, at bulk concentration [S]. Initially, the bulk product concentration is zero. Ficks' first law gives the total diffusion flux across a spherical surface of radius r towards the center as

$$J = 4\pi D_S r^2 \frac{dC_S}{dr} \tag{1}$$

At the stationary state, the flux J (moles/s) is independent of r and Equation (1) can be integrated to give the stationary concentration profile as

$$C_S(r) = [S] - \frac{J}{4\pi D_S r} \tag{2}$$

at distance $r \geqslant R$ from the center of the aggregate. D_S is the diffusion constant for the substrate. Equation (2) can be valid only if substrates are continuously removed at the surface $r = R$ with a rate J. This is accomplished by the substrate-to-product flux of the n_E enzymes. Also, for a stationary product profile, the product flux away from the interface is equal to J. If the bulk concentration of P is zero, this gives the stationary concentration profile for the product as

$$C_P(r) = \frac{J}{4\pi D_P r} \tag{3}$$

in analogy with Equation (2). In this way, the concentrations are determined by the, as yet unknown, flux J. The enzymes on the surface will 'see' the concentrations $C_S(R)$ and $C_P(R)$ so the local MM relation will give the flux per enzyme as

$$j_{ss} = N_{Av}\frac{J}{n_E} = \frac{k_{catS}C_S(R)/K_{MS}}{1 + \dfrac{C_S(R)}{K_{MS}} + \dfrac{C_P(R)}{K_{MP}}} \tag{4}$$

Avogadro's number N_{Av} enters in the relation between j_{ss} and J because j_{ss} is in the units of molecules per enzyme per second. Entering the expressions for C_S and C_P at $r = R$ from Equations (2) and (3) into Equation (4) gives a quadratic equation in j_{ss} that can be solved:

$$j_{ss} = 0.5 \left\{ \frac{(K_{MS} + [S])\,\beta + k_{catS}}{1 - K} \pm \sqrt{\left(\frac{(K_{MS} + [S])\,\beta + k_{catS}}{1 - K}\right)^2 - \frac{4k_{catS}[S]\beta}{1 - K}} \right\}$$

(5)

where

$$\beta = \frac{4\pi D_S R N_{Av}}{n_E}$$

(6)

$$K = \frac{K_{MS}}{K_{MP}} \frac{D_S}{D_P}$$

(7)

When $K = 1$, the result simplifies to

$$j_{ss} = \frac{k_{catS}[S]}{[S] + K_{MS} + k_{catS}/\beta}$$

(8)

In the extreme diffusion-limited case, $k_{catS}/\beta \gg [S] + K_{MS}$, the flux per aggregate is (cf. Equation 2 in Theory Box 2.1)

$$J = n_E j_{ss}/N_{Av} = 4\pi D_S R[S]$$

(9)

As a consequence, if enzyme concentration is increased at a constant aggregate concentration in the limit of Equation (9), there will be no change in overall product formed; overall enzyme activity is independent of the enzyme concentration in this limit (Nelsestuen and Martinez, 1997). As 'seen' by the substrate in solution, the whole aggregate surface appears as a catalytic site in this limit and the rate is determined by the diffusion. Equation (5) gives a maximum rate for large [S] of k_{catS}, while the apparent K_M may be different from the true K_{MS}, as obviously is the case in Equation (8). If the apparent K_M is identified as the concentration [S] for which the flux j_{ss} is half-maximal, one finds from Equation (5) that

$$K_M^{app} = K_{MS} + \frac{(1 + K)k_{catS}}{2\beta}$$

(10)

Very similar results can be valid also for interfacial enzymes. Consider the same setup as described above, except that the enzymes now access substrate from the interface. This introduces two more steps that must be balanced: substrates that have reached the interface will partition into it and products produced in the interface will leave. If these processes are diffusion-limited, they will be at local quasi-equilibrium such that the concentrations in the interface are at equilibrium with the concentrations just outside, i.e. $X_S^* = C_S(R)/K_S'$ and $X_P^* = C_P(R)/K_P'$. This will give the same relations as above, except that $k_{catS} \to k_{catS}^*$, $K_{MS} \to K_{MS}^* K_S'$ and $K_{MP} \to K_{MP}^* K_P'$ in Equations (4) to (8).

If the Michaelis–Menten equation holds in the microenvironment of the enzyme, the interpretation of the MM parameters in terms of the global concentrations may be dominated by contributions from events other than the catalytic cycle, in this case the replenishment. It can also be noted that the product inhibition influences the apparent initial rate, even though the bulk product concentration is zero. This happens when product removal from the microenvironment of the enzymes is not fast enough.

11.2 Exchange-Limited Kinetics of PLA2 in Detergent-Dispersed Mixed Micelles of Long Chain Phospholipids

Clear-looking solutions of the detergent-dispersed hydrophobic substrates are the obvious attractive choice for the assay of interfacial enzymes. Usually, the reaction progress under such assay conditions exhibits a linear initial region. Also, if the equilibrium constant of the reaction permits, virtually all the substrate is hydrolyzed at the end of the reaction progress. In fact, in most such assay conditions the shape of the reaction progress is rarely distinguishable from that predicted from the integrated reaction progress equation for soluble enzyme with a soluble substrate (Equation 3.8).

The appearance of the linear initial reaction progress with detergent-dispersed mixed micelles of long chain phospholipids is deceptive. The difficulty starts with the validity of the steady state assumption. Based on the apparently linear initial rate of hydrolysis it may be assumed that the steady state turnover is described by the interfacial kinetic path. However, for the reaction on small particles (Chapters 4 and 5) it is also possible that the kinetic path may include kinetic contributions from one or more steps involving the exchange of the enzyme or the substrate replenishment. As a first step towards understanding such possibilities, consider the accessibility and replenishment properties of mixed micelles that determine the microscopic steady state. Such dispersions usually consist of near equimolar ratios of substrate dispersed with a detergent. Typically, there are less than 100 substrate molecules per particle. Assuming that an enzyme molecule with an intrinsic turnover number of $100 \, s^{-1}$ is bound to such a particle, virtually half of the substrate will be hydrolyzed within less than 1 second if the product leaves the interface and if there is no product inhibition. Even with this assumption for an ensemble of mixed-micelle particles, the microscopic steady state for the interfacial turnover would last only a fraction of a second because X_S^* would decrease from 0.5 to 0.25 within 1 second. This is of course with the assumption that the substrate replenishment would not occur on the time scale of few turnover cycles in order to keep X_S^* constant in the microenvironment of the enzyme.

11.2.1 The intermicellar concentration (IMC)

The CMC of an amphiphile decreases in the presence of a hydrophobic solute or other amphiphiles in a mixed micelle. Typically, due to the partitioning effect, the intermicellar concentration (IMC) of an amphiphile in equilibrium with the ideally mixed micelle would also be lower than the CMC for that component. However, there is no good way to predict the behavior except in the simplest ideal-mixing model (Equation 3.15 and Theory Box 3.2). Intermolecular association of amphiphiles through ionic or hydrogen-bonding interactions can effectively decrease the IMC of each of the components to a much lower value. As a relatively simple example, both lysophosphatidylcholine and ionized long chain fatty acid form simple micelles over an extended concentration range, yet only the bilayer organization persists in the dispersion of an equimolar mixture of these two amphiphiles. Such a tendency is also expected to persist during the formation of premicellar aggregates.

11.2.2   The exchange-limited interfacial turnover

The exchange rates for the detergent monomers is rapid because the CMC for detergents is typically in the millimolar range (Table 2.3). Even with a significantly lower intermicellar concentration, the exchange rates for the detergent molecules are likely to remain rapid in mixed micelles. On the other hand, the exchange rate between bilayers for phospholipids through the monomer path occurs on the time scale of hours. For both the amphiphiles the exchange rates are unlikely to change significantly in mixed micelles because their monomer concentrations can only decrease in the aqueous phase in contact with mixed micelles. In short, the fact that the detergent exchange rate is rapid in mixed micelles is of little consequence for ascertaining what E^* 'sees' microscopically at the interface.

Evidence for lower rates of the enzyme-catalyzed hydrolysis in bilayers disrupted by micellar amphiphiles is provided by results in Figures 9.4 and 11.2. An abrupt decrease in the PLA2-catalyzed rate of hydrolysis of DMPM vesicles is seen as the vesicles are disrupted into mixed micelles by a neutral diluent. As analyzed further in Figure 11.2, at low mole fractions of the diluent the decrease in the rate is as predicted for a decrease in X_S^* and $K_{MS}^* = 0.35$. Note that the observed rate decreases significantly with the formation of mixed micelles. Another crucial result in Figure 11.2 is that the oxy/thio ratio also decreases abruptly to a value close to 1 at the same X_S^* where the rate decreases abruptly. Such a relationship is expected

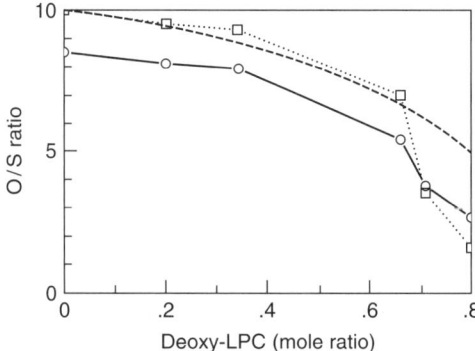

Figure 11.2   The oxy/thio ratio for the element effect (circles) and the normalized rate of hydrolysis of DMPM (squares) by pancreatic PLA2 as a function of added neutral diluent which is virtually completely partitioned in the vesicles. The dotted curve is the theoretical line for surface dilution with $K_{MS}^* = 0.35$ mole fraction. Adapted with permission from Jain *et al.* (1993a). Copyright (1993) American Chemical Society

if the rate-limiting step shifts from the chemical step to some other step for which the oxy/thio ratio is low. For a physical process, such as the exchange of substrate between the interface, the oxy/thio ratio is expected to be 1.

11.2.3 Replenishment by fusion–fission

The preferred mode for the transfer and exchange of solutes between mixed micelles is through the fusion–fission mechanism (Figure 2.7). Under such conditions the effective mixing rate for detergent-dispersed phospholipid is considerably more rapid (Figure 2.8 and Table 2.4) than the phospholipid monomer exchange rate. As a second-order collisional process it involves fusion of two micelles by collision followed by fission into two particles. The consequences of replenishment in the limit of fusion–fission of mixed micelles are analyzed in Theory Box 11.2.

THEORY BOX 11.2 Substrate replenishment through fusion–fission of mixed micelles

Consider the situation where substrate, S, and inhibitor, I, are mixed with a detergent, A, to form mixed micelles. Micelles are in large excess over enzymes. Furthermore, it is assumed that neither enzyme nor product leaves the micelles during the time course of the experiment. If the substrate concentration in solution is low, substrate replenishment in the enzyme-bearing micelles will be primarily through fusion–fission events with fresh micelles. Initially, the mole fraction of substrate in the micelles is X_S^{*0}. After an initial transient of very short duration, the system will enter a *stationary state* determined by the balance between enzyme activity and replenishment. This also determines the apparent initial rate.

In a fusion–fission event between a fresh micelle (with mole fraction X_S^{*0}) and a partially depleted one (with mole fractions X_S^{*f} and $X_P^{*f} = X_S^{*0} - X_S^{*f}$), substrates and products are assumed to be partitioned equally between the two new micelles. Just after the fission, the mole fractions are $(X_S^{*f} + X_S^{*0})/2$ and $X_P^{*f}/2$, and the substrate replenishment per event is $\Delta X^* = (X_S^{*0} - X_S^{*f})/2$. Thus, during the average time, τ, between these events, the enzyme must deplete the micelle by ΔX^* from $(X_S^{*f} + X_S^{*0})/2$ to X_S^{*f}. The enzyme rate is given by Equation (9.1) as

$$\frac{dX_S^*}{dt} = -\frac{1}{N_T}j = -\frac{k_{catS}^*}{N_T K_{MS}^*}\frac{X_S^*}{1 + \frac{X_S^*}{K_{MS}^*} + \frac{X_S^{*0} - X_S^*}{K_{MP}^*} + \frac{X_I^*}{K_I^*}} \tag{1}$$

Integrated over the time 0 to τ between two successive fusion–fission events, this gives

$$\frac{k_{catS}^*}{N_T K_{MS}^*}\tau = \left(1 + \frac{X_S^{*0}}{K_{MP}^*} + \frac{X_I^*}{K_I^*}\right)\ln\left(\frac{X_S^{*0} - \Delta X^*}{X_S^{*0} - 2\Delta X^*}\right) + \Delta X^*\left(\frac{1}{K_{MS}^*} - \frac{1}{K_{MP}^*}\right) \tag{2}$$

Equation (2) gives an implicit relation for X_S^{*f}, and thereby for ΔX^*, as a function of τ at the stationary state. $1/\tau$ is the rate of fusion–fission, which is proportional to the micelle concentration, $[m]$, as $1/\tau = k_f[m]$. Since $N_T \Delta X^*$ products are produced during time τ, the stationary flux per enzyme is

$$j_{ss} = N_T \Delta X^* / \tau \tag{3}$$

Together, Equations (2) and (3) give the apparent initial rate as a function of micelle concentration. The rate will decrease at later times when fusion–fission events are no longer most likely to occur with fresh undepleted micelles.

In the limit when the first-order term (the logarithmic term) dominates the right-hand side of Equation (2), ΔX^* can be solved explicitly and one finds

$$j_{ss} = k_f[S_0^*] \frac{e^{k_{cat}^{app}/k_f[S_0^*]} - 1}{2e^{k_{cat}^{app}/k_f[S_0^*]} - 1} \tag{4}$$

where $[S_0^*] = N_T X_S^{*0}[m]$ and

$$k_{cat}^{app} = \frac{k_{catS}^* X_S^{*0}}{K_{MS}^* \left(1 + X_S^{*0}/K_{MP}^* + X_I^*/K_I^*\right)} \tag{5}$$

is the apparent maximum rate at very high substrate concentration. This is exactly the same as the relaxation rate, $k_i N_S^0$, given in Equation (9.17). However, here it appears in the apparent initial rate because the integrated MM equation (Equation 2) is of very short duration and is repeated over and over. The stationary flux in Equation (4) is indistinguishable (with a deviation of at most 1% of the apparent maximum rate) from the MM result

$$j_{ss} = \frac{k_{cat}^{app} [S_0^*]}{K_M^{app} + [S_0^*]} \tag{6}$$

where

$$K_M^{app} \approx 1.6 \frac{k_{cat}^{app}}{k_f} \tag{7}$$

It can be noted that, although I is a competitive inhibitor that binds the catalytic site, it affects both k_{cat}^{app} and K_M^{app} equally and therefore appears uncompetitive. This is because the apparent MM relation is in terms of the bulk substrate concentration, $[S_0^*]$, while the competition with inhibitor is in terms of the local concentrations X_I^* and X_S^*. From Equation (5) one also finds that the apparent k_{cat} is reduced by 50% at an inhibitor mole fraction

$$X_I^*(50) = K_I^* \left(1 + \frac{X_S^{*0}}{K_{MP}^*}\right) \tag{8}$$

This resembles the result in Equation (9.20), which was also based on the integrated MM equation. Even if there is no product present in the bulk, the apparent initial rate can be strongly influenced by product inhibition if product accumulates in the microenvironment of the enzyme.

> If fusion–fission is limited by the diffusional encounters of micelles, the rate constant in water at 20 °C is
>
> $$k_f = 16\pi DRN_{Av} = \frac{8R_0 T}{3\eta} \approx 7 \times 10^9 \ \text{M}^{-1}\text{s}^{-1} \tag{9}$$
>
> Here R is the radius of a micelle, D is the diffusion rate, R_0 is the gas constant, T is the absolute temperature and η is the viscosity. This result follows from the Smoluchowski expression (Equation 2 in Theory Box 2.1), where the relative diffusion rate of two micelles is $2D$ and the encounter radius is $2R$. Furthermore, from Stokes' law, $D = k_B T/(6\pi\eta R)$. Thus the diffusion-limited enzyme rate from Equation (4) would be
>
> $$j_{ss} \approx 3.3 \times 10^9 \ [S_0^*] \quad \text{s}^{-1} \tag{10}$$
>
> if $[S_0^*]$ is in M. If fusion–fission is the mode of replenishment, the enzyme rate could never be faster than this. There is no evidence that a diffusion-limited fusion-fission occurs in any known case.

11.3 Effect of Bile Salts on the Pancreatic PLA2 Catalyzed Hydrolysis of Phosphatidylcholines

As shown in Figure 1.7, phospholipids and other dietary lipids emulsified with bile salts provide the natural substrate interface for pancreatic PLA2. Also, detergent-dispersed mixed micelles of substrates have been used for assays of interfacial enzymes. From the theoretical considerations developed in the preceding sections, it is clear that the equilibrium and exchange behavior of the mixed micelles of long chain phospholipids would mask the kinetic contributions of the events of the turnover cycle during the reaction progress. For examining the feasibility of the fusion–fission limited reaction progress consider the effect of bile salts on the hydrolysis of phosphatidylcholines. As summarized in Figure 11.3, the rate of hydrolysis increases dramatically at low mole fraction and then decreases to <10 % of the peak value as the bile salt concentration exceeds its CMC. These results are not consistent with a simple model in which micellization increases the observed rate; however, several effects can be dissected.

The effect of detergents on the hydrolysis of phosphatidylcholines catalyzed by PLA2 can be rationalized, but not analyzed, in terms of the interfacial kinetic paradigm. As shown in Figure 11.3 the rate of hydrolysis of the bee venom catalyzed reaction is enhanced by Triton X-100. However, such an effect of uncharged Triton is not seen for the pancreatic PLA2 catalyzed hydrolysis which is dramatically affected by anionic bile salts. In both cases the effect of the increasing detergent mole ratio is biphasic, i.e. the maximum effect is seen only at a certain optimum range of the detergent-to-substrate mole ratio. At least the initial bile–salt-dependent increase in the rate is due to enhanced binding of the enzyme. For example, as shown in Figure 11.4, the fluorescence emission intensity of PLA2 increases with the increasing concentration of deoxycholate in the presence of the ether analog of phosphatidylcholine. A maximum is reached at the mole ratio of about 1:1 phospholipid to bile salt. These results are consistent with the results in Figure 11.3,

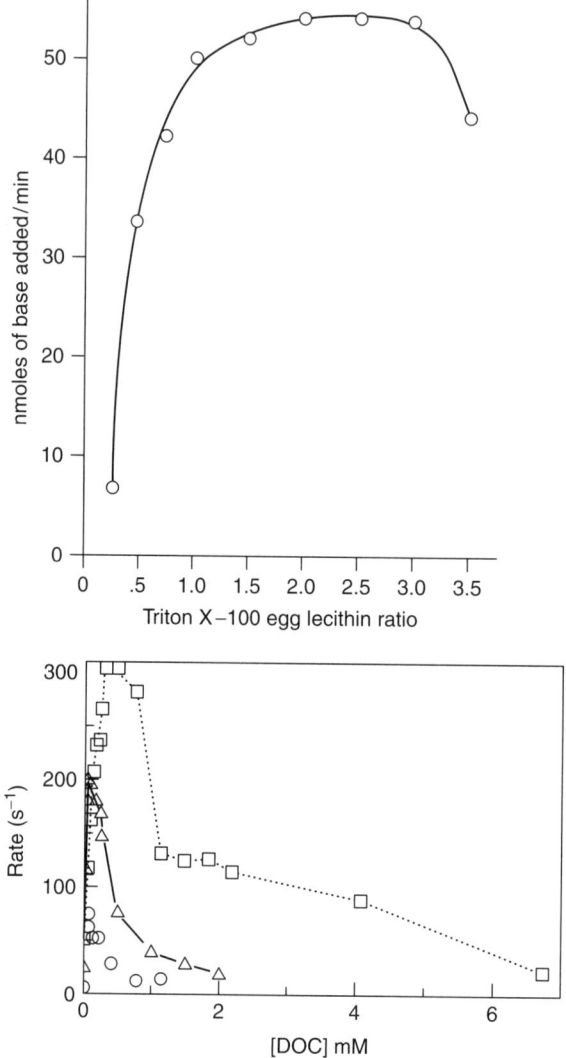

Figure 11.3 (Top) Initial rate of hydrolysis by bee venom PLA2 of 10 mM egg phosphatidylcholine containing the varying mole ratio of Triton X-100 detergent. From Upreti and Jain (1978). Reproduced by permission of Academic Press. (Bottom) Dependence of the initial rate of hydrolysis of 10 mM (squares), 1 mM (triangles) and 0.3 mM (circles) $DC_{14}PC$ vesicles in the presence of deoxycholate (DOC) by pig pancreatic PLA2. The CMC of DOC is 3 mM and IMC would be about 1.6 mM. Thus under these conditions, the maximum rate is observed when the mole fraction of DOC partitioned into vesicles is <0.1. Reprinted with permission from Jain *et al.* (1993a). Copyright (1993) American Chemical Society

and together they show that a rate increase at low mole fractions of the bile salt is attributed to enhanced substrate accessibility due to an increase in the binding of the enzyme to the substrate interface.

A striking feature of results in Figure 11.3 is that the rate of hydrolysis decreases dramatically at higher deoxycholate concentrations. For example, the maximum in the observed rate with 1 mM substrate is seen at about 0.3 mM deoxycholate,

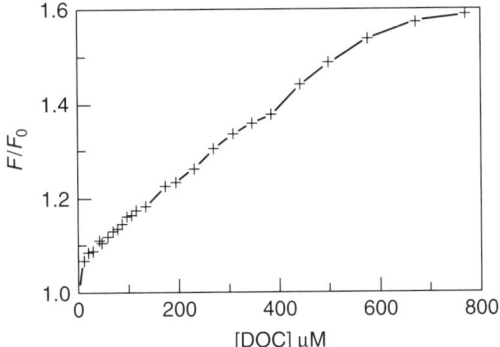

Figure 11.4 Relative fluorescence intensity for the emission from Trp-3 of pig pancreatic PLA2 and 1 mM ditetradecylphosphatidylcholine vesicles as a function of total DOC (deoxycholate) concentration. Reprinted with permission from Jain *et al.* (1993a). Copyright (1993) American Chemical Society

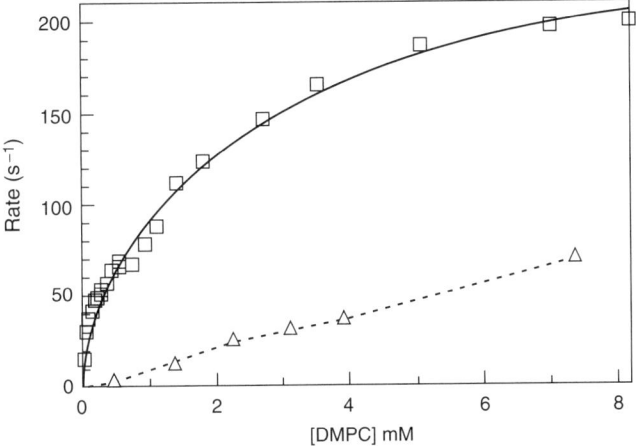

Figure 11.5 Dependence of the initial rate of hydrolysis of $DC_{14}PC$ vesicles by pig pancreatic PLA2 in the presence of 0.15 mM (squares) and 2 mM (triangles) DOC. Reprinted permission from Jain *et al.* (1993a). Copyright (1993) American Chemical Society

whereas enhanced binding is seen even at 0.8 mM deoxycholate (Figure 11.4). These results show that the observed rate decreases as the deoxycholate concentration reaches a value high enough to disrupt the bilayer organization. A comparison with the calculated curve in Figure 11.2 shows that beyond 0.3 mole fraction deoxycholate does not act as a diluent. The decrease in the rate of hydrolysis at a higher mole fraction of deoxycholate is due to the micellization of phosphatidylcholine.

The adverse effect of micellization on the observed rate is also shown in Figure 11.5. In the presence of a constant deoxycholate concentration, the rate of hydrolysis increases with the increasing substrate concentration. This is expected because the fraction of the enzyme bound to the interface increases with the interface concentration. However, as developed in Theory Box 11.2, an interpretation of these results is far from trivial. Note that the observed rate is lower at high deoxycholate concentrations where the mixed micelles are most likely to be present over the whole concentration range. Together these results suggest that in this assay system the observed rate does not change monotonically or ideally with the increasing concentration of the components.

11.4 Kinetic Concerns for Interfaces of the Dispersed Phase

The efficiency of the collisional fusion–fission depends on the bulk concentration of the particles, and the underlying rate constant will depend on the intrinsic properties of the interface of the particle. Thus the kinetics of the fusion–fission process depends on the composition and phase behavior of the particles, which in turn depends on the detergent to lipid ratio in the micelles (Fullington *et al.*, 1990). In this section we outline some of the unresolved concerns to meet the necessary conditions that allow analysis of the kinetic information obtained with assays on interfaces of dispersed phases:

(a) Even under ideal conditions X_S^* is <1 with the assumption that the components are mixed and dispersed only as micelles. From the example above it is clear that the dispersions do not behave ideally, and X_S^* values are not easily predictable as a function of the mole ratio and the bulk concentration.

(b) The idealized kinetics dominated by fusion–fission is based on the assumption that the replenishment rate depends on the micelle concentration; i.e. only the number of micelles and not their properties change with the changing concentration of the components. This appears to be virtually impossible for the dispersions of naturally occurring phospholipids in detergents with millimolar CMC.

(c) The nature of the interface of mixed micelles is not characterized. Indications are that the size and composition dispersity of the dispersed particles change with the composition and the concentration of the components.

11.4.1 Surface dilution in detergent dispersions

Surface dilution studies require that the diluent amphiphile should not have an affinity for the active site. Although this is a reasonable assumption, it has not been rigorously tested for the detergents. In addition to the concerns about the possible impurities in the commercially available detergents, many of the synthetic polymeric detergents are sold in heterodisperse mixtures. It has been argued that changing mole ratios of the dispersed components faithfully reflect the mole fraction of the substrate at the interface. Even ignoring the problem of intermicellar exchange rates and narrow dispersity, carefully constructed phase diagrams (Hofman and Mysels, 1988; Lasch, 1995) show the tendency of codispersions to undergo a phase transition between different structural forms. This makes it difficult to predict the interfacial behavior of mixtures, even under equilibrium conditions (Kamrath and Frances, 1986). The problem becomes acute if codispersions consist of an amphiphile with the micelle-forming tendency and the other with the bilayer propensity. Thus not only the composition of mixed micelles changes with the detergent to phospholipid composition, but the changes in shape, size and dispersity of the particles add up to a complex phase behavior (Helenius and Simons, 1975; Carey *et al.*, 1983; Lichtenberg *et al.*, 1983). In short, to abide by the phase constraints of mixed micelles, only a window of conditions can be established that may be suited to take advantage of the linear rates observed with mixed micelles.

11.4.2 Polymorphism, metastability and lyotropic behavior

Simple micelles in dilute solutions do not interact with each other and their shapes are approximately spherical. The aggregation number changes and elongated rod- or disc-shaped micelles are formed in mixtures of amphiphiles to balance the packing constraints in the head group and the nonpolar region dictated by the effective polarity of the head group and the size and stiffness of the chain. The stiffness of the alkyl chain arises from trans–gauche conformational change in which the trans conformer is favored by about 0.5 kcal/mole with an activation energy of about 3 kcal/mole. The gauche–trans conformational entropy means that the all-trans chain becomes less favorable. Such conformational factors effectively reduce the length and increase the cross-sectional area of an alkyl chain relative to that of an all-trans conformer.

A dramatic change in the solution properties is often seen at the higher concentrations of micelles, and more so for the mixed micelles. Such changes are often accompanied by the formation of liquid crystalline phases (mesophases) in which the order persists over a significantly longer range, i.e. the aggregation number may increase from less than 100 to well over a million. Examples of such mesophases include egg yolk, butter and liquid soaps. In general, the motional properties of individual amphiphiles in micelles and mesophases are not very different. The gross organization and long-range order of the metastable mesophases of such lyotrophic dispersions may approach that of bilayers. In this wide range of the organizational morphism, simple micelles of short chain phospholipids and bilayer vesicles of long chain phospholipids represent the two extremes, with the other polymorphs in the middle range of stability, i.e. submilliseconds for micelles to weeks for bilayer vesicles.

Organized structures assumed by an amphiphile in the aqueous phase are often related through phase boundaries in the phase diagrams as a function of the water content, the presence of other amphiphiles, ionic environment or temperature. Besides the concentration dependence of the equilibrium between the phases, the phase boundaries also depend on the experimental conditions. Additional difficulties come from metastability and complex kinetics of the phase change which can result in long hysteresis. For example, n-octadecylphosphocholine forms a stable bilayer in which the acyl chains are interdigitated in the hydrophobic interior. On heating, the bilayer undergoes a sharp and rapid transition to a micelle at 24 °C, yet in the cooling cycle the transition occurs below 15 °C after several days (Jain *et al.*, 1985). Such phase changes are also seen in aqueous dispersions of lysophospholipid, but their contribution can often be ignored by taking advantage of the long hysteresis during the micelle to bilayer transition. On the other hand, such problems may be exaggerated under the conditions relevant for monitoring the reaction progress in complex mixtures like dispersions.

11.5 The Nonideality Factor

In Chapter 3 we discussed the simplest possible partitioning as a perfectly ideal process. Several sources of nonideality in the 'interfacial solution behavior' of a solute partitioned or codispersed in the interface are expected. Sufficiently dilute solutions are ideal, and the activity factor accounts for the problems arising from the compensating or repulsive solute–solute and solute–solvent interactions at increasing concentrations. In this case we have chosen the activity factor on the form $1/(1 + \varphi X_S^*)$, which gives the nonideality correction to first order in the concentration X_S^* (Theory Box 11.3). For an ideal mixture of solute in the interface, $\varphi = 0$ and the activity factor is 1.

In terms of the surface density, an ideal solution means that φ equals the area correction factor (Theory Boxes 3.2, 7.2, and 11.3). Positive φ values, resulting from surface compensation or the solute-enhanced binding of the solute, will give an activity coefficient of <1. On the other hand, negative φ, from repulsive solute–solute interactions that decrease the partitioning of additional solute, will give an activity coefficient >1. The estimated $\varphi = -1$ for PNPB (Table 7.1) suggests that its partitioning into POPG becomes less favorable with increasing X_S^*. In general, as discussed above, the partitioning of certain solutes could also induce major organizational changes in the aggregates, creating discontinuities in the substrate concentration dependence of enzyme activity.

Partitioning of a solute in the interface is related to K_S' and thus determines the mole fraction of the substrate that the bound enzyme 'sees' during the steady state. When the substrate is a very different kind of molecule from the diluent, we expect the partitioning to show nonideal behavior (Theory Box 11.3). Effectively, this shows up as a concentration-dependent dissociation constant and the partitioning equilibrium can be expressed as

$$[S] = X_S^* \gamma_S K_S' \approx X_S^* \frac{K_S'}{1 + \varphi X_S^*} \tag{11.1}$$

where $\gamma_S = 1/(1 + \varphi X_S^*)$ is the activity coefficient. Furthermore, the total concentration of substrate, $[S_T]$, added to the system is

$$[S_T] = [S] + [S^*] = X_S^* \frac{K_S'}{1 + \varphi X_S^*} + \frac{X_S^*}{1 - X_S^*}[A_T] \tag{11.2}$$

assuming for simplicity that all of the diluent A is in the interface. X_S^* is determined from Equation (9) in Theory Box 3.2 (with L replaced by S).

The main problem is to determine K_S' and φ separately. If it is possible to measure the amount of S in the interface, $[S^*]$, one can study $[S]/[S^*]$ as a function of $[S]$. From Equation (11.2) this can be written as

$$\frac{[S]}{[S^*]} = \frac{K_S'}{[A_T]} \frac{1 - X_S^*}{1 + \varphi X_S^*} = \frac{K_S' - (1 + \varphi)[S]}{[A_T]} \tag{11.3}$$

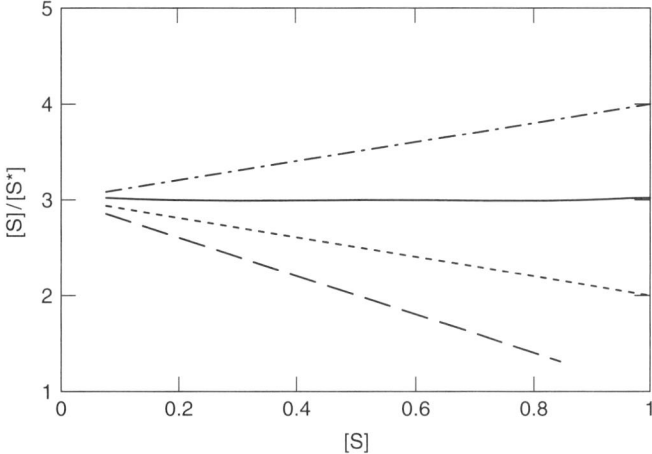

Figure 11.6 The ratio of the concentrations of S in the solution and in the interface, $[S]/[S^*]$, as a function of $[S]$, from Equation (11.3). $K'_S = 3$ and $[A_T] = 1$. From upper to lower curves, $\varphi = -2$, $-1, 0, 1$

The intercept when $[S] \to 0$ equals $K'_S/[A_T]$, which can be used to determine K_S'. When $\varphi = -1$, the slope is zero and added S goes into the solution and the interface in the same proportions throughout. When $\varphi < -1$, the slope is positive and added S goes preferentially into solution at higher S. When $\varphi > -1$, the slope is negative and added S goes preferentially into the interface at high S, corresponding to favorable S^* to S^* interactions (cf. Figure 11.6).

11.5.1 Effect of nonideal partitioning on substrate–enzyme binding

Binding of the substrate to the catalytic site of the enzyme removes the substrate from the interface. Thus, nonidealities associated with the partitioning into the interface will also affect the effective binding of substrate to enzyme in the same way as substrate dissociation from the interface. Thus we expect the equilibrium relation $K_S^* [E^*S] = \gamma_S X_S^* [E^*]$ to hold, and

$$\frac{[E^*S]}{[E^*]} = \frac{\gamma_S X_S^*}{K_S^*} = \frac{X_S^*}{K_S^* \left(1 + \varphi X_S^*\right)} = \frac{[S]}{K'_S K_S^*} \tag{11.4}$$

In the second equality we have introduced the form of the activity coefficient discussed in Theory Box 11.3, $\gamma_S = 1/\left(1 + \varphi X_S^*\right)$, and in the last equality we have entered the equilibrium partitioning from Equation (11.1). Thus, the nonideality for the partitioning disappears when the equilibrium is considered in terms of the aqueous concentration $[S]$. This is a consequence of the fact that an equilibrium concentration is independent of the path; i.e. in agreement with the detailed balance condition, we always get exactly the same equilibrium concentration of complex, $[E^*S]$, regardless of whether the substrate binds from the interface or from the solution. Thus, the nonideality of partitioning cannot influence complex formation at equilibrium. The necessary assumption for this result is that S, when complexed with the enzyme, does not retain its interactions with the membrane; as discussed further in Chapter 12, this may not always be the case.

Under catalytic conditions, the E^*S complex formation is not equilibrated and the situation may be different. With ideal partitioning, the steady state concentration is determined by $[E^*S]/[E^*] = X_S^*/K_{MS}^*$, where $K_{MS}^* = (k_{-1}^* + k_2^*)/k_1^*$. When $k_2^* < k_{-1}^*$, one finds that $K_{MS}^* \approx k_{-1}^*/k_1^* = K_S^*$; in this limit, the binding reaction is equilibrated and the activity coefficient can be included exactly as in Equation (11.4), $[E^*S]/[E^*] = \gamma_S X_S^*/K_{MS}^*$. In the other limit, however, when $k_2^* > k_{-1}^*$, the binding is not equilibrated, $K_{MS}^* \approx k_2^*/k_1^*$, and only k_1^* is present from the equilibrium constant K_S^*. Then the question arises as to whether the nonideality of partitioning should be associated primarily with the process of S entering the interface, i.e. with k_{-1}^*, or with the process of substrate leaving the interface, i.e. with k_1^*. In the first case, there is no activity coefficient in the complex formation, $[E^*S]/[E^*] = X_S^*/K_{MS}^*$. In the second case, it is the same as in Equation (11.4), $[E^*S]/[E^*] = \gamma_S X_S^*/K_{MS}^*$. This uncertainty is difficult to remove.

THEORY BOX 11.3 Thermodynamics of nonideal partitioning

Assuming that [S] is sufficiently dilute, the chemical potential per mole of S in solution can be expressed as

$$\mu_S^* = \mu_S^{*0} + R_0 T \ln([S]) \qquad (1)$$

where R_0 is the gas constant and T is the absolute temperature. The interface can be considered as a solution of S^* in A^* (if S is the minor species), and the chemical potential can be defined as

$$\mu_S^* = \mu_S^{*0} + R_0 T \ln\left(\gamma_S X_S^*\right) \qquad (2)$$

Here the standard potential for S refers to the limit $X_S^* \to 0$, where the corresponding activity coefficient, γ_S, by definition approaches 1. The equilibrium partitioning is determined by the requirement that the chemical potentials are equal: $\mu_S = \mu_S^*$. This gives

$$[S] = X_S^* \gamma_S e^{-\left(\mu_S^0 - \mu_S^{*0}\right)/R_0 T} = X_S^* \gamma_S K_S' \qquad (3)$$

where the true equilibrium constant, K_S', has been defined from the standard potentials in the usual way. This relation differs from Equation (3.11) through the activity coefficient which depends on X_S^* in the general (nonideal) case. Thus, the combination $\gamma_S K_S'$ could be considered as an apparent (concentration-dependent) dissociation constant.

Rather than using the mole fraction X_S^* as concentration units in the interface, we could have chosen the number density, i.e. moles per unit surface area. If a_A and a_S denote the molar surface areas of A^* and S^*, respectively, the surface density of S is

$$\rho_S = \frac{X_S^*}{a_A X_A^* + a_S X_S^*} = \frac{X_S^*}{a_A \left(1 + \varphi X_S^*\right)} \qquad (4)$$

In the last equality, the *area correction factor*, $\varphi = a_S/a_A - 1$, has been introduced (cf. Theory Boxes 3.2 and 7.2). The chemical potential can be defined from the number density as

$$\mu_S^* = \mu_{S,\rho}^{*0} + R_0 T \ln \left(f_S \rho_S \right) = \mu_{S,\rho}^{*0} - R_0 T \ln (a_A) + R_0 T \ln \left(\frac{f_S X_S^*}{1 + \varphi X_S^*} \right) \quad (5)$$

where $\mu_{S,\rho}^{*0}$ is the standard potential when using ρ_S. In the last equality, Equation (4) was used. Here f_S is the activity coefficient in terms of the surface density and $f_S \to 1$ when $\rho_S \to 0$. The chemical potentials in Equations (2) and (5) are the same by definition. Thus, the standard potentials are related as

$$\mu_{S,\rho}^{*0} = \mu_S^{*0} + R_0 T \ln(a_A) \quad (6)$$

which just refers to the change in units of the equilibrium constant when the concentration is a surface density rather than a mole fraction. The activity coefficients are related as

$$f_S = \left(1 + \varphi X_S^* \right) \gamma_S \quad (7)$$

It can be argued that the number density is a more fundamental quantity to describe the entropy of dilution than is the mole fraction (Ben-Naim, 1987; see also Chapter 12). If so, $f_S = 1$ for an ideal solution and

$$\gamma_S = 1 / \left(1 + \varphi X_S^* \right) \quad (8)$$

where φ is simply the area correction. On the other hand, if there are nonidealities other than the size difference between A* and S*, the activity coefficients are unknown and it may be more convenient to use mole fraction units and the activity coefficient in Equation (8). In this case, Equation (8) should be considered as a first-order expansion valid for small X_S^*. The best situation is when A and S are very similar molecules so that there is little size difference and little difference in their interactions, in which case the parameter φ accounts for solute–solute interactions.

11.6 Summary: Nonidealities for Replenishment and Binding

Although mixed micelles and microemulsions are widely used for assaying interfacial enzymes, the intrinsic properties of the dispersed interfaces are such that it is nearly impossible to establish the values of the variables for the events of the interfacial turnover cycle in the microenvironment of the bound enzyme. Even if such variables could be characterized independently, it is quite likely that the rate-limiting step in the kinetic path for the processive turnover would be the substrate replenishment rate governed by the fusion–fission-dependent exchange from the excess particles.

The origins of the problem of analysis lie in the fact that typically 10 to 100 phospholipid molecules are present in a mixed micelle. If the number remains the same on the enzyme containing particle, the local substrate mole fraction on the enzyme-containing micelle would decrease by 50 % within 0.1 second if the intrinsic turnover rate is of the order of a few hundred per second, as is the case for PLA2. The replenishment rate for long chain phospholipids in mixed micelles is expected to be, and is, exceedingly slow via the monomer exchange mechanism because their aqueous concentration is immeasurably low. The fact that the detergent exchange rate is rapid in mixed micelles, because of their high CMC, is of little consequence for determining the X_S^* the E^* 'sees'. In short, it is crucial that the substrate and product exchange rates on the enzyme-containing mixed micelle be rapid on the time scale of the intrinsic catalytic turnover rate. Otherwise, a change amounts to a change in the kinetic mechanism.

Dilution of the substrate at the interface to vary X_L^* is fundamental to the measurement of interfacial equilibria and turnover. Although the detergent-dispersed phases can be used for the measurement of the equilibrium binding and distribution parameters, the surface dilution for the interpretation of the turnover cycle requires that the interface remain intact. The substrate replenishment by fusion–fission events becomes rate-limiting for the reaction progress on the detergent-dispersed substrate.

Ideal miscibility of solutes in the interface is probably an exception rather than a rule. Certainly high mole fractions of solutes codispersed with amphiphiles give rise to phase boundaries indicative of the formation of different dispersed phases. The nonideality factor formulated above is only for the partitioning situation with small deviations from ideal miscibility, where the gross organization of the interface is retained. Its basic origin is from a difference in the interfacial area of the substrate and the diluent amphiphiles, but it can be used as a first-order correction also for differences in interactions.

12 Effects of Reduction of Dimensionality

It is in vain that we say what we see; what we see never resides in what we say.
Michael Focault

Consensus science can only provide an illusion of certainty.
Roger Pielke

Interfacial enzyme kinetics differs in many ways from solution kinetics. Nevertheless, as seen in previous chapters, it is possible to set up and apply the standard Michaelis–Menten scheme when due account is taken of the constraints on concentrations in the interface. In this way, the Michaelis constant in the interface will be determined by the units used for interface concentration, usually the mole fraction. However, to compare properties between enzymes in solution and in the interface, it is necessary to set the concentration units to the same scale.

Interfacial enzymes access their substrate in an essentially two-dimensional space. The reduction of dimensionality has two major kinetic consequences: first on the thermodynamics of the equilibrium binding and second on the kinetics of the diffusion processes involved. In this chapter we will consider the equilibrium and diffusion processes within the constraints of interfaces, i.e. issues related to the association and diffusion of molecules within a confined space. By calculating the entropy loss for a molecule that is confined to the interface it is possible to define an accessible volume that can serve as the basis for the local concentrations in the interface. Thus, the loss of one translational degree of freedom of a solute on binding to the interface can best be described as a reduction in accessible volume rather than as a reduction of dimensionality. Consequently, an enzyme that binds to the interface will 'see' a high local concentration of ligands (substrate and/or inhibitor) if they are partitioned in the interface. Under some conditions this will lead to a correspondingly large increase in ligand–enzyme complex formation in the interface. Perhaps surprisingly, under other fairly general conditions, the high local concentration in the interface is expected to lead to no or only a moderate increase in complex formation.

The process of transmembrane signal transduction in cells often involves the interfacial interaction of two proteins confined to the membrane. For example, rhodopsin kinase, which is anchored to rod outer segment membranes via a covalently attached 15-carbon farnesyl group, binds to the transmembrane protein rhodopsin after a flash of light, and this leads to termination of the visual cycle signal transduction cascade. To what extent does membrane confinement dictate the specificity of the kinase–target interaction? Given a very large number of protein kinases present in cells, this specificity issue is an important matter. It seems reasonable to think that such confinement will increase the concentration of one component that the other component 'sees', and this local concentration effect will favor binding of the two molecules in the interface. It is important to determine if this is always the case. This requires that specific models describing the bimolecular interaction be hypothesized, which enables quantitative expressions of binding energy to be derived. Such expressions guide the design of experiments to test the model and ultimately lead to an understanding of the interfacial recognition event.

The kinetics of ligand association in the interface will be influenced by the surface diffusion of the reactants. Diffusion in two dimensions (2D) has properties that are quite different from three dimensions (3D). Adam and Delbrück (1968) suggested that confining a reaction to a surface would considerably speed up the association rate. We will explore the limitations of this suggestion as it pertains to enzyme–ligand association in the interface. The corresponding one-dimensional case has been studied in great detail, both theoretically and experimentally (for reviews see Berg and von Hippel, 1985; von Hippel and Berg, 1989). Under optimal conditions in dilute solution, one-dimensional diffusion along the DNA can increase the association rate for a gene-regulatory protein to its DNA binding site by several orders of magnitude. Just as is the case in one dimension (1D), however, due to the high concentrations of most constituents inside a living cell, at best only moderate rate increases can be achieved from a reduction in dimensionality (Berg, 1988).

12.1 Dissection of the Entropy Loss on Interface Binding

A binding equilibrium is the result of a competition between favorable interactions between the reaction partners and the favorable entropy of keeping them unbound and free to roam the solution. The favorable binding-free energy includes the lower enthalpy caused by van der Waal's interactions, hydrogen bonds, etc., as well as the higher entropy from the hydrophobic effect. The entropy loss on binding is due to the loss of freedom when the six translational and six rotational degrees of freedom of two reaction partners together are reduced to three translational and three rotational degrees of freedom for the complex. When binding is very tight, the maximum entropy loss per degree of freedom lost can be calculated from quantum mechanics as given by Equation (10) in Theory Box 12.1. This is the limit considered by Page and Jencks (1971) and Jencks (1975).

12.1.1 Motional constraints and accessible volume in the interface

An amphiphile partitioning into the interface or an enzyme binding to the interface has to give up one translational and two rotational degrees of freedom. Two

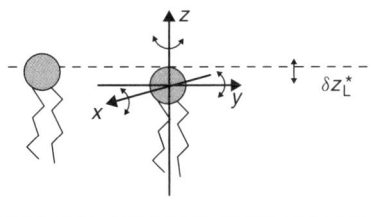

Figure 12.1 Degrees of freedom for an amphiphile in the bilayer interface (dotted lines). The amphiphile molecule has free translational motion in the plane of the interface along the x and y coordinates. It has restricted motion perpendicular to the plane in the z coordinate. It has free rotations around its main axis, the z axis, and very restricted rotations around the x and y axes

translational degrees of freedom in the plane of the interface remain while the motion perpendicular to the interface is severely constrained (Figure 12.1). Similarly, one rotational degree of freedom involving rotations around the main molecular axis remains, while two involving rotations 'head over tail' are severely constrained. However, confinement is not likely to be very tight (Jähnig, 1983; Ben-Tal *et al.*, 2000). In fact, the molecules retain considerable freedom to 'bob' up and down perpendicular to the interface. Simulations of lipid motions suggest that this perpendicular motion could be several Å (Berger *et al.*, 1997).

Consider the consequences of confinement of a solute in a membrane–water interface (Chapter 3). Molecules L can partition either into water solution, L, or the interface, L*. The membrane is built up from molecules A. The numbers of L molecules and A molecules are, n_L^* and n_A^* respectively, in the interface. Each molecule contributes the surface area a_L and a_A.

An L molecule that is bound in the interface has an accessible volume that is determined by

$$V^* = \left(n_A^* a_A + n_L^* a_L\right) \delta z_L^* = n_M^* a_A \left(1 + \varphi X_L^*\right) \delta z_L^* \tag{12.1}$$

where φ is the area correction factor (Theory Boxes 3.2, 7.2, and 11.3) and $n_M^* = n_L^* + n_A^*$; δz_L^* is the allowed freedom of movement perpendicular to the surface. Based on the calculated entropy loss (see Theory Box 12.1), δz_L^* is related to the allowed fluctuations, the variance σ_z^2, as

$$\delta z_L^* = \sqrt{2\pi e \sigma_z^2} \tag{12.2}$$

where e is the base of the natural logarithms. However, if the confinement is very tight so that $\sigma_z^2 \to 0$, quantum mechanical effects set in and δz_L^* must be chosen as $\delta z_L^* = \delta z_L^{*\,\mathrm{min}} \approx 0.6/\sqrt{MW_L}$ Å (cf. Equations 9 and 10 in Theory Box 12.1) where MW_L is the molecular weight of the L molecules.

The local concentration per unit accessible volume in the interface (here and below indicated by curly brackets) can be defined as

$$\{L^*\} = \frac{n_L^*}{V^* N_{Av}} \tag{12.3}$$

THEORY Box 12.1 Entropy change on translational confinement

Consider a particle of mass m in a one-dimensional 'box' of length Z. The canonical distribution function in coordinate $z(0 < z < Z)$ and momentum p is

$$\rho(z,p) = \frac{1}{Z}\frac{1}{\sqrt{2\pi mkT}}\exp\left(-\frac{p^2}{2mkT}\right) \tag{1}$$

and the entropy is

$$S_{\text{box}} = -k\iint \rho\ln(\rho h)\,dp\,dz = k\ln(Z) + \frac{k}{2} + \frac{k}{2}\ln\left(\frac{2\pi mkT}{h^2}\right) \tag{2}$$

where h is Planck's constant and k is Boltzmann's constant.

If the particle is held in a smaller region of space at $z = z_0$ by a harmonic potential with force constant κ, the canonical distribution function is

$$\rho(z,p) = \sqrt{\frac{\kappa}{2\pi kT}}\frac{1}{\sqrt{2\pi mkT}}\exp\left(-\frac{\kappa(z-z_0)^2}{2kT}\right)\exp\left(-\frac{p^2}{2mkT}\right) \tag{3}$$

and the entropy is

$$S_{\text{harm}} = -k\iint \rho\ln(\rho h)\,dp\,dz = \frac{k}{2}\ln\left(\frac{2\pi kT}{\kappa}\right) + k + \frac{k}{2}\ln\left(\frac{2\pi mkT}{h^2}\right) \tag{4}$$

The variance in the position of the particle is given by

$$\sigma_z^2 = \iint (z-z_0)^2\rho\,dp\,dz = \frac{kT}{\kappa} \tag{5}$$

The frequency of the harmonic oscillator is determined by

$$\nu = \sqrt{\frac{\kappa}{m}}\frac{1}{2\pi} \tag{6}$$

Equations (3) to (5) hold in the classical limit where $h\nu \ll kT$. Using Equations (5) and (6), one finds

$$\frac{h\nu}{kT} = \sqrt{\frac{h^2}{2\pi mkT}}\frac{1}{\sqrt{2\pi\sigma_z^2}} \tag{7}$$

The first factor, $\sqrt{h^2/(2\pi mkT)}$, is the thermal wavelength of the particle and equals $1.0/\sqrt{\text{MW}}$ Å at temperature 300 K if the mass corresponds to mole weight MW. Thus, the classical limit, $h\nu \ll kT$, holds as long as the constraints are not more severe than allowing fluctuations in position, $\sqrt{\sigma_z^2}$, that are larger than $1.0/\sqrt{2\pi\text{MW}} = 0.40/\sqrt{\text{MW}}$ Å. In this limit, the entropy change on confinement is determined by the difference

$$\Delta S_c = S_{\text{harm}} - S_{\text{box}} = \frac{k}{2} + k\ln\left(\frac{\sqrt{2\pi\sigma_z^2}}{Z}\right) = k\ln\left(\frac{\sqrt{2\pi e\sigma_z^2}}{Z}\right) \tag{8}$$

The entropy change discussed in the main text corresponds to choosing the confinement δz as $\sqrt{2\pi e \sigma_z^2} = \delta z$, where e is the base of the natural logarithms.

If the confinement is more severe than the classical limit allows, the particle in the potential must be treated quantum mechanically, and the classical entropy calculation in Equation (8) is no longer applicable. In the most severe limit, where $h\nu \gg kT$, the entropy approaches zero, and the entropy change on the loss of one degree of translational freedom is determined by

$$\Delta S_c^{\mathrm{max}} = 0 - S_{\mathrm{box}} = -k\ln(Z) - \frac{k}{2} - \frac{k}{2}\ln\left(\frac{2\pi mkT}{h^2}\right)$$

$$= k\ln\left[\frac{\sqrt{h^2/(2\pi emkT)}}{Z}\right] \tag{9}$$

A complete quantum mechanical entropy calculation shows that the classical approximation, Equation (8), holds quite well for decreasing σ_z^2 all the way to the limiting result, Equation (9), for complete loss of freedom. For a molecule with molecular weight MW at temperature 300 K, this gives

$$\frac{\Delta S_c^{\mathrm{max}}}{k} = \ln\left(\frac{0.6}{Z\sqrt{\mathrm{MW}}}\right) \tag{10}$$

if Z is given in units of Å. For a standard solution of 1 M, the accessible volume per solute particle corresponds to a cube with side length $Z = 11.86$ Å. For MW = 100 one finds from Equation (10) that $\Delta S_c^{\mathrm{max}}/k = -5.3$, which corresponds to about 10 e.u. lost per translational degree of freedom. The corresponding free energy increase is $-T\Delta S_c^{\mathrm{max}} = 13$ kJ/mole at 300 K. This is the estimate used by Jencks and coworkers (Page and Jencks, 1971; Jencks, 1975), and it corresponds to the maximal change which is valid for strong confinement. However, note that the entropy loss for a weak confinement, Equation (8), could be significantly smaller.

12.1.2 Free energy differences

The chemical potential of molecules L can be defined from the concentration in terms of the number of molecules (or moles) per unit accessible volume as (Ben-Naim, 1987)

$$\mu_L = \mu_L^0 + kT\ln([L]) \tag{12.4a}$$

$$\mu_L^* = \mu_L^{*0} + kT\ln\left(\{L^*\}\right) \tag{12.4b}$$

in the water phase and interface, respectively. By using Boltzmann's constant, k, rather than the gas constant in these expressions, the free energy is counted per molecule rather than per mole (cf. Theory Box 11.3). At equilibrium, the chemical potential of an L molecule in the two regions must be the same, $\mu_L = \mu_L^*$, and the standard free energy of binding the interface can be calculated as

$$\Delta G_L^0 = \mu_L^{*0} - \mu_L^0 = -kT \ln\left(\frac{\{L^*\}}{[L]}\right) = kT \ln\left(K_L' a_A \delta z_L^* N_{Av}\right) \tag{12.5a}$$

For the last equality, the equilibrium partitioning of L (Equation 11.1), $[L] = K_L' X_L^*/(1 + \varphi X_L^*)$, together with Equations (12.1) and (12.3) were used. On the assumption that the perpendicular freedom of motion, δz_L^*, is the same for A and L molecules, this can be written as

$$\Delta G_L^0 = kT \ln\left(\frac{K_L'}{\{A^*\}^0}\right) \tag{12.5b}$$

where $\{A^*\}^0$ is the local solvent concentration in the absence of solute L. This standard free energy difference contains everything in terms of molecular interactions, conformational freedom, etc. Since the standard states in the two phases are the same, one molecule per unit accessible volume, there is no contribution from the choice of concentration units, and ΔG^0 is a 'true' free energy difference.

The translational confinement in the membrane is accounted for by the δz_L^* factor. The rotational confinement, where a molecule in the membrane can only take up certain orientations for its main axis, is included in ΔG^0. This leads to an entropy loss in the bound state that increases the free energy (makes it less favorable). It may be more convenient to separate out this part as well. If it is assumed that the allowed orientations cover only a fraction $\delta \Omega_L^*$ of all possible solid angles, then

$$\Delta G_L^0 = \Delta G_L^i - kT \ln\left(\delta \Omega_L^*\right) \tag{12.6}$$

What remains, ΔG^i, is the *intrinsic binding free energy* for a molecule placed in a particular allowed position and orientation in the membrane. This contains all the interactions, including differences in conformational freedom–e.g. segmental motions in the hydrophobic tails–that hold the molecule in the interface. (Actually, ΔG^i is the weighted average over all such allowed positions and orientations.) Thus the intrinsic binding free energy contains all the effects that hold the molecule in the interface, which must be broken if the molecule is moved back into solution without giving it back its rotational and translational degrees of freedom. From Equations (12.5) and (12.6) we can write

$$\Delta G_L^i = kT \ln\left(\frac{K_L'}{\{A^*\}^0} \delta \Omega_L^*\right) \tag{12.7a}$$

or, equivalently,

$$K_L' = \frac{\{A^*\}^0}{\delta \Omega_L^*} \exp\left(\frac{\Delta G_L^i}{kT}\right) \tag{12.7b}$$

The entropy loss due to confinement in the interface is

$$\Delta S_c = k \ln\left(\frac{V^*}{V^w} \delta \Omega_L^*\right) \tag{12.8}$$

where V^w is the volume of the water phase, while V^* is the accessible volume of the interface.

Equation (12.8) describes the loss of entropy in relation to the strength of confinement of the center of gravity of a molecule and the orientation of its main axis. Each molecule on binding to the interface gives up one degree of translational freedom and–possibly–two degrees of rotational freedom. This is described here as a reduction in accessible volume in spatial and orientational coordinates. The separation of this confinement and other conformation restrictions cannot in general be made strictly, and the resolution of the terms in Equation (12.6) is only approximate. If the confinement is not too tight–as may be expected for the interface binding–the momentum distribution in each constrained coordinate is the same as in the unconstrained case (see Theory Box 12.1). In this case, the entropy loss is due only to the constraint of the spatial coordinates, as described above, and the reduction of dimensionality is in effect a reduction of accessible volume. It should be stressed, however, that the reduction in accessible volume has been calculated from the entropy loss (see Theory Box 12.1) rather than the other way around. In this way, the free energy differences in Equations (12.5) and (12.7a) are calculated from the appropriate thermodynamic definitions.

If the confinement is very tight and δz_L^* from Equation (12.2) approaches zero, Equation (12.8) would predict that the entropy loss becomes infinite. However, in this limit, quantum mechanical effects will provide a lower limit. In the equations above, a very tight confinement can be accounted for by using a minimum value for $\delta z_L^{min} \approx 0.6/\sqrt{MW_L}$ Å (where MW_L is the molecular weight of L molecules) corresponding to the thermal wave length (see Theory Box 12.1). This very tight limit corresponds to total loss of freedom and may be applicable, for instance, if the molecule is held in position by covalent forces. Similar considerations also apply to the maximal loss of orientational entropy; i.e. $\delta\Omega_L^*$ would not be zero even if the orientational constraints are very tight.

The local solvent concentration was defined as

$$\{A^*\}^0 = \frac{n_A^*}{a_A \delta z_L^* N_{Av}} \tag{12.9}$$

The local concentration is not known unless some estimate of δz_L^* can be made. For lipid molecules $a_A \approx 50$ Å^2, $\sqrt{\sigma_z^2} \approx 1$ Å which gives $\delta z_L^* \approx 4$ Å and $\{A^*\}^0 \approx 8$ M. This is the local concentration of solvent in the interface which corresponds to the standard 55 M for the concentration of water in the aqueous phase. If the freedom of movement of the lipids perpendicular to the surface is as large as suggested by simulations (Berger *et al.*, 1997), $\sqrt{\sigma_z^2} \approx 5$ Å, the estimate of δz_L^* would be correspondingly larger and $\{A^*\}^0$ correspondingly smaller. These numerical estimates assume that L is an amphiphile of the same nature as the membrane-forming molecules, A. If L is a small hydrophobic molecule that partitions into the membrane, δz_L^* would more closely correspond to the displacements of the trans-gauche kinks formed by the rotation of the methylene residues. Large hydrophobic solutes would probably sequester in the space in the middle of the bilayer. If, on the other hand, L is a protein that binds at the surface, δz_L^* would have some other very small value because of specific close-range interactions along the i-face. In either case, δz_L^* is defined from Equation (12.2) and determined by the shape of the interaction potential.

To find a free energy difference that is independent of the choice of concentration units (Equation 12.5), it is imperative that the units are the same in both phases. A conventional choice is to use mole fractions in both phases. This defines the unitary standard free energy difference

$$\Delta G_L^U = -kT \ln \left(\frac{X_L^*}{X_L^w} \right) = kT \ln \left(\frac{K_L'}{[W]} \right) \tag{12.10}$$

where $X_L^w \approx [L]/[W]$ and $[W]$ (≈ 55 M) is the concentration of water in the aqueous phase. This is quite different, by a term $kT \ln(\{A^*\}^0/[W])$, from the free energy difference given in Equation (12.6). The unitary free energy has been used to calculate a units-invariant free energy of interaction (Gurney, 1953; Cantor and Schimmel, 1980; Tanford, 1980; Peitzsch and McLaughlin, 1993). However, since the solvents are not the same, the mole fraction units in the two phases are not commensurate and the meaning of the unitary free energy difference remains obscure. The chemical potential is defined from the increase in Gibbs' free energy of the system when the number of solute molecules is increased by one at constant T and p. The solute molecule can be placed anywhere within the confines of the solution and the solvent molecules will have to accommodate equally regardless of where it is placed. This freedom affects the entropy as determined by the volume and therefore introduces the number density as the fundamental concentration unit for the calculation of the chemical potential (Ben-Naim, 1987; see also Theory Box 12.1 where the entropy effect in one dimension has been calculated from statistical mechanics). In contrast, the assertion that the mole fraction is the fundamental concentration variable that describes the entropy of dilution is based on a lattice model where solvent and solute molecules occupy equivalent positions and can freely be exchanged for one another. Clearly, this cannot describe all the entropy effects at an interface or for the contacts along the i-face (next section), particularly not when solute and solvent molecules are different in size. Although the mole fraction is a useful concentration measure in many cases, its relation to the number density depends on the nature of the solvent and it is therefore not suitable for the calculation of a free energy difference between different solvents.

12.2 Synergistic Effects of the Interface on Enzyme–Substrate Binding, Local Concentration and Scaffolding

In Chapter 6 we considered the detailed balance condition for the enzyme–substrate binding versus the binding of enzyme and substrate to the interface (Figure 6.1). From this relationship (Equation 6.25), we can define a cooperativity factor

$$\omega_{\text{coop}} = \frac{K_L}{K_L' K_L^*} = \frac{K_d}{K_d^L} \tag{12.11}$$

The first ratio expresses how much better the free ligand binds to the interface-bound enzyme than to the free enzyme and the second ratio expresses how much better the EL complex binds the interface than does E. The detailed balance

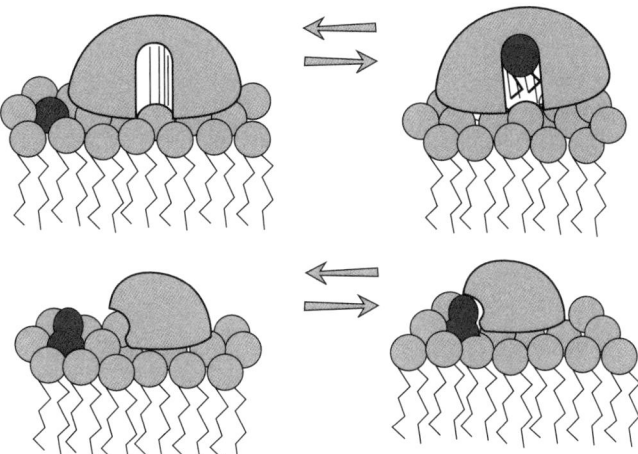

Figure 12.2 Enzyme–substrate interactions in the interface. (Top) Total engulfment where the enzyme pulls the substrate out of the interface leaving no substrate–interface interactions. (Bottom) Marginal contacts where the enzyme–substrate binding does not disturb the interface interactions either for the substrate or the enzyme

condition implies that these two effects must be the same. The cooperativity, or synergism, is most easily discussed on the basis of the second ratio. Two extreme cases (Figure 12.2) can be distinguished: total engulfment and marginal contact. In the first case, the ligand L does not retain any contacts with the interface once it has entered the catalytic site. In this case, the presence of L in the catalytic site does not influence the binding interactions of EL to the interface, giving $K_d^L = K_d$ and $\omega_{coop} = 1$. As discussed in Chapters 7 and 8, there could still be an allosteric effect whereby the presence of the ligand changes the conformation of the enzyme to one that binds the interface better. Thus, $\omega_{coop} = \omega_{allostery}$ is the expected interface effect in the case of total engulfment. In this case, the high local concentration of ligand in the interface is of no help since the interactions, ΔG_L^i, used to increase the concentration must be broken again on entry to the catalytic site.

If the enzyme contacts the ligand in the interface without disrupting the ligand–interface interactions or tightening the orientational constraints, the local concentration effect will be maximal. The dissociation of E^*L to EL will require not only the breaking of the enzyme–interface contacts, represented in the dissociation constant K_d, but also the breaking of all the favorable interactions between L and the interface, ΔG_L^i. L by itself does not gain any rotational or translational freedom in the dissociation process $E^*L \rightarrow EL$, other than what the complex gains, and this part is included equally in K_d and K_d^L. Therefore by the definition of ΔG_L^i, this is just the contribution of L to the binding of E^*L in the interface so that $K_d^L = K_d \exp[\Delta G_L^i/(kT)]$. To this we should also add the possibility of an allosteric effect and

$$\omega_{coop} = \frac{K_d}{K_d^L} = \omega_{allostery} \exp\left[-\Delta G_L^i/(kT)\right] = \omega_{allostery} \frac{\{A^*\}^0}{K_L'} \frac{1}{\delta\Omega_L^*} \tag{12.12}$$

The last equality is from Equation (12.7a). In this limit, the interface holds the reaction partners in optimal configurations to form the E^*L complex. This scaffolding effect has two contributions, the local concentration factor $\{A^*\}^0/K_L' = \{L^*\}/[L]$ and the orientational factor, $1/\delta\Omega_L^*$. Considering the synergism from a comparison of $E + L \rightarrow EL$ versus $E^* + L^* \rightarrow E^*L$, part of the price, the free-energy cost, of constraining the reactants has been paid already in their association to the interface and does not have to be paid again when the E^*L complex is formed. Equation (12.12) is the maximum effect of the interface on the binding of ligand to the enzyme. As discussed above, the local concentration of the membrane-forming molecules (lipids), $\{A^*\}^0$, is ca. 1 to 8 M, so the effect can be very large if K_L' is small.

The cases depicted in Figure 12.2 are the two extreme limits. Real situations for the various interfacial enzymes are expected to be somewhere in between, requiring partial removal of the substrate from the interface, or at least some distortion in the substrate–interface interactions. Nevertheless, the range of possible synergism effects is huge, estimated to be about six orders of magnitude.

12.3 Diffusion Times in 1D, 2D and 3D

The increased complex formation when reactants are constrained to an interface can be ascribed primarily to the reduction in accessible volume and the consequent increase in local concentrations. Similarly, the rate of complex formation is expected to increase in the interface relative to solution due to increased local concentrations. However, since diffusion effects in 2D and 3D are significantly different, this must also be considered for the kinetic description.

The mean time for a molecule to find a target can be calculated as outlined in Theory Box 12.2. For a centrally located target with reaction radius b in a spherical region of radius $R(\gg b)$, the mean association time via diffusion is

$$\tau_3 = \frac{R^2}{3D_3}\frac{R}{b} = \frac{V}{4\pi D_3 b} \tag{12.13}$$

where D_3 is the diffusion constant and V is the volume of the spherical region. Similarly, the mean time to find a target in a circular region with radius R via two-dimensional diffusion is

$$\tau_2 = \frac{R^2}{2D_2}\left[\ln\left(\frac{R}{b}\right) - \frac{3}{4}\right] \tag{12.14}$$

where D_2 is the diffusion constant in the plane. In 1D, the corresponding result is

$$\tau_1 = \frac{R^2}{3D_1} \tag{12.15}$$

Diffusion is a volume-filling process. In the mean-time expressions, $R^2/2D$ or $R^2/3D$ corresponds to the time required for a molecule to diffuse by a random walk so that its trajectory fills the whole empty region. In 1D, the probability of finding the target in this process is one–the molecule must pass the point where the target is–so that the size of the target does not matter. In 3D, the probability of hitting the target is only $b/R \ll 1$ and the particle must diffuse through the empty region many times before the target is located. In 2D, the probability of hitting the target is fairly large, $1/[\ln(R/b) - 3/4]$ and the size of the target matters very little, only through the logarithmic term. These mean times were calculated for a molecule that starts with equal probability anywhere in the empty region outside the target and requires that $R \gg b$.

A comparison between τ_3 and τ_2 (Equations 12.13 and 12.14) shows that, for the same radius R, the encounter or association in the plane is much faster. This led Adam and Delbrück (1968) to suggest that reduction of dimensionality would be employed in nature as a mechanism to speed up reaction rates. However, it should be stressed that the reduction of dimensionality described above is accompanied by a substantial reduction in the volume that needs to be searched. If one were to compress the sphere to a thin circular disc with thickness $2b$, while retaining the original volume V, the time to find the target would actually increase: the compression would give a disc radius R determined by $V = 2b\pi R^2$ so that the mean association time from Equation (12.14) would be $\tau_2 = (V/4\pi D_2 b)[\ln(R/b) - 3/4]$. This is approximately a factor $\ln(R/b)$ larger than the association time in the spherical region of the same volume (Equation 12.13), if $D_2 = D_3$. Thus it is not the change in dimensionality *per se* that decreases the association time, it is the reduction in the total volume that needs to be searched. If the target is on a membrane it is usually, though not always, as discussed further below, more efficient to search processively only the membrane surface.

THEORY BOX 12.2 Mean-time calculations

Consider the situation depicted in Figure 12.3 where a target of radial extension b is located at the center of a region of radial extension R; a sphere in 3D, a circular disc in 2D and a line in 1D. If a molecule starts from the perimeter, what is the mean time before it has reached the target? If $p(r, t)$ denotes the probability density of finding the molecule at a certain position r at time t, then the probability that the molecule has not bound at time t is

$$P(t) = \int_b^R p(r, t)\, dV \tag{1}$$

where the integral is over the line, the disc or the volume. The absorbing boundary condition is $p(r = b, t) = 0$. The mean time before capture can be defined as

$$\tau = \int_0^\infty P(t)\, dt \tag{2}$$

Let us introduce the time-integrated probability density

$$\tilde{p}(r) = \int_0^\infty p(r, t)\, dt \tag{3}$$

At infinite time, the molecule has been captured and $p(r, t = \infty) = 0$ for $b < r < R$. Thus, the molecule starting at the perimeter must with probability one pass across any 'surface' between the perimeter and the target. This probability can be expressed as the time-integrated diffusion

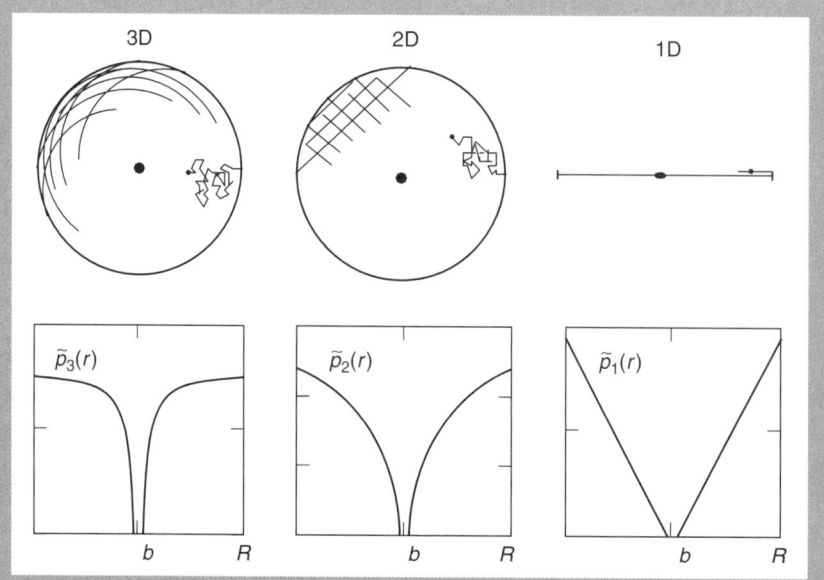

Figure 12.3 (Top) Geometries for the mean-time calculations. (Bottom) The time-integrated probability densities or concentration profiles as defined in Theory Section 12.2, per unit volume (3D), per unit surface area (2D) and per unit length (1D)

flux across a surface with radius r centered on the target

$$2D_1 \frac{d\tilde{p}_1}{dr} = 1, \qquad \text{in 1D} \tag{4a}$$

$$2\pi r D_2 \frac{d\tilde{p}_2}{dr} = 1, \qquad \text{in 2D} \tag{4b}$$

$$4\pi r^2 D_3 \frac{d\tilde{p}_3}{dr} = 1, \qquad \text{in 3D} \tag{4c}$$

With the absorbing boundary condition, $\tilde{p}(r = b) = 0$, the solutions are

$$\tilde{p}_1(r) = \frac{1}{2}\frac{r - b}{D_1}, \qquad \text{in 1D} \tag{5a}$$

$$\tilde{p}_2(r) = \frac{1}{2\pi D_2} \ln\left(\frac{r}{b}\right), \qquad \text{in 2D} \tag{5b}$$

$$\tilde{p}_3(r) = \frac{1}{4\pi D_3}\left(\frac{1}{b} - \frac{1}{r}\right), \qquad \text{in 3D} \tag{5c}$$

These densities, or concentration profiles, have been plotted in Figure 12.3. In the three-dimensional case, the concentration levels off and becomes constant at large distance from the target. In the two-dimensional and, even more so, in the one-dimensional case, the density continues to increase at larger distances. Alternatively, one can say that the depletion zone surrounding each target is finite in 3D but reaches to large distances in 2D and 1D. This is the reason why the diffusion reaction quickly reaches a steady state in 3D, while no steady state exists in 2D or 1D. From

Equations (1) to (5) one finds the mean times

$$\tau_1 = 2 \int_b^R \tilde{p}_1(r)\, \mathrm{d}r \approx \frac{R^2}{2D_1} \tag{6a}$$

$$\tau_2 = \int_b^R \tilde{p}_2(r) 2\pi r\, \mathrm{d}r \approx \frac{R^2}{2D_2}\left[\ln\left(\frac{R}{b}\right) - \frac{1}{2}\right] \tag{6b}$$

$$\tau_3 = \int_b^R \tilde{p}_3(r) 4\pi r^2\, \mathrm{d}r \approx \frac{R^3}{3D_3 b} \tag{6c}$$

The approximations are for the neglect of terms of the order b^2/R^2. Equation (6a) is the same as given in Equation (6) in Theory Box 10.2. The results in the main text (Equations 12.13 to 12.15) were calculated in a similar way from the initial condition that the molecule starts with equal probability anywhere in the empty region between b and R. Equations (6b) and (12.14) can also be compared with Equation (3) in Theory Box 5.2 where τ_2 was calculated for a molecule starting anywhere on a spherical surface rather than a circular disc.

12.3.1 Diffusion with mixed dimensionality

Consider the three situations depicted in Figure 12.4 with a single spherical target with radius b at the center of a right circular cylinder of radius R and height $2H$. A molecule that starts in the vessel finds the target after mean time

$$\tau_A = \frac{V}{4\pi D_3 b} = \frac{R^2 H}{2D_3 b} \tag{12.16}$$

For the three-dimensional diffusion-reaction process the shape of the vessel does not matter much, unless it is very deformed ($H \gg R$ or $H \ll R$), and the association time

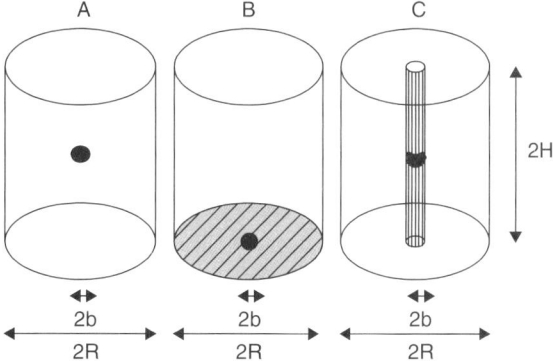

Figure 12.4 Diffusion geometries for the three cases discussed in the text. A: free three-dimensional diffusion to the central target. B: free diffusion on to the plane in the center, thereafter two-dimensional diffusion on the plane to the central target. C: free diffusion on to the thin cylinder in the middle, thereafter one-dimensional diffusion to the central target

is determined by the volume (Equation 12.13). In Figure 12.4, case B, the target is embedded in a surface to which the molecule can bind and diffuse along. Similarly, in case C, the target is embedded in a thin (radius b) cylinder that can serve as an essentially one-dimensional diffusion guide for the molecule. This model has been used to describe the association of gene-regulatory protein to a specific binding site on DNA (Berg and Blomberg, 1976). The presence of these diffusion guides will subdivide the three-dimensional association process into the sum of 1D and 2D. In case B, the first step is to bind the surface. This process takes place in 1D with mean time given by Equation (12.15) and $\tau_1 = 4H^2/(3D_3)$; the diffusion rate through solution is assumed to be D_3. Thereafter, the diffusion takes place along the surface with mean time from Equation (12.14). Thus the total association time in case B is

$$\tau_B = \frac{4H^2}{3D_3} + \frac{R^2}{2D_2}\left[\ln\left(\frac{R}{b}\right) - \frac{3}{4}\right] \tag{12.17}$$

Similarly, in case C the first step is a two-dimensional diffusion through the solution to the diffusion guide followed by a one-dimensional diffusion along it, giving

$$\tau_C = \frac{R^2}{2D_3}\left[\ln\left(\frac{R}{b}\right) - \frac{3}{4}\right] + \frac{H^2}{3D_1} \tag{12.18}$$

When $H \approx R$ and if $D_2(D_1)$ is not too different from D_3, both τ_B and τ_C can be expected to be smaller than τ_A, and there is a rate enhancement from the surfaces introduced. However, if there is too much surface, $R \gg H$ in case B or $H \gg R$ in case C, the introduction of the lower-dimensional diffusion guides will slow down the process, particularly if $D_1, D_2 \ll D_3$ as expected.

12.3.2 Effective association rates in 2D and 3D

The mean times are related to the rate constants in the following way. The targets are assumed distributed evenly throughout the solution. Around each target we imagine a cylinder like those in Figure 12.4 such that the whole solution is taken up by these cylinders packed together. Molecules diffuse freely through the solution, but when they move from one imagined cylinder into a neighboring one, they move between equivalent positions in the two cylinders and could just as well have been reflected by the cylinder surface. Thus, the diffusion-reaction process can be considered as taking place in a single cylinder around a single target. The dimensions of the cylinders are chosen such that they accommodate one target and the amount of interface that on average is associated with each target.

Since the target concentration–the number of targets per unit volume–by the definition of V is $C = 1/V$, the diffusion-limited association rate constant can be defined from the mean time in Equation (12.16) as

$$k_1 = N_{Av}/(\tau_A C) = 4\pi D_3 b N_{Av} \tag{12.19}$$

which is the usual Smoluchowski expression (cf. Theory Boxes 2.1 and 11.1). For $D_3 = 10^{-6}$ cm^2/s and $b = 4$ Å, Equation (12.19) gives $k_1 = 3 \times 10^8$ M^{-1}s^{-1}. Similarly, in the planar surface, the target concentration is $X_L^* = b^2/R^2$ if the target is a ligand

and ligands have the same size as the other membrane-forming molecules with area $a_M = \pi b^2$. The diffusion-time expressions are valid only for $R \gg b$, i.e. for $X_L^* \ll 1$. The effective diffusion-limited bimolecular association rate constant in 2D can be expressed from the mean time in Equation (12.14) as

$$k_1^* = \frac{1}{\tau_2 X_L^*} = \frac{4\pi D_2}{\left[-\ln\left(X_L^*\right) - 3/2\right] a_M} \tag{12.20}$$

This is the rate constant if ligand concentration in the interface is counted as the mole fraction (cf. Equation 12.22 below). It is not a proper rate constant since it depends on the target concentration. For $a_M = 50$ Å^2, $D_2 = 10^{-8}$ cm^2/s and $X_L^* = 10^{-3}$, Equation (12.20) gives $k_1^* = 5 \times 10^6$ s^{-1}. Thus, the turnover rate can be limited by the diffusion rates only at very low substrate (or ligand) concentrations, such that $k_1[S] < k_{catS}$ or $k_1^* X_S^* < k_{catS}^*$. For k_{catS} or $k_{catS}^* = 500$ s^{-1}, this requires $[S] < 10^{-6}$M or $X_S^* < 10^{-4}$.

Let us compare two situations where enzymes (targets) E and ligand L can be either in solution or in the interface of a membrane. Which arrangement will lead to the fastest association rates? The rate of complex formation in solution is determined by

$$r_3 = \frac{d[EL]}{dt} = 4\pi D_3 b N_{Av}[E][L] \tag{12.21}$$

If the ligand is in the mole fraction $X_L^* = [L^*]/[M^*]$ in the interface, the rate of complex formation is

$$r_2 = \frac{d\,[E^*L]}{dt} = k_1^* X_L^* \,[E^*] = \frac{4\pi D_2}{\left[-\ln\left(X_L^*\right) - 3/2\right] a_M} \frac{[E^*]\,[L^*]}{[M^*]} \tag{12.22}$$

Here, the mole fraction of ligand in the interface is $X_L^* = [L^*]/[M^*]$. The ratio of the two rates, in 2D and 3D, is

$$\frac{r_2}{r_3} = \frac{k_1^*}{k_1\,[M^*]} = \frac{D_2}{D_3 b a_M N_{Av}\left[-\ln\left(X_L^*\right) - 3/2\right]} \frac{1}{[M^*]} \tag{12.23}$$

The comparison is for the same number of molecules in solution as in the interface so that the concentrations are $[E] = [E^*]$ and $[L] = [L^*]$. The ratio r_2/r_3 is larger than one only when the concentration of interface molecules is smaller than

$$[M^*] < \frac{1}{b a_M N_{Av}} \frac{D_2}{D_3\left[-\ln\left(X_L^*\right) - 3/2\right]} \tag{12.24}$$

Thus, confinement of the reactants to the interface can be rate-enhancing only as long as the interface is not too large. For some reasonable choices of parameter values, the requirement is that $[M^*] < 1$ to 10 mM. This estimate is based on $b = 4$ Å, $a_M = 50$ Å^2, $X_L^* = 10^{-4}$ and $D_2/D_3 = 10^{-2}$ to 10^{-3}, which is a rough estimate of the ratio of the diffusion rates for phospholipids in a membrane and in aqueous solution (Jain, 1988). However, both D_2 and D_3 should be the sum of the diffusion rates for the substrate and the enzyme so that the ratio D_2/D_3 also depends on the enzyme. The 1 to 10 mM limit is of the order of magnitude of the size of the inner membrane of a typical eukaryotic cell and smaller than that of a typical prokaryot. The limit is also much smaller than the phospholipid concentration in tissue, ca. 50 mM (Chapter 1). Thus, enzymes *in vivo* are not likely to benefit much kinetically from this effect. To the extent that the enzyme kinetics is limited by the substrate association rate, some interfacial activation could be caused by interfacial confinement. However, if the interface area is too large and/or the diffusion rate in the interface too slow, confining the reaction to 2D will only decrease the rate.

This comparison is for the rates of complex formation, on the assumption that they are diffusion limited both in 3D and 2D. If complex formation is not diffusion-limited, it is impossible to formulate any general comparison between two-dimensional and three-dimensional kinetics. The calculations assume that the molecules in the interface are in orientations that are favorable for complex formation; this is inherent in the assumption of diffusion limitation. If orientations are not favorable, association may be limited by an activation step that requires a major reorientation. Similarly, the association step could be limited by the rate of pulling the substrate out of the interface, as may be the case for PLA2. For complex formation in solution, finding the correct orientation is part of the diffusion process and should be included in the effective reaction radius b used in Equations (12.13) and (12.19). As a consequence, the value of b may be larger in the interface if the substrate is already in a favorable orientation, causing some increase in the two-dimensional rate. Such an increase could also be caused by a relief of some activation step required for binding. This could be the case for the 'lid' opening in lipase (Chapter 7), although there is no experimental information about the binding rates in this system.

This comparison of the two-dimensional and three-dimensional rates is valid when the targets and ligands are dilute in the interface ($X_L^* \ll 1$). The situation will be quite different for an enzyme that acts on an interface that contains a large mole fraction of substrate. As discussed in Chapter 5, such an enzyme is not likely to be limited by two-dimensional diffusion. However, if the interface is spread out over a large number of aggregates and if the enzymes are strongly bound to the interface (scooting kinetics), the overall rate may become low because of exchange limitations. The enzymes may remain on the aggregates even after all substrates are gone without being able to access new aggregates. The exchange rate should not be too small even if interface binding is favored, $K_d < [M^*]$. Similarly, if substrates are localized only in some interfaces, it is important that the enzymes show sufficient specificity in their interface binding and not spend too much time searching the wrong interface. Thus, for interfacial enzymes, too strong or indiscriminate interface binding may be more of a problem rather than creating the potential for a rate increase due to two-dimensional diffusion.

12.4 Rate Enhancement by Facilitated Diffusion in 2D

Consider a situation where the enzymes, or receptors, are in the interface while ligands are in solution. Is the rate of ligand–receptor association increased if ligands confine their search to the interface? If so, how strongly should they be bound to the interface during the search? The ligand–receptor association can be described in a two-step scheme (Berg, 1988):

$$
\begin{array}{c}
\mathrm{L} + \mathrm{E}^* \xrightarrow{k_1} \\[4pt]
k_\mathrm{d} \uparrow \downarrow k_\mathrm{a} \qquad\qquad \mathrm{E}^*\mathrm{L} \\[4pt]
\mathrm{L}^* + \mathrm{E}^* \xrightarrow{k_1^*}
\end{array}
\tag{12.25}
$$

The rate constants are defined from the rate equations as

$$
\frac{\mathrm{d}\,[\mathrm{L}^*]}{\mathrm{d}t} = k_\mathrm{a}\,[\mathrm{L}]\,[\mathrm{M}^*] - k_\mathrm{d}\,[\mathrm{L}^*] - k_1^* X_\mathrm{L}^*\,[\mathrm{E}^*]
\tag{12.26}
$$

$$
\frac{\mathrm{d}\,[\mathrm{E}^*\mathrm{L}]}{\mathrm{d}t} = k_1^* X_\mathrm{L}^*\,[\mathrm{E}^*] + k_1\,[\mathrm{L}]\,[\mathrm{E}^*]
\tag{12.27}
$$

where $X_\mathrm{L}^* = [\mathrm{L}^*]/[\mathrm{M}^*]$. The effective association rate constant in this scheme can be defined from

$$
\frac{\mathrm{d}\,[\mathrm{E}^*\mathrm{L}]}{\mathrm{d}t} = k_\mathrm{a}^{\mathrm{eff}}\,[\mathrm{L_T}]\,[\mathrm{E}^*]
\tag{12.28}
$$

A standard steady state analysis, setting $\mathrm{d}[\mathrm{L}^*]/\mathrm{d}t = 0$ and solving $[\mathrm{L}^*]$ from Equation (12.26), allows identification of the effective association rate constant, $k_\mathrm{a}^{\mathrm{eff}}$, for formation of $\mathrm{E}^*\mathrm{L}$. Relative to the rate of direct association, k_1, it can be expressed as

$$
\frac{k_\mathrm{a}^{\mathrm{eff}}}{k_1} = \frac{k_1^*/k_1 + K_\mathrm{L}' + k_1^*\,[\mathrm{E}^*]\,/\,(k_\mathrm{a}\,[\mathrm{M}^*])}{[\mathrm{M}^*] + K_\mathrm{L}' + k_1^*\,[\mathrm{E}^*]\,/\,(k_\mathrm{a}\,[\mathrm{M}^*])}
\tag{12.29}
$$

For simplicity, in this calculation $[\mathrm{M}^*]$ has been considered constant. From Equation (12.29) it is obvious that $k_\mathrm{a}^{\mathrm{eff}} > k_1$ only when $k_1^* > k_1[\mathrm{M}^*]$, as expected from Equation (12.23). The direct association from solution to a membrane-bound receptor (half-sphere) has the diffusion-limited rate constant

$$
k_1 = 2\pi D_3 b N_{\mathrm{Av}}
\tag{12.30}
$$

The diffusion-limited association rate constant in the interface, k_1^*, is given by Equation (12.20). Ligand binding and dissociation from the interface has rate constants k_a and k_d, which satisfy $k_\mathrm{d}/k_\mathrm{a} = K_\mathrm{L}'$ (cf. Equation 2.7). From Equation (12.29) it can be seen that if $k_1^* > k_1[\mathrm{M}^*]$, the effective association rate constant will be maximal when $K_\mathrm{L}' \to 0$.

Thus for a two-dimensional diffusion guide, the fastest association is achieved when all ligands are confined to the interface, as long as the condition given by Equation (12.24) is satisfied. Otherwise, when the concentration of the interface is too large, association through solution is faster. This either/or situation is quite different from the case where the target is in a one-dimensional diffusion guide, e.g. for *sliding* along a DNA chain. In this case the maximal association rate is achieved when the ligand spends 50 % of the time free in solution and the rest searching by sliding along the one-dimensional 'surface' where the target is located (Berg *et al.*, 1981, 1988). This difference is due to the probability of locating a target, which in 1D increases only as the square root of the time spent in 1D, whereas the probability increases in direct proportion to the time spent in 2D (Berg, 1985, 1988).

To calculate the rate enhancement we also need to know the rate constant, k_a, of binding the interface. This will depend on the geometrical arrangements in solution. $1/(k_a[M^*])$ corresponds to the mean time it takes to bind the interface from solution. If membrane and targets are distributed evenly throughout the solution, with concentrations $[M^*]$ and $[E^*]$, respectively, e.g. in an arrangement like Figure 12.3B, where $2\pi R^2 H N_{Av} = 1/[E^*]$ and $R^2/b^2 = [M^*]/[E^*]$, one finds

$$k_a[M^*] = 3D_3/(4H^2) = 3D_3\pi^2 b^4 N_{Av}^2[M^*]^2 \tag{12.31}$$

As discussed above, this thought construction is not as useful for small values of $[M^*]$ when there is little interface surface in the system. In this case, a homogenous distribution of a continuous interface throughout the solution volume is not possible. If the interface instead is divided up into a large number of smaller aggregates, the limit discussed in Theory Box 2.1 or 11.1 would be more appropriate: $k_a = 2\pi D b N_{Av}/\sqrt{N_T}$, where $N_T = [M^*]/[E^*]$ is the number of interface molecules associated with each target. In this limit, the rate enhancement does not decrease as rapidly with decreasing $[M^*]$ as depicted in Figure 12.5.

Figure 12.5 shows the expected rate enhancement based on Equations (12.29) and (12.31). Only for very low enzyme (target) concentration, below the nanomolar range, could there be any substantial rate enhancement from the 2D-diffusion. Also, if the interface is at too high a concentration, large $[M^*]$, dilution in the interface becomes a problem. The main limitation comes from the slowdown of the lateral diffusion rate in the interface relative to that in solution. Lipid molecules in the interface may have a diffusion rate constant that is 100- to 1000-fold slower than in solution (Jain, 1988).

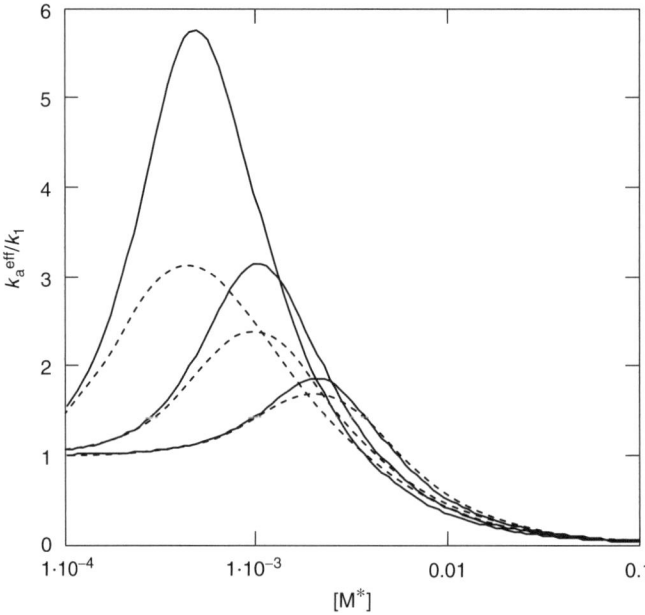

Figure 12.5 Rate enhancement from Equations (12.29) and (12.31) using $D_2/D_3 = 0.003$, $b = 4$ Å, and $a_M = 50$ Å^2. Solid curves, $K'_L = 0$ and dotted curves, $K'_L = 10^{-3}$ M. For each set of curves, $[E^*] = 10^{-9}$, 10^{-8}, 10^{-7} M (from upper to lower curves)

12.5 Summary: Dimensionality Effects on the Equilibrium and Diffusion

As a first approximation, the main effects of confinement and orientation of reactants in the interface can be described in terms of high local concentrations. However, it is important to define the concentration units in the interface and solution properly so that the consequences of the change in dimensionality can be compared directly. From classical statistical mechanics it can be argued that the basic concentration variable that describes the entropy of dilution is the number density (molecules per unit of accessible volume). Note that the mole fraction units for a solute in the interface versus the aqueous phase are not directly comparable since their properties must depend on the local concentration of solvent.

The entropy of confinement in the interface can be calculated from statistical mechanics, and this entropy loss can be directly interpreted as a reduction in accessible volume with a concomitant increase in local concentration. It should be stressed that the local concentration in this way can be defined from thermodynamics (entropy) and will not depend on an *ad hoc* estimate of interface thickness. This makes it possible to estimate the interaction free energies involved in interface binding and to discuss the relationship between complex formation in solution relative to that in the interface.

This relationship is intrinsic in the detailed balance conditions developed in Chapters 6 through 9. The high local concentration of substrate that an enzyme 'sees' in the interface could lead to a substantial increase in substrate–enzyme binding, i.e. an interfacial K_S^* activation of the enzyme. However, if the enzyme must dislodge the substrate from the interface before it can bind the catalytic site, little or no increase in enzyme–substrate binding would be expected from a confinement of the reactants to the interface. It is of course also possible that confinement in the interface actually reduces accessibility, and therefore the enzyme–substrate binding, if the molecules are in configurations that are not favorable for interaction.

The high local concentrations and accessibility in the interface influence not only the stability of the Michaelis complex but also the replenishment, i.e. the rate with which complexes can form during the steady state turnover. No enzyme–substrate reaction can ever be faster than the rate with which diffusion brings the reactants together. However, the rate can be limited by diffusion only at very low concentrations of both substrate and enzyme. For an interfacial enzyme, the two-dimensional diffusion in the interface is not likely to be limiting, particularly if substrate is a major component of the interface, in which case the local concentration is high. As discussed throughout this book, the main problem in this case may be to move substrates or enzymes between aggregates. At very high concentrations of interface, notably corresponding to the membrane concentration present in living tissue, two-dimensional association may become slower than if the reaction partners were free in solution. However, the interfacial enzymes are in the interface because that is where their substrates are. For such enzymes, the main problem is not the dimensionality of the search process but rather its specificity and processivity. The heterogeneity of the distribution of substrate is more important than the dimensionality.

References

Adam, G. and Delbrück, M. (1968). Reduction of dimensionality in biological diffusion processes, in *Structural Chemistry and Molecular Biology*, A. Rich and N. Davidson (Eds.), W.H. Freeman, San Francisco. pp. 198–215.

Aniansson, E.A.G., Wall, S.N., Almgren, M., Hoffman, H., Kleimann, I., Ulbricht, W., Zana, R., Lang, J. and Tondre, C. (1976). Theory of the kinetics of micellar equilibria and quantitative interpretation of chemical relaxation studies of micellar solutions of ionic surfactants. *J. Phys. Chem.*, **80**, 905–922

Apitz-Castro, R.J., Jain, M.K. and de Haas, G.H. (1982). Origin of the latency phase during the action of phospholipase A_2 on unmodified phosphatidylcholine vesicles. *Biochim. Biophys. Acta*, **688**, 349–56.

Araujo, P.S., Rosseneu, M.Y., Kremer, J.M.H., van Zolen, E.J.J. and de Haas, G.H. (1979). Structure and thermodynamic properties of the complexes between phospholipase A_2 and lipid micelles. *Biochemistry* **18**, 580–6.

Baybert, T., Yu, B.Z., Lin, H., Browning, J., Jain, M.K. and Gelb, M.H. (1993). Human nonpancreatic secreted phospholipase A_2: interfacila parameters, substrate specificities, and competitive inhibitors. *Biochemistry*, **32**, 573–82.

Ben-Naim, A. (1987). *Solvation Thermodynamics*, Plenum, New York, 243.

Ben-Tal, N., Honig, B., Bagdassarian, C.K. and Ben-Shaul, A. (2000). Association entropy in adsorption processes. *Biophys. J.* **79**, 1180–7.

Berg, O.G. (1985). Orientation constraints in diffusion-limited macromolecular association. The role of surface diffusion as a rate-enhancing mechanism. *Biophys. J.* **47**, 1–14.

Berg, O.G. (1988). Surface diffusion as a rate-enhancing mechanism: the effectiveness of one- and two-dimensional surface sliding. *Makromol. Chem. Macromol. Symp.* **17**, 161–75.

Berg, O.G. and Blomberg, C. (1976). Association kinetics with coupled diffusional flows. Special application to the *lac* repressor–operator system. *Biophys. Chem.* **4**, 367–81.

Berg, O.G. and von Hippel, P.H. (1985). Diffusion-controlled macromolecular interactions. *Ann. Rev. Biophys. Biophys. Chem.* **14**, 131–60.

Berg, O.G., Winter, R.B. and von Hippel, P.H. (1981). Diffusion-driven mechanisms for protein translocation on nucleic acids. I: models and theory. *Biochemistry*, **20**, 6929–48.

Berg, O.G., Yu, B.-Z., Rogers, J. and Jain, M.K. (1991). Interfacial catalysis by phospholipase A_2: determination of the interfacial kinetic rate constants. *Biochemistry* **30**, 7283–97.

Berg, O.G., Rogers, J., Yu, B-Z., Yao, J., Romsted, L.S. and Jain, M.K. (1997). Thermodynamic and kinetic basis of interfacial activation: resolution of binding and allosteric effects on pancreatic phospholipase A_2 at zwitterionic interfaces. *Biochemistry*, **36**, 14512–30.

Berg, O.G., Cajal, Y., Butterfoss, G.L., Grey, R.L., Alsina, M.A., Yu, B.-Z. and Jain, M.K. (1998). Interfacial activation of triglyceride lipase from *Thermomyces* (Humicola) *lanuginosa*: kinetic parameters and a basis for control of the lid. *Biochemistry* **37**, 6615–27.

Berg, O.G., Tsai, M.D., Gelb, M.H. and Jain, M.K. (2001). Interfacial enzymology: the secreted phospholipase A_2 paradigm. *Chem. Rev.,* **101**, 2613–2653.

Berger, O., Edholm, O. and Jähnig F. (1997). Molecular dynamics simulation of a fluid bilayer of dipalmitoylphosphatidylcholine at full hydration, constant pressure, and constant temperature. *Biophys. J.* **72**, 2002–13.

Bian, J. and Roberts, M.F. (1991). Thermodynamic properties of short chain phosphatidyl-choline isomers: micellization and adsorption at an air/water interface. *J. Phys. Chem.,* **95**, 2572–7.

Bonsen, P.P.M., de Haas, G.H., Pieterson, W.A. and Van Deenen, L.L.M. (1972). Studies on phospholipase and its zymogen from pancreas, IV. The influence of chemical modification of the lecithin structure on substrate properties. *Biochim. Biophys. Acta* **270**, 364–82.

Brzozowski, A.M., Savage, H., Verma, C.S., Turkenburg, J.P., Lawson, D.M., Svendsen, A. and Patkar, S. (2000). Structural origins of the interfacial activation in *Thermomyces* (Humicola) *lanuginosa* lipase. *Biochemistry,* **39**, 15071–82.

Buser, C.A., Sigal, C.T., Resh, M.D. and McLaughlin, S. (1994). Membrane binding of myris-toylated peptides corresponding to the NH_2 terminus of Src. *Biochemistry* **33**, 13093–101.

Cajal, Y. and Jain, M.K. (1997). Synergism between mellitin and phospholipase A_2 from bee venom: apparent activation by intervesicle exchange of phospholipids. *Biochemistry,* **36** 3882–93.

Cajal, Y., Berg, O.G. and Jain, M.K. (1995). Direct vesicle–vesicle exchange of phospholipids mediated by polymyxin B. *Biochem. Biophys. Res. Comm.* **210**, 746–52.

Cajal, Y., Ghanta, J., Surolia, A.K., Easwaran, E. and Jain, M.K. (1996a). Specificity for the transfer of phospholipids through polymyxin B mediated intermembrane molecular contacts. *Biochemistry,* **35**, 5684–95.

Cajal, Y., Rogers, J., Berg, O.G. and Jain, M.K. (1996b). Intermembrane molecular contacts by polymyxin B mediate exchange of phospholipids. *Biochemistry,* **35**, 299–308.

Cajal, Y., Berg, O.G. and Jain, M.K. (2000a) Product accumulation during the lag phase as the basis for the activation of phospholipase A_2 on monolayers. *Langmuir* **16**, 252–7.

Cajal, Y., Svendsen, A., Girona, V., Patker, S.A. and Alsina, M.A. (2000b). Interfacial control of lid opening in *Thermomyces lanuginosa* lipase. *Biochemistry,* **39**, 413–23.

Cantor, C.R. and Schimmel, P.R. (1980). *Biophysical Chemistry, Part 1: The Conformation of Biological Macromolecules,* W.H. Freeman, San Francisco.

Carey, M.C., Small, D.M. and Bliss, C.M. (1983). Lipid digestion and absorption. *Ann. Rev. Physiol.* **45**, 651–77.

Cevc, G. (1990). Membrane electrostatics. *Biochim. Biophys. Acta,* **1031**, 311–82.

Cevc, G. and Marsh, D. (1987). *Phospholipid Bilayers: Physical Principles and Models,* Wiley, New York, p. 442.

Chakrabarti, P. (1993). Anion binding sites in protein structures. *J. Mol. Biol.* **234**, 463–82.

Collins, K.D. and Washabaugh, M.W. (1985). The Hoffmeister effect and the behaviour of water. *Quart. Rev. Biophys.* **18**, 323–422.

Dam-Mieras, M.C.E., Slotboom, A.J., Pieterson, W.A. and de Haas G.H. (1975). The interaction of phospholipase A$_2$ with micellar interfaces. The role of the N-terminal region. *Biochemistry*, **14**, 5387–94.

de Haas, G.H., Bonsen, P.P.M., Pieterson, W.A. and van Deenen, L.L.M. (1971). Studies on phospholipase A$_2$ and its zymogen from porcine pancreas. III. Action of the enzyme on short-chain lecithins. *Biochim. Biophys. Acta*, **239**, 252–66.

de Haas, G.H., Dijkman, R., Ransac, S. and Verger, R. (1990). Competitive inhibition of lipolytic enzymes. IV. Structural details of acylamino phospholipid analogues important for the potent inhibitory effects on pancreatic phospholipase A$_2$. *Biochim. Biophys. Acta*, **1046**, 249–57.

de Haas, G.H., Dijkman, R., Lugtigeid, R.B., Dekker, N., Van den Berg, L., Egmond, M.R. and Verheij, H.M. (1993). Competitive inhibition of lipolytic enzymes. IX. A comparative study on the inhibition of pancreatic phospholipase A$_2$ from different sources by (R)-2-acylamino phospholipid analogues. *Biochim. Biophys. Acta*, **1167**, 282–88.

de Haas, G.H., Dijkman, R., Boots, J.W.P. and Verheij, H.M. (1995). Competitive inhibition of lipolytic enzymes. XI. Estimation of the interfacial dissociation constants of porcine pancreatic phospholipase A$_2$ for substrate and inhibitor in the absence of detergents. *Biochim. Biophys. Acta*, **1257**, 87–95.

Dennis, E.A. (1973). Phospholipase A$_2$ activity towards phosphatidylcholine in mixed micelles: surface dilution kinetics and the effect of thermotropic phase transition. *Arch. Biochem. Biophys.* **158**, 485–93.

Dennis, E.A. (1983). Phospholipases. *The Enzymes* **16**, 307–53.

Derwenda, Z.S. (1994). Lipases. *Adv. Protein Chem.*, **45**, 1–52.

Dijkstra, B.W., Drenth, J. and Kalk, K.H. (1981). The active site and the catalytic mechanism of phospholipase A$_2$. *nature*, **289**, 604–6.

Engelman, D.M., Steitz, T.A. and Goldman, A. (1986). Identifying non-polar transbilayer helices in amino acid sequences of membrane proteins. *Ann. Rev. Biophys. Chem.*, **15**, 321–53.

Fullington, D.A., Shoemaker, D.G. and Nichols, J.W. (1990). Characterization of phospholipid transfer between mixed phospholipid-bile salt micelles. *Biochemistry* **29**, 879–86.

Gelb, M.H., Jain, M.K., Hanel, A.M. and Berg, O.G. (1995). Interfacial enzymology of glycerolipid hydrolases: lessons from phospholipase A$_2$. *Ann. Rev Biochem.* **64**, 653–88.

Ghomashchi, F., O'Hare T., Clary, D. and Gelb, M.H. (1991a). Interfacial catalysis by phospholipase A$_2$: evaluation of the interfacial rate constants by steady-state isotope effect studies *Biochemistry*, **30**, 7298–305.

Ghomashchi, F., Yu, B.-Z., Berg, O.G., Jain, M.K. and Gelb, M.H. (1991b). Interfacial catalysis by phospholipase A$_2$: substrate specificity in vesicles. *Biochemistry*, **30**, 7318–29.

Griffith, O.H. and Rijan, M. (1999). Bacterial phosphatidylinositol- specific phospholipase C: structure, function, and interaction with lipids. *Biochim. Biophys. acta*, **1441**, 237.

Gurney, R.W. (1953). *Ionic Processes in Solution*, Dover, New York, 275 pp.

Hauser, H. (1984). Some aspects of the phase behavior of charged lipids. *Biochim. Biophys. Acta*, **772**, 37–50.

Hauser, H., Pascher, I., Pearson, R.H. and Sundell, S. (1981). Preferred conformation and molecular packing of phosphatidylethanolamine and phosphatidylcholine. *Biochim. Biophys. Acta*, **650**, 21–51.

Helenius, A. and Simons, K. (1975). Solubilization of membranes by detergents. *Biochim. Biophys. Acta*, **415**, 29–79.

Hille, J.D.R., Op den Kelder, D., Sauve, M., de Haas, G.H. and Egmond, M.R. (1981). Physicochemical studies on the interaction of pancreatic phospholipase A_2 with a micellar substrate analog. *Biochemistry*, **20**, 4068–73.

Hofman, A.F. and Mysels, K.J. (1988). Bile salts as biological surfactants. *Colloids Surf. Sci.*, **30**, 134–73.

Homan, R. and Jain, M.K (2000) Biology, pathology and interfacial enzymology of pancreatic phospholipase A_2, In *Intestinal Lipid Metabolism* C.M. Mansbach, P. Tso and A. Kuksis (Eds.), Kluwer Academic Press and Plenum, New York, pp. 81–104.

Homan, R. and Krause, B.R. (1997). Established and emerging strategies for inhibition of cholesterol absorption. *Curr. Pharm. Des* **3**, 29–44.

Honig, B. and Nicholls, A. (1995). Classical electrostatics in biology and chemistry. *Science* **268**, 1144–9.

Hunter, R.J. (Ed.) (1989). *Foundations of Colloid Science*, Vols I and II, Clarendon Press, 1089.

Hurley, J.H. and Misra, S. (2000) Signaling and subcellular targeting by membrane-binding domains: *Ann. Rev. Biophys. Biomol. Struct.* **29**, 49–79.

Jähnig F. (1983). Thermodynamics and kinetics of protein incorporation in membranes. *Proc. Natl Acad. Sci. USA*, **80**, 3691–5.

Jain, M.K. (1973). Enzymatic hydrolysis of various components in membranes and related systems. *Curr. Topics in Memb. Transport*, **4**, 175–254.

Jain, M.K. (1988). *Introduction to Biomembranes*, II edition, Wiley, New York, p. 423.

Jain, M.K., Berg, O.G. (1989). The kinetics of interfacial catalysis by phospholipase A_2 and regulation of interfacial activation: hopping versus scooting. *Biochim. Biophys. Acta*, **1002**, 127–56.

Jain, M.K. and Cordes, E.H. (1973a) Phospholipases I. Effect of n-alkanols on the rate of hydrolysis of egg phosphatidylcholine. *J. Membrane Biol.* **14**, 101–18.

Jain, M.K. and Cordes, E.H. (1973b). Phospholipases II. Effect of sonication and addition of cholesterol to the rate of hydrolysis of various lecithins. *J. Membrane Biol.* **14**, 119–34.

Jain, M.K. and de Haas, G.H. (1983). Activation of phospholipase A_2 by freshly added lysophospholipids. *Biochim. Biophys. Acta*, **736**, 157–62.

Jain, M.K. and Jahagirdar, D.V. (1985a). Action of phospholipase A_2 on bilayers: effect of fatty acid and lysophospholipid additives on the kinetic parameters. *Biochim. Biophys. Acta*, **814**, 313–18.

Jain, M.K. and Jahagirdar, D.V. (1985b) Action of phospholipase A_2 on bilayers: effect of inhibitors. *Biochim. Biophys. Acta*, **814**, 319–26.

Jain, M.K. and Maliwal, B.P. (1993). Spectroscopic properties of the states of pig pancreatic phospholipase A_2 at interaces and their possible molecular origin. *Biochemistry*, **32**, 11838–46.

Jain, M.K. and Rogers, J. (1989). Substrate specificity for interfacial catalysis by phospholipase A_2 in the scooting mode. *Biochim. Biophys. Acta*, **1003**, 91–7.

Jain, M.K. and Vaz, V.L.C. 1987, Dehydration of the lipid-protein microinterface on binding of phospholipase A_2 to lipid bilayers. *Biochim. Biophys. Acta*, **905**, 1–8.

Jain, M.K., Egmond, M.R., Verheij, H.M., Apitz-Castro, R.J., Dijkman, R. and de Haas G.H. (1982). Interaction of phospholipase A_2 and phospholipid bilayers. *Biochim. Biophys. Acta*, **688**, 341–8.

Jain, M.K., Crecely, R., Hille, J.D.R., de Haas G.H. and Grunner, S. (1985). Phase properties of the aqueous dispersions of n-octadecylphosphocholine. *Biochim. Biophys. Acta*, **813**, 68–76.

Jain, M.K., de Haas, G.H., Marecek, J.F. and Ramirez, F. (1986a). The affinity of phospholipase A_2 for the interface of the substrate and analogous. *Biochim. Biophys. Acta*, **860**, 475–83.

Jain, M.K., Maliwal, B.P., de Haas, G.H. and Slotboom, A.J. (1986b). Anchoring of phospholipase A_2: the effect of anions and deuterated water, and the role of N-terminus region. *Biochim. Biophys. Acta*, **860**, 448–61.

Jain, M.K., Rogers, J., Jahagirdar, D.V., Marecek, J.F. and Ramirez, F. (1986c). Kinetics of interfacial catalysis by phospholipase A_2 in intravesicular scooting mode, and heterodiffusion of anionic zwitterionic vesicles. *Biochim. Biophys. Acta*, **860**, 435–47.

Jain, M.K., Rogers, J., Marecek, J.F., Ramirez, F. and Eibl, H. (1986d). Effect of the structure of phospholipid on the kinetics of intravesicle scooting of phospholipase A_2. *Biochim. Biophys. Acta*, **860**, 462–74.

Jain, M.K., Rogers, J. and de Haas G.H. (1988). Kinetics of binding of phospholipase A_2 to lipid–water interface and its relationship to interfacial activation. *Biochim. Biophys. Acta*, **940**, 51–62.

Jain, M.K., Yuan, W. and Gelb, M.H. (1989). Competitive inhibition of phospholipase A_2 in vesicles. *Biochemistry*, **28**, 4135–9.

Jain, M.K., Ranadive, G.N., Rogers, J., Yu, B.-Z. and Berg, O.G. (1991a). Interfacial catalysis by phospholipase A_2: dissociation constants for calcium, substrate, products, and competitive inhibitors. *Biochemistry*, **30**, 7306–17.

Jain, M.K., Ranadive, G.N., Yu, B.-Z. and Verheij, H.M. (1991b). Interfacial catalysis by phospholipase A_2: monomeric enzyme is fully catalytically active at the bilayer interface. *Biochemistry*, **30**, 7330–40.

Jain, M.K., Rogers, J., Berg, O.G. and Gelb, M.H. (1991c). Interfacial catalysis by phospholipase A_2: activation by substrate replenishment. *Biochemistry*, **30**, 7340–48.

Jain, M.K., Tao, W., Rogers, J., Arenson, C., Eibl, H. and Yu, B.-Z. (1991d). Active-site directed specific competitive inhibitors of phospholipase A_2: novel transition state analogues. *Biochemistry*, **30**, 10256–268.

Jain, M.K., Ghomashchi, G., Yu, B.-Z., Bayburt, T., Murphy, D., Houck, D., Brownell, J., Reid, J.C., Solowiej, J.E., Jarrell, R., Sasser, M. and Gelb, M.H. (1992a). Fatty acid amides: scooting mode-based discovery of tight-binding competitive inhibitors of secreted phospholipases A_2. *J. Med. Chem.*, **35**, 3584–6.

Jain, M.K., Yu, B.-Z., Rogers, J., Gelb, M.H., Tsai, M.D., Hendrickson, E.K. and Hendrickson, S. (1992b). Interfacial catalysis by phospholipase A_2: The rate-limiting step for enzymatic turnover. *Biochemistry*, **31**, 7841–7.

Jain, M.K., Rogers, J., Hendrickson, H.S. and Berg, O.G. (1993a). The chemical step is not rate-limiting during the hydrolysis by phospholipase A_2 of mixed micelles of phospholipid and detergent. *Biochemistry*, **32**, 8360–67.

Jain, M.K., Yu, B.Z. and Berg, O.G. (1993b). Relationship of interfacial equilibria to interfacial activation of phospholipase A_2. *Biochemistry*, **32**, 11319–29.

Jain, M.K., Gelb, M.H., Rogers, J. and Berg, O.G. (1995). The kinetic basis of interfacial catalysis and activation of phospholipase A_2. *Methods Enzymol.*, **249**, 567–614.

Janssen, M.J.W., van de Wiel, W.A.E.C., Beiboer, S.H.W., van Kampen, M.D., Verheij, H.M., Slotboom, A.J. and Egmond, M.R. (1999). Catalytic role of the active site histidine of porcine pancreatic phospholipase A_2 probed by the variants H48Q, H48N and H48K. *Protein Eng.*, **12**, 497–503.

Jencks, W.P. (1975). Binding energy, specificity, and enzymatic catalysis: the Circe effect. *Adv. Enzymol.*, **43**, 219–410.

Kamrath, R.F. and Frances, E.I. (1986). Phenomena in mixed surfactant systems, in *ACS Symposium Series No. 311*, J.F. Scamehorn (Ed), ACS, Washington, DC, Ch. 3.

Lasch, J. (1995). Interaction of detergents with lipid vesicles. *Biochim. Biophys. Acta*, **1241**, 269–92.

Lebowitz, J.L., Helfand, E., Praestgaard, E. (1965). Scaled particle theory of fluid mixtures. *J. Chem. Phys.*, **43**, 774–9.

Leslie, C.C. (1997). Properties and regulation of cytosolic phospholipase A_2. *J. Biol. Chem.*, **272**, 16709–12.

Lichtenberg, D., Robson, R.J., Dennis, E.A. (1983). Solubilization of phospholipids by detergents. Structural and kinetic aspects. *Biochim. Biophys. Acta*, **737**, 285–304.

Lin, H.K. and Gelb, M.H. (1993). Competitive inhibition of interfacial catalysis by phospholipase A_2. Differential interaction of inhibitors with the vesicle interface as a factor of inhibitor potency. *J. am. chem. Soc.*, **115**, 3932–42.

Low, M.G. (1989). The glycosylphosphatidylinositol anchor of membrane proteins. *Biochim. Biophys. Acta*, **988**, 427–54.

McLaughlin, S. (1989) The electrostatic properties of membranes. *Ann. Rev. Biophys and Biophys. Chem.* **18**, 113–36.

Min, J.H., Jain, M.K., Wilder, C., Paul, L., Apitz-Castro, R., Aspleaf, D.C. and Gelb, M.H. (1999). Membrane-bound plasma platelet activating factor acetylhydrolase acts on substrate in the aqueous phase. *Biochemistry*, **38**, 12935–42.

Min, J.H., Jain, M.K. and Gelb, M.H. (2000). Do membrane bound enzymes access their substrates from the membrane or aqueous phase: interfacial versus non-interfacial enzymes. *Biochim. Biophys. Acta*, **1488**, 20–27.

Mukerjee, P. (1967). The nature of the association equilibria and hydrophobic bonding in aqueous solutions of association colloids. *Advan. Colloid Interface Sci.*, **1**, 241–75.

Murray, D., Arbuzova, A., Honig, B. and McLaughlin, S. (2001). The role of electrostatic and nonpolar interactions in the association of peripheral proteins with membranes, in *Current Topics in Membranes: Peptide–Lipid Interactions*, S. Simin and T. McIntosh (Eds.), Academic Press, in press.

Nelsestuen, G.L. and Martinez, M.B. (1997). Steady state enzyme velocities that are independent of [enzyme]: an important behavior in many membrane and particle-bound states. *Biochemistry*, **36**, 9081–6.

Nichols, J.W. (1985). Thermodynamics and kinetics of phospholipid monomer-vesicle interaction. *Biochemistry*, **24**, 6390–8.

Nichols, J.W. (1988). Phospholipid transfer between phosphatidylcholine-taurocholate mixed micelles. *Biochemistry*, **27**, 879–86.

Nichols, J.W. and Pagano, R.E. (1982). Use of resonance energy transfer to study the kinetics of amphiphile transfer between vesicles. *Biochemistry* **21**, 1720–6.

Op den Kamp, J.A.F., Kaurez, M. Th. and Van Deenen, L.L.M. (1975) Action of pancreatic phospholipase A_2 on phosphatidylcholine bilayers in different physical states. *Biochim. Biophys. Acta*, **406**, 169–77.

Pascher, I. and Sundell, S. (1986). Membrane lipids: preferred conformational states and their interplay. The crystal structure of dilauroylphosphatidyl-*N*,*N*-dimethylethanolamine. *Biochim. Biophys. Acta*, **855**, 68–78.

Page, M.I. and Jencks, W.P. (1971). Entropic contributions to rate acceleration in enzymic and intramolecular reactions and the chelate effect. *Proc. Natl. Acad. Sci. USA*, **68**, 1678–83.

Pan, Y.H., Epstein, T.M., Jain, M.K. and Bahnson, B.J. (2001). Five coplanar anion binding sites on one face of phospholipase A_2: relationship to interface binding. *Biochemistry*, **40**, 609–17.

Panaiotov, I., Ivanova, M. and Verger, R. (1997). Interfacial and temporal organization of enzymatic lipolysis. *Curr. Opin. Colloid Interface Sci.*, **2**, 517–24.

Peitzsch, R.M. and McLaughlin, S. (1993). Binding of acylated peptides and fatty acids to phospholipid vesicles: pertinence to myristoylated proteins. *Biochemistry*, **32**, 10436–43.

Pieterson, W.A., Vidal, J.C., Volwerk, J.J. and de Haas G.H. (1974). Zymogen-catalyzed hydrolysis of monomeric substrates and the presence of a recognition site for lipid-water interfaces in phospholipase A_2. *Biochemistry*, **13**, 1455–60.

Quinn, D.M. and Sutton, L.D. (1991). Theoretical basis and mechanistic utility of solvent isotope effects, in *Enzyme Mechanism from Isotope Effects*, P.D. Cook. (Ed.), CRC Press, Boca Raton, p. 73–126.

Ramirez, F. and Jain, M.K. (1991). Phospholipase A_2 at the bilayer interface. *Proteins*, **9**, 229–39.

Ransac, S., Riviere, C., Soulie, J.M., Gancet, C., Verger, R. and de Haas, G.H. (1990). Competitive inhibition of lipolytic enzymes. I. A kinetic model applicable to water-insoluble competitive inhibitors. *Biochim. Biophys. Acta*, **1043**, 57–66.

Ransac, S., Iavanova, M., Verger, R. and Panaiotov, I. (1997). Monolayer techniques for studying lipase kinetics. *Methods Enzymol*, **286**, 263–92.

Rogers, J., Yu, B.-Z. and Jain, M.K. (1992). Basis for the anomalous effect of competitive inhibitors on the kinetics of hydrolysis of short chain phosphatidylcholines by phospholipase A_2. *Biochemistry*, **31**, 6056–62.

Rogers, J., Yu, B.-Z., Serves, S.V., Tsivgoulis, G.M., Sotiropoulos, D.N., Ioannou, P.V. and Jain, M.K. (1996). Kinetic basis for the substrate specificity during hydrolysis of phospholipids by secreted phospholipase A_2. *Biochemistry*, **35**, 9375–84.

Rogers, J., Yu, B.-Z., Tsai, M.D., Berg, O.G. and Jain, M.K. (1998). Cationic residues 53 and 56 control the anion-induced interfacial k^*_{cat}-activation of pancreatic phospholipase A_2. *Biochemistry*, **37**, 9549–56.

Romero, G., Thompson, K. and Biltonen, R.L. (1987). The activation of porcine pancreatic phospholipase A_2 by dipalmitoylphosphatidylcholine large unilamellar vesicles. *J. Biol. Chem.* **262**, 13476–82.

Sarda, L. and Desnuelle, P. (1958). Action de la lipase pancreatique sur les esters en emulsion. *Biochim. Biophys. Acta*, **30**, 513–21.

Schurer, G., Lanig, H. and Clark, T. (2000). The mode of action of phospholipase A_2: semiemperical MO calculations including the protein environment. *J. Phys. Chem.*, **B104**, 1349–61.

Scott, D.L. and Sigler, P.B. (1994). Structure and catalytic mechanism of secretory phospholipase A_2. *Adv. Protein Chem.*, **45**, 53–88.

Scott, D.L., Mandel, A.M., Sigler, P.B. and Honig, B. (1994). The electrostatic basis for the interfacial binding of secretory phospholipase A_2. *Biophys. J.* **67**, 493–504.

Segel, I.H. (1975) *Enzyme Kinetics: Behavior and Analysis of Rapid Equilibrium and Steady-State Enzyme Systems*, Wiley, New York, 957 pp.

Segrest, J.P., DeLoof, H., Dohlman, J.G., Brouillette, C.G. and Anantharamaiah, G.M. (1990). Amphipathic helix motif: classes and properties. *Proteins*, **8**, 103–117.

Sekar, K., Eswaramoorthy, S., Jain, M.K. and Sundaralingam, M. (1997a). Crystal structure of the complex of bovine pancreatic phospholipase A_2 with the inhibitor 1-hexadecyl-3-(trifluoroethyl)-*sn*-glycero-2-phosphomethanol. *Biochemistry*, **36**, 14186–91.

Sekar, K., Yu, B.-Z., Rogers, J., Lutton, J., Liu, X., Chen, X., Tsai, M.D., Jain, M.K. and Sundaralingam, M. (1997b). Phospholipase A_2 engineering. Structural and functional roles of the highly conserved active site residue aspartate-99. *Biochemistry*, **36**, 3104–14.

Seshadri, K., Vishveshwara, S. and Jain, M.K. (1994). Binding of active site directed ligands to phospholipase A_2: implications on the molecular constraints and catalytic mechanism. *Proc. Indian Acad. Sci.* **106**, 1177–89.

Sharp, K.A., Nicholls, A., Friedman, R. and Honig, B. (1991). Extracting hydrophobic free energies from experimental data: relationship to protein folding and theoretical models. *Biochemistry*, **30**, 9686–97.

Smith, W.L. (1989). The eicosanoids and their biochemical mechanism of action. *Biochem. J.* **259**, 315–24.

Smoluchowski, M. von, Z. (1917). Versuch einer mathematischen Theorie der Koagulationskinetik kolloider Lösungen. *Z. Phys. Chem.* **92**, 129–68.

Soltys, C.E. and Roberts, M.F. (1994). Fluorescence studies of phosphatidylcholine micelle mixing: relevance to phospholipase kinetics. *Biochemistry*, **33**, 11608–17.

Tanford, C. (1980). *The Hydrophobic Effect: Formation of Micelles and Biological Membranes*, 2nd edition, John Wiley, New York, 233 pp.

Tatulian, S.A. (1983). Effect of lipid phase transition on the binding of anions to dimyristoylphosphatidylcholine liposomes. *Biochim. Biophys. Acta*, **736**, 189–95.

Tausk, R.J.M., Karmiggelt, J., Oudshoorn, C. and Overbeek, J. Th. G. (1974). Physical chemical studies of short-chain lecithin homologues. I. Influence of the chain length of the fatty acid ester and of electrolytes on the critical micelle concentration. *Biophys. Chem.*, **1**, 175–83.

Thunnissen, M.M.G.M., Eiso, A.B., Kalk, K.H., Drenth, J., Dijkstra, B.W., Kuipers, O.P., Dijkman, R., de Haas, G.H. and Verheij, H.M. (1990). X-ray structure of phospholipase A_2 complexed with a substrate-derived inhibitor. *Nature*, **347**, 689–691.

Tilbeurgh, H. van Bezzine, S., Cambillau, C., Verger, R. and Carriere, F. (1999). Colipase: structure and interaction with pancreatic lipase. *Biochim. Biophys. Acta*, **1441**, 173–84.

Upreti, G.C. and Jain, M.K. (1978). Effect of the state of phosphatidylcholine on the rate of its hydrolysis by phospholipase A_2 (bee venom). *Arch. Biochem. Biophys.*, **188**, 364–75.

Upreti, G.C. and Jain, M.K. (1980). Action of phospholipase A_2 on unmodified bilayers: organizational defects are preferred sites of action. *J. Membrane Biol.* **55**, 113–23.

Upreti, G.C., Rainier, S. and Jain, M.K. (1980). Intrinsic differences in the perturbing ability of alkanols in bilayer: action of phospholipase A_2 on the alkanol modified phospholipid bilayers. *J. Membrane Biol.*, **55**, 97–112.

Valentin, E. and Lambeau, G. (2000). Increasing molecular diversity of secreted phospholipases A$_2$ and their receptors and binding proteins. *Biochim. Biophys. Acta.* **1488**, 59–70.

Van, Eijk, J.H., Verheij, H.M., Dijkman, R. and de Haas G.H. (1983). Interaction of phospholipase A$_2$ from *Naja melanoleuca* snake venom with monomeric substrate analogs. Activation of the enzyme by protein–protein or lipid–protein interactions. *Eur. J. Biochem.*, **132**, 183–8.

Van, Oort, M.G., Dijkman, R., Hille, J.D.R. and de Haas, G.H. (1985a). Kinetic behavior of porcine pancreatic phospholipase A$_2$ on zwitterionic and negatively charged single chain substrates. *Biochemistry*, **24**, 7987–93.

Van Oort, M.G., Dijkman, R., Hille, J.D.R. and de Haas, G.H. (1985b). Kinetic behavior of porcine pancreatic phospholipase A$_2$ on zwitterionic and negatively charged double chain substrates. *Biochemistry*, **24**, 7993–9.

Verger, R. and de Haas, G.H. (1976). Interfacial enzyme kinetics of lipolysis. *Ann. Rev. Biophys. Bioeng.* **5**, 77–117.

Verger, R., Mieras, M.C.E. and de Haas, G.H. (1973). Action of phospholipase A at interfaces. *J. Biol. Chem.*, **248**, 4023–34.

Verheij, H.M. (1995). Structure and mechanism of pancreatic phospholipase A$_2$ – a molecular biology approach. in *Phospholipase A$_2$ in Clinical Inflammation, Molecular Approaches to Pathophysiology*, K.B. Glaser, P. Vadas (Eds.), CRC Press, Boca Raton, pp. 3–24.

Verheij, H.M., Volwerk, J.J., Jensen, E.H.J.M., Puijk, W.C., Dijkstra, B.W., Drenth, J. and de Haas, G.H. (1980). Methylation of histidine-48 in pancreatic phospholipase A$_2$. Role of histidine and calcium ion in the catalytic mechanism. *Biochemistry*, **19**, 743–50.

Verheij, H.M., Slotboom, A.J. and de Haas G.H. (1981). Structure and function of phospholipase A$_2$. *Rev. Physiol. Biochem. Pharmacol.*, **91**, 91–203.

von Hippel, P.H. and Berg, O.G. (1989). Facilitated target location in biological systems. *J. Biol. Chem.*, **264**, 675–8.

Waite, M. (1987). *The Phospholipases*, Plenum, New York.

Wilshut, J.C., Regts, J., Westenberg, H. and Scherphof, G. (1978). Action of phospholipases A$_2$ on phosphatidylcholine bilayers. Effects of the phase transition, bilayer curvature and structural defects. *Biochim. Biophys. Acta*, **508**, 185–96.

Wilschut, J.C., Regts, J. and Scherphof, G. (1979). Action of phospholipase A$_2$ on phospholipid vesicles. *FEBS Lett.*, **98**, 181–6.

Wimpley, W.C. and White, S.H. (1996). Experimentally determined hydrophobicity scale for proteins at membrane interfaces. *Nature Str. Biol.* **3**, 842–8.

Woolley, P. and Petersen, S.B. (Eds.) (1994). *Lipases: Their Structure, Biochemistry and Application*, Cambridge University Press, Cambridge, p. 363.

Yau, W., Wimpley, W., Gawrisch, K. and White, S.H. (1998). The preference of tryptophan for membrane interfaces. *Biochemistry*, **37**, 14713–18.

Yu, B.-Z., Berg, O.G. and Jain, M.K. (1993). The divalent cation is obligatory for the binding of ligands to the catalytic site of secreted phospholipase A$_2$. *Biochemistry*, **32**, 6485–92.

Yu, B.-Z., Ghomashchi, F., Cajal, Y., Annand, R.R., Berg, O.G., Gelb, M.H. and Jain, M.K. (1997a). Use of an imperfect neutral diluent and outer vesicle layer scooting mode hydrolysis to analyze the interfacial kinetics, inhibition, and substrate preferences of bee venom phospholipase A$_2$. *Biochemistry*, **36**, 3870–81.

Yu, B.-Z., Rogers, J., Ranadive, G.N., Baker, S., Wilton, D.C., Apitz-Castro, R. and Jain, M.K. (1997b). Gossypol modification of Ala-1 of secreted phospholipase A_2: a probe for the kinetic effects of sulfate glycoconjugates. *Biochemistry*, **36**, 12400–11.

Yu, B.-Z., Rogers, J., Nicol, G.R., Theopold, K.H., Seshadri, K., Vishweshwara, S. and Jain, M.K. (1998). Catalytic significance of the specificity of divalent cations as K_s^* and k^*cat cofactor for secreted phospholipase A_2. *Biochemistry*, **37**, 12576–87.

Yu, B.-Z., Berg, O.G. and Jain, M.K. (1999a). Hydrolysis of monodisperse substrate by phospholipase A_2 occurs at vessel walls and air bubbles. *Biochemistry*, **38**, 10449–56.

Yu, B.-Z., Rogers, J., Tsai, M.D., Pidgeon, C. and Jain, M.K. (1999b). Contributions of residues of pancreatic phospholipase A_2 to interfacial binding, catalysis and activation. *Biochemistry*, **38**, 4875–84.

Yu, B.Z., Janssen, M.J.W., Verheij, H.M. and Jain, M.K. (2000a). Control of the chemical step by leucine-31 of pancreatic phospholipase A_2. *Biochemistry*, **39**, 5702–11.

Yu, B.Z., Poi, M.J., Ramagopal, U.A., Jain, R., Ramakumar, S., Berg, O.G., Tsai, M.D., Sekar, K. and Jain, M.K. (2000b). Structural basis of the anionic interface preference and k_{cat}^*-activation of pancreatic phospholipase A_2. *Biochemistry*, **39**, 12312–23.

Yuan, C. and Tsai, M.D. (1999). Pancreatic phospholipase A_2: new views on old issues. *Biochim. Biophys. Acta*, **1441**, 215–22.

Zhou, F. and Schulten, K. (1996). Molecular dynamics study of phospholipase A_2 on a membrane surface. *Proteins: Structure, Function, and Genetics*, **25**, 12–27.

Zographi, G., Verger, R. and de, Haas, G.H. (1971). Kinetic analysis of the hydrolysis of lecithin monolayers by phospholipase A_2. *Chem. Phys. Lipids*, **7**, 185–206.

Index

Abbreviations, xxiii–xxvi, 8
Accessibility of substrate, 12–15, 26, 44, 47, 73
Accessible volume, 268, 269
Accumulated products, 179, 185, 228, 232, 235, 242
Activation, *see also* allosteric effect
   allosteric, 147, 173
   apparent, 94–95
   by cofactor, 213
   by enzyme exchange, 117–21
   by NaCl, 186
   by product desorption, 177
   by substrate replenishment, 94, 179
   factor, 146
   model for k_{cat}^*-, 190
   model for K_L^*-, 146–7
Active site
   sPLA2, 112–14
   tl-lipase, 164–7
Active site directed ligands, 47, 202
Acyl chain, *see* Polymethylene
Acylated protein, 128
Acyltransferase, 11, 12, 18, 89
Affinity, *see also* binding
   enzyme-substrate, 53, 138
   enzyme-interface, 138
Aggregates, *see also* Micelles, Vesicles
   dispersed, 24–7, 41–2
   organized, 26, 30
   metastability, 88, 261
   polymorphism, 88, 261
   premicellar, 59, 182
Alkyl chain, *see* Polymethylene
Allergenic effects, 16
Allosteric effect, 117, 138, 249, 275
   interfacial, 164, 167, 173
   k_{catS}^*-, 164, 184, 186–91, 229
   K_S^*-, 146, 164, 167–8, 183
Allostery, *see* Allosteric effect
Amphipathic helix, 170
Amphiphile, aggregated, 23

Amphiphiles, 3, 7, 8, 23, 28
Amphiphilic solutes, *see* amphiphiles
Anchoring, 128, 138, 170
Anion binding to protein, 19, 193–4
Anion partitioning in the interface, 128
Apparent MM-parameters, 122, 153, 177, 179
   with inhibitors, 210, 217, 223
Arachidonate, 17
Arachidonic acid, 17
Area correction factor, 70, 157

Bilayer, 2, 9, 26, 38, 39
   gel–fluid transition, 88, 114, 230
   intervesicle exchange of phospholipid, 39
   phospholipid flip-flop, 39, 88
   phospholipid lateral diffusion, 39, 88, 115
   transbilayer movement of phospholipid 39, 88
Bile salts, 13, 187
   effect on PLA2 kinetics, 187, 257–9
Binding,
   cofactor, 135–6
   enzyme to interface, 91, 125, 129, 133, 139–42, 144, 157
   solute to interface, *see* partitioning
   substrate to enzyme, 134, 137–8, 263, 274
Bipolar lipids, 4

Carbonyl isotope effect, 110
Cardiolipin, 8, 9
Catalytic mechanism of sPLA2, 112–14
   active site, 13, 18, 19, 112, 192
   calcium-coordinated oxyanion, 112
   catalytic triad, 112
   kinetic parameters, 108, 180
   kinetic-isotope effect, 106
   oxy/thio effect, 108, 110, 114, 117, 254
   quality of interface effects, 114
   solvent-isotope effect, 110
   substrate specificity, 111
   temperature dependence, 115

Catalytic mechanism of sPLA2 (*Continued*)
    tetrahedral intermediate, 112
    transition state, 112
Catalytic parameters, uses of, 110
Catalytic turnover path, 77, 175
    aqueous phase, 47, 57
    interface, 77, 98
Cell, 2
Cerebroside, 9, 39
Charge compensation, 186, 191, 196
Charge neutralization, 188–90
Charge preference, 184–7
Charge reversal, 196
Chemical potential, 264, 271
Chemical step, 50, 106, 110
Cholesterol, 5, 12, 13, 18, 42
Cholesterol esters, 12, 13, 15
CL, 8
Clatharate water, 6, 11
Cloud point, 33
CMC, xxii, 27–36, 63, 173
Cofactor
    binding, 135–6
    partitioning, 136
    kinetic effects, 107, 213
Colipases, 171
Concentration
    in the interface, 13, 69, 82, 175
    local, 137, 269, 273
    of interface, 82
    of substrate, 81–2
    total, 24, 81–2
    units in the interface, 69
Conformational reciprocity, 196
Constraints, kinetic, 50
Critical micelle concentration, *see* CMC
Critical micellization concentration, *see* CMC
Cyclooxygenase, 17
Cytoplasmic space, 2

Delay to steady state, 227–30, 232
    effect of the phase transition temperature,
        230
    enzyme concentration dependence, 229, 238
    in monolayer kinetics, 235, 243–5
Desorption, *see* Dissociation rate
Detailed balance condition, 130, 136–8,
        146–47, 183, 191, 216, 274
Detailed balance condition and non-ideal
        partitioning, 263
Dielectric constant, 2
Diffusion
    lateral, 39, 88, 115

    limit, 36, 115, 252, 280
    times, 276–80
Diluent, *see* Neutral diluent
Dilution, 212, 229, 260
Dimensionality, 48, 267
Dispersed phases, 24–7, 41, 249
    as substrate, concerns for kinetics, 260
Dispersions, *see* Emulsion, Micelles, Vesicles
Dissociation constant, 37, 47
    enzyme-interface, 37, 125–6, 133, 144
    solute-interface, 35, 36, 126
    substrate-enzyme, 47, 134, 145
    two-dimensional, 126
Dissociation rate
    enzyme-interface, 36–7
    solute-interface, 36–7, 65
Dixon plot, 207, 210

Eicosanoids, 5, 11, 12, 17
Electrical double layer, 127
Electrophoretic mobility, 127, 129
Electrostatic potential, 127
Emulsion, 13, 25, 41
Ensemble averaging, 79, 83, 85
Ensemble behavior, 52, 92, 95
Entropy, 268, 270–1
Enzyme state, statistical weight, 131
Enzymes
    interfacial, 3, 160
    matrix, 3, 161
    soluble, 3
Equilibrium constants, effective, 132
Exchange, 36–40
    between interfaces, 36, 88
    between micelles, 36, 40
    between vesicles, 39, 88, 90
    mechanisms, 38
    of enzymes, 117–21
    of monomers, 33, 36, 38, 251
    protein, 38
    micelle to monomer, 33, 36
Exchange-limited kinetics, 177–8, 249–50,
        253–4
Excluded-area effect, 140
Excluded-surface effect, 140

Facilitated diffusion, 282
Flip-flop, 39, 88
Flux per enzyme, 53, 98
Free-energy for transfer of peptide residues,
        169
Fusion, 38
Fusion–fission, 40–1, 183, 251, 255
Fusion–fission in mixed micelles, 38, 40–1, 255

Ganglioside, 8, 9, 12
Gel−fluid phase transition, 88, 114, 230
Glyceride, 4, 12, 13, 15
Glycoconjugates, sulfated, 203
Glycolipids, 4, 7, 9, 44
Glycosyl diacylglycerol, 8
GM, 8

Hemifusion, 38
Heparin, 203
Hoffmeister series, 33, 129
Hopping mode, 79, 120, 123
Hydration layer, 129
Hydropathy plot, 169
Hydrophobic, 2, 6
    area, 196
    drive of methylene residues, 86
    drive of amino acid residues, 168
    effect, 10
    residues, 168
    segments, 169

i-face 19, 126, 168, 191−5
Inhibition, *see also* Inhibitor
    allosteric, 199
    competitive, 108, 199−203
    covalent modification, 204
    noncompetitive 199, 225
    nonspecific, 199, 206−7
    product, 103, 109, 177, 229
    specific, *see* competitive inhibition
    uncompetitive 199, 256
Inhibitor, *see also* Inhibition
    interface-based, 208
    partitioning, 215
    potency, 212, 219
    solution based, 222−4
    sPLA2, 200−7
Initial rate, 53, 98, 101, 104
Interface, *see also* Substrate
    accessibility, 249
    binding region, *see* i-face
    composition, 7−13, 42
    concentration, 82
    density, concentration in the interface, 13, 69, 82, 175
    diluent, *see* Neutral diluent
    equilibrium constants, 125−148
    organization, 7−9, 25−7, 34, 42, 205−6, 261
    phenomena, 9
    preference, 184−187
    recognition region of PLA2, *see* i-face
    solvent, *see* Neutral diluent
Interfacial allostery, *see* Allosteric effect

Interfacial enzymes, criteria for assays, 43
Interfacial mechanism, 3, 60
Interfacial reaction loop, 47−8, 98
Intermicellar concentration, 40, 69, 253
k^*_{catS}-activation, two-state model, 190−1

Kinase, 11, 12
Kinetic paradigm, 47−52
    Interfacial, 98−100
    Michaelis−Menten, 50−2, 98−101
Kinetic parameters for sPLA2, 108, 180
Kinetic path,
    kinetic mechanism, 14, 49, 56−9, 110, 175−6
    hopping mode, 120−123
    quasi-scooting mode, 149−166
    scooting mode, 77−117
    solution, 47−73
    with rapid substrate replenishment 117−120, 149−166
Krafft point, 33

Lateral phase−separation effect, 114
Leukotrienes, 17
Lid, *see* lipase
Lineweaver−Burk plot, 54, 62, 73, 151, 210, 217
Lipase, 11, 12, 13, 15
    activation, 165−7
    lid on, 164−7
    tl-, 160−6
    triglyceride hydrolase, 57, 158
Lipid−protein interaction, 129, 168−171
Lipolytic enzymes, *see* lipase
Lipoproteins, 42
Lipoxygenase, 17
Local concentration, 276
Lysophospholipid, 12, 17

Matrix mechanism, 3, 60−2
Membrane anchors, 128, 138, 170
Membrane−associated enzymes, 168, 206
Membrane−binding domains, 170
Membranes of cell, 2
Metastability, 88, 261
Micellar rate, 174
Micelle−monomer equilibrium, *see* monomer−micelle equilibrium
Micelles, 9, 26
    aggregation number, 28, 32
Micellization, 27−36, *see also* CMC
    cooperativity, 34
    equilibrium, 32−5
    free energy, 32
    kinetic-order, 34
    monomer in equilibrium, 35

Micellization (*Continued*)
  nonindealities, 31
  structural requirements, 31
  thermodynamic parameters, 32–3
Michaelis constants, 53, 100, 107
Michaelis–Menten equation, 47, 52, 61, 98
  integrated 54, 101–4, 118, 121, 123, 214
Michaelis–Menten parameters, 52–3, 99–100, 107,
  see also Apparent MM-parameters
Mixed-micelles, 38
MJ33, competitive inhibitor of sPLA2, 112, 146, 193, 195, 202, 220
MM-equation, *see* Michaelis–Menten equation
Mole fraction in the interface, 67, 69, 175
Monolayer, 26, 28
  at air–water interface, 29, 236
  kinetics, 243–245
  product accumulation, 242
  reaction progress, 228, 235–7, 245
  trough, 40, 236
  unstirred layer, 40, 236, 240
Monomer aggregate equilibrium, 34
Monomer exchange, 33, 36, 38, 251
Monomer–micelle equilibrium, 32–5
Monomer rate, 56–59, 174, 180–1

Neutral diluent, 67, 73, 75, 99, 126, 133, 143, 205
  area, 157
  binding to the active site, 126, 133, 145
Non-ideal partitioning, 68, 70, 262–5
Non-idealities in mixed-micelles, 68, 249–264
Non-ideality factor, 262
Nonpolar solutes, 3, 5, 6
Number density, *see* concentration

Organelles of cell, 2
Orientational confinement, 268–9, 272
Oxy/thio effect, 108, 110, 114, 117, 254
O/S ratio, *see* oxy/thio effect

PA, 8
PAF analogs, 60
PAF, *see* platelet activating factor
PAF-acetylhydrolase, kinetic mechanism, 60, 63
Partitioning, 7, 47, 63, 66, 73
  effect on reaction rate, 158
  equilibria, 68, 70, 125, 221
  equilibrium dissociation constant, 67
  in micelles, 68–9
  molecular area effects, 157–158
  non-ideality, 68, 70, 154, 157, 264–265

PC, 8
PE, 8
Penetrating power of enzyme, 237, 247
Peptide–mediated exchange, 38, 93–4
PG, 8
Phases, dispersed, 24–7, 41–2, 249
Phases, organized, 24–7, 41–2
Phosphatidic acid, 8, 12
Phosphatidylcholine, 8, 12, 13, 39, 59, 60, 257
Phosphatidylethanolamine, 8, 41
Phosphatidylinositol, 8, 11, 12, 89
Phosphatidylserine, 8, 128
Phospholipase, 12, 16
Phospholipase A$_2$, *see* sPLA2
Phospholipid, 2, 8, 9, 12, 27, 42
  hydration, 87
  preferred conformations, 87
  properties 39, 86, 88
PI, 8
Plasmalogen, 8
Platelet activating factor (PAF), 12, 17
Poisson distribution, 79, 82–5, 92
Poisson law, 79, 84
Poisson–Boltzmann equation, 127
Polymethylene chain, 6–9, 31–2, 86–8, 111, 169, 192, 219
Polymorfism, 88, 261
Polymyxin B, 93, 106
Premicellar aggregates, *see* Aggregates
Primary rate constants, *see* Rate constants
Processivity, 48, 77–81, 123, 125, 249
Product accumulation, 179, 185, 228, 232
  in monolayer, 235, 242
Product dissociation from interface, 175
Product inhibition, *see* Inhibition
Prostacyclin, 17
Prostaglandins, 17
Protection method, 142–5
PS, 8

Quality of interface, 48, 82, 114, 167
Quasi-equilibrium, 183
Quasi-scooting mode, 121, 150, 173, 175

Rate
  enhancement, 282, 284
  constants, 35–7, 47–8, 50, 56, 98
  parameter, k_{cat}, 52
Reaction path, *see* kinetic path
Reaction progress, 54, 101–4, 118, 121, 123, 214
Reduction of dimensionality, 267
Replenishment, 44, 73, 93, 105, 205–6
  by fusion–fission, 255
  by monomer exchange, 35–6, 65, 153, 251

Residence time, 26, 35, 36
  of enzyme, 229

Salting-out effect, 32–3, 129
Scaffolding, 147, 183, 274–6
Scooting mode, 77–9, 83, 86, 89, 149
  catalytic parameters, 108
  constraints, 83–84
  turnover processivity, 98–102
Secreted phospholipase A$_2$, *see* sPLA2
Signals, transmembrane, 11, 267
SM, 8
Smoluchowski expression, 36, 257, 280
Solubility, 7, 9
  limit, 150, 154
  monomer in equilibrium, 35
Solute concentration, total, 24, 81
Solvent isotope effect, 110
Sparingly soluble substrate, 150, 154–6
Sphingolipid, 9, 11, 44
Sphingomyelin, 8, 12, 39, 89
sPLA2, 13, 16, 18, 48, *see also* Catalytic
    mechanism of sPLA2
Standard free energy of binding, 189, 272
Steady-state
  microenvironment, 149
  assumption, 55
  condition, 51, 101, 175
  flux, 52, 99
  rate
  effect of partitioning, 71, 158
  effect of desorption, 65
Substrate
  accessibility, 12–15, 26, 44, 47, 73
  concentration dependence of flux, 53, 162–3,
    183
  concentration variables, 81–83
  depletion in dispersed particles, 250
  replenishment, 44, 65, 73, 93, 153, 175, 251,
    255
  sparingly soluble, 154–156
  specificity, 111
Sulfolipid, 9
Surface area, 2, 69, 210
Surface charge, 127
Surface potential, 127

Surface pressure, 40, 235, 242
Surface tension, 29, 33, 235
Symbol list, xxii–xxvi

Target search by lateral diffusion, 115–17,
    276–82
Thermodynamic box, 130, 146–7, 203
Thromboxane, 17
Toxic effects, 16
Translational confinement, 268–271, 272
Transmembrane propensity, 169
Transmembrane signals, *see* signals
Triglyceride, 4, 12, 42
Triglyceride lipase, *see* Lipase
Triton X-100, effect on PLA2 kinetics, 258
Turnover
  cycle, 47–8, 52, 98, 121, 125, 175
  number, 53, 98
  per enzyme, *see* flux per enzyme
  processive, 97
Two-dimensional solvent, *see* Neutral diluent

Unitary standard free energy, 274
Unstirred layer
  and trough geometry, 236, 239
  diffusion through, 40, 239–41
  equilibration through, 241
Water interactions, *see* Hydrophobic
Water-structure breakers, 33
Water-structure makers, 33

Vesicles 26, 27, 160
  dispersity of size, 84, 87
  exchange between, 38–9, 88, 90
  integrity during reaction progress, 83, 86, 90
  population, 83
  properties, 27, 88
  size, 27, 87
  stability, 39, 86
Vesicle-to-vesicle contacts, protein-mediated,
    38, 93, 149
Vesicle-to-vesicle exchange, protein-mediated,
    38, 93, 149

Zeta potential, 128
Z-factor, 212